MW00509832

INTRODUCTION TO ENGINEERING DESIGN

Book 12, 2nd Edition

ENGINEERING SKILLS AND ROBOTIC CHALLENGES

James W. Dally

University of Maryland, College Park

Stanley J. Reeves and Thaddeus Roppel

Auburn University

College House Enterprises, LLC
Knoxville, Tennessee

Copyright © 2019 by College House Enterprises, LLC.

Reproduction or translation of any part of this work beyond that permitted by Sections 107 and 108 of the 1976 United States Copyright Act without permission of the copyright owner is unlawful. Requests for permission or for further information should be addressed to the Publisher at the address provided below.

This textbook is intended to provide accurate and authoritative information regarding the various topics covered. It is distributed and sold with the understanding that the publisher is not engaged in providing legal, accounting, engineering or other professional services. If legal advice or other expertise advice is required, the services of a recognized professional should be retained.

The manuscript was prepared using MS Word with 11 point Times New Roman font and converted to pdf using Adobe Acrobat Pro DC.

College House Enterprises, LLC.
5713 Glen Cove Drive
Knoxville, TN 37919, U. S. A.
Phone and/or FAX (865) 558 6111
Email   jwd@collegehousebooks.com
http://www.collegehousebooks.com

ISBN-978-1-935673-44-6

PREFACE

This book is the second edition of the twelfth textbook in this series: **Introduction to Engineering Design**. Jim Dally, working with College House Enterprises, LLC and faculty members at the University of Maryland, the University of Nevada @ Reno, University of California @ Irvine and Auburn University, has prepared eleven previous textbooks in this series—a new one almost every 18 months. These books are written for the first-year engineering students at one of the universities listed above. Several Colleges of Engineering have adopted one or more of the books in this series to introduce design and engineering skills for their first or second year students. The design, build and testing of a hovercraft model, described in Book 9, was such an interesting and challenging project that it was used for twelve semesters with approximately 8,000 students at the University of Maryland. The project is also assigned for the first year Introduction to Engineering Design at the University of Nevada @ Reno, where it is in its ninth year. A project involving a quadcopter in Book 11 has been used by the University of California @ Irvine, for the past four years. Book 10 prepared for the University of Maryland involves the design, build and testing of an autonomous Over -Sand Vehicle (OSV) is in its fifth year. This second edition of Book 12 prepared for Auburn University involves Robotic Challenges in addition to several chapters dealing with content necessary for entering first year students to understand.

This textbook is used to support the students during the entire semester. Some of the material may be covered in lecture, recitation or in a computer laboratory or a model shop. Additional material is covered with reading assignments. In other instances, the students use the text as a reference document for independent study. Exercises, provided at the end of each chapter, may be used for assignments. The book contains 15 chapters presenting many topics that first year engineering students should understand as they proceed through the engineering curriculum.

Three chapters dealing with engineering and society are included. A historical perspective on the role engineering played in developing civilization and on improving the lives of the masses is presented in Chapter 1. In this chapter, we move from the past into the present and indicate the current relationship between business, consumers and society. The twenty greatest engineering achievements of the 20th century are briefly described. Chapter 5 discusses the balance between safety and performance. Methods to evaluate and recognize risky environments are discussed. The chapter includes a listing of hazards, which is important in identifying the many different ways users of a product can be injured. Chapter 4 on ethics, character and engineering includes a large number of topics so the instructor can select from among them. A description of the Challenger and Columbia accidents is also given, because both of these fatal crashes provide excellent case histories covering safety related conflicts between management and engineers. A new reference to a video produced by the New York Times has been added. We strongly recommend viewing it.

An introduction to the engineering profession, described in Chapter 2, covers engineering disciplines, on-the-job activities, salary statistics and registration information for your PE license. A useful student survival guide is also included in Chapter 3. We strongly encourage you to read it as it contains information for managing time that will be useful throughout your life.

The very important topic of communications is covered in two chapters. Chapter 8, on technical reports, describes many aspects of technical writing and library research. The most important lesson here is that a technical report is different than a term paper for the History or English Departments. An effective professional report is written for a predefined audience with specific objectives. The technical writing process is described, and many suggestions to facilitate composing, revision, editing and proofreading are given. Chapter 7, on design briefings, shows the distinction among speeches, presentations and group discussions. Emphasis is placed on the technical presentation and the importance of preparing excellent visual aids. PowerPoint slides are usually employed as visual aids in a design briefing.

Information on team skills is covered in Chapter 6. An updated version of an introduction to MATLAB is provided in Chapter 11. Chapter 12 describes basic electric circuits and batteries to provide technical background helpful for the design and control of robots. Sensors that may be used to complete the task assigned as part of a robotic challenge are described in detail in Chapter 13. Concepts of statics and dynamics are introduced in Chapter 14 to enable the students to understand the forces and moments and their effect on controlling the motion of their robot. Finally, an introduction to programming an Arduino μprocessor in a version of C is included in Chapter 15.

Two chapters on engineering graphics using computer programs are included. A discussion of preparing engineering drawings is described in Chapter 9. Several CAD/CAE programs are available on the college's computers and instructional videos are available to assist you in mastering the software. The use of tables and graphs in communicating engineering information using Microsoft Excel is presented in Chapter 10.

ABET AND EDUCATIONAL OUTCOMES

ABET's Criteria 2000 is considered in writing each edition of these books on Introduction to Engineering Design. It is believed that a first course in engineering design should include many of the educational outcomes listed below:

Communication Skills:

1. Engineering Graphics
    *   Understand the role of graphics in engineering design.
    *   Understand orthographic projection in producing multi-view drawings.
    *   Understand three-dimensional representation with pictorial drawings.
    *   Understand dimensioning and section views.
    *   Demonstrate capability of preparing drawings using computer methods.
    *   Demonstrate understanding of engineering graphics by incorporating appropriate high-quality drawings in the design documentation.
2. Engineering Briefings:
    *   Within a team format, present an engineering review for the class using appropriate visual aids.
    *   Each team member demonstrates briefing skills.
3. Engineering Reports:
    *   The team's work is documented in a professional style report incorporating time schedules, costs, parts list, drawings and an analysis.

Team Experience:

- Develop an awareness of the challenges arising in teamwork.
- Demonstrate planning from conceptualization to the evaluation of several programs for robotic tasks.
- Understand and demonstrate sharing responsibility among team members.
- Demonstrate teamwork in preparing reports and presenting briefings.

Software Applications:

- Demonstrate entry-level skills in using spreadsheets for calculations and data analysis.
- Show a capability to prepare graphs and charts with a spreadsheet.
- Show a capability to prepare professional quality visual aids.
- Understand entry-level skills in a feature-based solid modeling program.
- Demonstrate these computer skills in preparing appropriate materials for briefings and reports.

ACKNOWLEDGEMENTS

Acknowledgments are always necessary in preparing a textbook, because many people are involved in helping improve the presentation. First, it is important to recognize the contributions of many dedicated faculty members and teaching fellows at the University of Maryland. To date more than 100 different faculty members have taught this introductory engineering course on at least one occasion over the past 20 years. Many of them have made suggestions for improving the material in this book. In particular, thanks are due to Dr. William L. Fourney for his commitment in providing the leadership necessary to operate a multi-section offering of this course and in providing support for the many different instructors and teaching fellows involved each semester.

In Book 12, Stanley Reeves updated the Chapter 11 on MATLAB and Thaddeus Roppel provided essential information on the sequencing of the 15 chapters.

Finally, we appreciate our families who tolerate our absences as we teach, perform research and spend countless hours at the keyboard writing, editing and drawing.

James W. Dally and the Keystone Faculty including:
Vince Brannigan, Kevin Calabro, William L. Fourney, Chris Cadou, Chris Davis, Mel Gomez, Ken Kiger, Dave Lovell, Tom Murphy, Peter Sunderland, Guangming Zhang, and many others.
University of Maryland at College Park

And from Auburn University
Stanley Reeves and Thaddeus Roppel

August, 2018
Knoxville, TN

CONTENTS

CHAPTER 9 ENGINEERING GRAPHICS

CHAPTER 10 MICROSOFT EXCEL

CHAPTER 11 BEGINNING MATLAB® FOR ENGINEERS

CHAPTER 12 BASIC ELECTRIC CIRCUITS AND SENSORS

CHAPTER 13 SENSORS

CHAPTER 14 ROVER DYNAMICS AND CONTROL

CHAPTER 15 BUILDING AN AUTONOMOUS VEHICLE CONTROL SYSTEM USING THE ARDUINO PLATFORM

CHAPTER 1

ENGINEERING AND SOCIETY

1.1   INTRODUCTION

It should be made clear from the beginning of this chapter that engineering and society are closely joined. Engineering is an integral part of society, and consequently, engineering must be treated as a social enterprise as we continue to develop sociotechnical systems. When engineers develop a new product or improve an existing system, we change society. The magnitude of the change differs from one product to another, and the effect on society may be trivial or enormous, but be assured that there is an impact. For example, when transistors were first developed from new ultra-pure semiconductor crystals, the impact was revolutionary. While the first transistor was developed in the mid 1950s, the integrated circuit introduced in 1959, and the microprocessor released in 1971, the technological revolution resulting from introducing these three products in the microelectronics industry is still underway. Micro-electronic based products are markedly affecting the way in which society lives today. The media refers to the current period as the "information age". New products, which are introduced almost daily, are driven by technology. Products based on new advances in microelectronics will continue to change the way everyone works and plays for the foreseeable future [1].

You often hear about new products and changing life styles. Television, radio, Internet, smart phones and even newspaper advertising keep everyone well informed of the availability and prices of both old and new products. However, you rarely hear much about the actual technology involved even when the product is new and revolutionary. The media and the public are usually not interested in learning about the technology employed. This is an unfortunate fact, because it would be easier to introduce new or modified systems if the public were more technologically literate.

Engineers are, in part, responsible for this apparent wall between the public and the understanding of technology. Engineers often speak in such technical terms that those few individuals interested in our work become confused. As engineers, we must clearly recognize the need to effectively communicate with society. We can begin by learning how to explain why and how products work in terms more easily understood by the public. We must also be more enterprising in understanding society and their changing views relative to technology. Lastly, we need to think on a more global scale. Engineers develop sociotechnical systems that are often large and complex. These systems rely on hundreds of products and many different types of services to function efficiently. Even though, as an engineer, you may be dedicated to developing a product that is only a small part in a much larger system, you should understand the entire system. Those development teams that have a thorough understanding of all aspects of the complete sociotechnical system develop superior designs.

Let's consider an example of a sociotechnical system—the commercial air transport system. Do we have technical products involved in this system? Yes, literally hundreds of advanced state-of-the-art products. Do we provide technical services for the airlines? Yes. Does an engineer working for General Electric in the development of a new jet engine need to understand what Boeing is doing in the design of a new aircraft or modifications of an existing aircraft? Does that same engineer need to know about the sound levels that the public will accept as the plane lands and takes off? Is the engineer working for Boeing required to understand the causes of each and every airline accident? Again, Yes.

From these simple questions, it should be evident that sociotechnical systems are extremely complex. It will require significant effort for you to understand the detail of the interaction of the many products incorporated in these systems with the public. Nevertheless, engineers that understand and appreciate the improvements, which both the customer and society seek, will develop the most successful and safe products.

Finally, as an engineer, you must understand that systems are adaptive. They are dynamic and change with time in response to several societal and economic forces. Recognize that society changes. The requirements of society and the customer in the 1950s were much different than those that exist today. The requirements acceptable a generation ago are not applicable today. The customers, an important segment of society, typically expect that the new model of the product will be much improved over the older model and be lower in price. Others in society, not necessarily the customers, will expect (perhaps demand) that you manufacture the product without degrading the environment.

An example is useful to distinguish between the customer and society. Let's consider the automobile as the product. Customers of automobiles demand style, comfort, mobility, convenience, affordability, reliability, performance and value. Society requires (through federal regulation) safety, fuel efficiency, reduced emissions and less pollution. The fact that the requirements from the customer and society often conflict should be recognized. Conflicting requirements challenge engineers to create improved designs that advance the state of the art in all areas of technology.

The governments (Federal, State, County and City) are also involved in the relationship between engineers and society. Various government agencies frequently issue regulations that affect both the sociotechnical systems and the product. Consider the commercial air transport system again. Obviously, safety is a very serious issue. The Federal Aviation Administration (FAA) is responsible for issuing regulations to insure safe operations of commercial aircraft. Consider 1996, which was not a good year for the FAA or the commercial airlines, because several very serious accidents occurred with significant loss of life. As a consequence of a fire in the cargo bay, which caused the crash of a Boeing 737 and the loss of all aboard, the FAA now requires the airlines to have a fire suppressant system on all planes. Regulations from government agencies drive a portion of the changes to any sociotechnical system.

This understanding of the relationship between engineering and society is relatively new. If you examine the five-volume tome, **A History of Technology**, [2] published by Oxford University Press over the period from 1954-58, you will find a chronological, deterministic catalog of technological development. However, the editors largely ignore the relationship between engineering and society. In his review of this tome, Hughes [3] states that the editors belatedly conclude their five-volume treatise with an essay on technology and the social consequences. Today, the relationship is more clearly recognized although educators have not yet managed to include adequate treatment of this very complex relationship in engineering curricula. A few examples of the relationships between engineering and society will be described in this chapter. For a more complete treatment, read reference [4].

1.2   ENGINEERING IN EARLY WESTERN HISTORY

Technology has impacted the way that we live since the beginning of recorded history. Long before the first engineering college was founded, intelligent, hard-working men[1] worked as engineers developing new products and processes. You may study history and cite many engineering developments that changed, and for the most part, benefited society.

Prior to about 6,000 BC, there was little evidence of technology. Men and women were hunters and gatherers. They scrounged for food and lived in caves if they were lucky and shacks or huts if they were not. The significant developments in the period from 6,000 to 3,000 BC were in agriculture with the domestication of animals and the cultivation of grains. With time, it was possible to insure a relatively stable, if limited, food supply and people began to congregate together in villages and small cities.

---

[1] Historical accounts of engineering achievements indicate that the profession was male dominated. The author is sensitive to the gender issue in engineering education, but it must be recognized that the presence of women in the profession is very recent, when considered relative to a historical span of at least 8,000 years.

The first technology involved the building of roads, bridges and boats for local transportation. People clustered in villages and small cities wanted to transport food, water and other materials over relatively short distances. The geographic regions where technology was evident were extremely local. Early Western history involved only Mesopotamia, Egypt, Greece and then Rome. In a few cities, there were aqueducts to transport water and a few primitive sewerage systems. In most regions, only the ruling class lived well. The great monuments of Egypt and Greece were built with slave labor. In fact, slaves (people who were on the losing side in a war or those born into slavery) provided most of the power required for building, mining and agriculture for many thousand years. The rise of religion is often credited with the reduction in the prevalence of slavery. However, the steam engine developed by Thomas Savery, Thomas Newcomen and James Watt in the early 18th century signaled the beginning of the end of the need for power provided by humans (slaves and other unfortunate souls in bondage). As society enters the 21st century, the development of the low-cost, high-performance computers, a wide variety of specialized software programs and effective communication 24×7 by the Internet, is another beginning of the end for many employees performing repetitive tasks or clerical work.

The Egyptians

The Egyptians built many huge monuments and were masters in moving large blocks of stone from distant quarries to a construction site and then elevating these blocks to their respective positions in gigantic structures. Some estimates indicate that 20,000 to 50,000 men were required to drag 20 to 50 ton blocks of stone used in the construction of these monuments. In spite of the majestic grandeur and durability of these monuments, the new innovations introduced by the Egyptian engineers were not outstanding. They constructed temples in a simple fashion using only uprights (columns) and cross pieces (short, deep beams). While the arch was known during this period, it was constructed with mud bricks and never adapted to stone construction. Classical treatments of the history of technology tell us very little about the impact of these developments on society. However, it is apparent that the construction of a pyramid or a temple required a significant segment of the working population for many decades. This effort probably did little to improve the welfare of even the **free** segment of the population. It is also evident that the effort required of the slaves to move the countless large, heavy stones must have been brutal.

The Greeks

Although they had very capable engineers, the Greeks are better known for their contributions to science. Thales, Euclid and Aristotle are recognized for their contributions to the development of the basics of geometry and physics. This new knowledge of geometric relations was applied almost immediately in architectural engineering. However, scientific understanding during this period was so limited that nearly 2,500 years elapsed before this knowledge was expanded sufficiently to be useful to the engineering community.

Greek engineers built cities, roads, water (hydraulic) systems and some machinery. Archimedes and Hero were innovators who used the pulley, screw, lever and hydraulic pressure to develop cranes, catapults, pumps, and several hydraulic devices. The Hellenistic period brought better buildings and local roads that improved living conditions for the upper class. However, the construction of the city infrastructure was still performed largely by slaves because other forms of power were nonexistent. The Hellenistic world was divided into extremely independent city-states. Politics of the period did not foster cooperation between these governmental entities; hence, very few long roads were built. Transportation was largely by sea in boats equipped with oars powered by a galley of slaves. Some of the boats were fitted with a square sail, but these were so poorly designed that they were effective only on down-wind tacks.

The Romans

The Roman engineers followed the Greeks, and while they were considered by some historians to be less inventive, they were masters of detail. As the Roman Empire spread from the Middle East to Scotland, the Roman engineers built long roads connecting the distant cities. The roads were so well constructed that they were in service for many centuries. The large cities they constructed in the conquered lands had paved streets and heated homes with water supplies and sewer systems. The Roman engineers had little or no theoretical knowledge, but they had developed practical methods of construction that served their purposes well. During this period, the subjects of statics and mechanics of materials were not well understood. The theory used to design beams would not be developed for another 1,700 years. The Roman engineers based their construction on experience, and they used very large safety factors. These empirical methods, while crude by today's standards, sufficed to produce many bridges, roads and aqueducts. A great bridge over the Tagus River in Spain supported a road nearly 200 m long with six spans 60 m above the river. Considering it was completed at about 100 AD, this bridge was a remarkable engineering feat. The durability of the Roman construction is evident in their structures, some of which are still in use today, almost 2,000 years after they were commissioned.

The Roman engineers deviated from the Egyptian and the Greeks in their construction techniques by using much smaller building stones and bricks. They were able to produce massive structures with the smaller blocks by utilizing mortar and cement. The use of smaller stones, bricks and cement clearly impacted society. Buildings could be constructed with much less effort because thousands of slaves were not necessary to move a single block of material. Life for the slaves clearly improved although the construction work was still very demanding. Also, the upper classes benefited from the more rapid construction of buildings and houses.

While the Roman engineers were outstanding in the construction of civil infrastructure, they did very little to advance the use of power in driving the few machines available. Humans were still used in most instances to power pumps, mills and cranes. In rare cases, water wheels were used to grind grain. These wheels were of the undershoot design and much less effective than the overshoot design that was developed later. While references on the history of technology are quiet on the issue, the life of a worker powering a pump or a mill must have been very difficult.

Eventually, the vast Roman Empire crumbled, and the world entered a period that the historians classify as medieval civilization from 325 to 1,300 AD. The fact that barbarians stripped and destroyed the cities does not imply that progress stopped, but it slowed to a snail's pace. When Rome collapsed, the Byzantine Empire was the next civilization to follow. Significant engineering achievements were made in this period in developing dome structures and curved dams. The author has omitted discussion of these developments to keep this historical treatment brief.

Early Agriculture and the Use of Animals

Regardless of the slow development of engineering and the absence constructing monuments during the Middle Ages, an early agricultural revolution occurred. White [5] has described the changes that occurred as agriculture moved from the dry sandy soil of Italy into the heavy alluvial and wet soil of northern Europe. The plows that were effective in the light sandy soils would not turn the sod in the heavy, moisture-laden soil of the north. A new, heavy plow was invented that would cut through the grass sod and turn over the soil. However, this plow required eight yoked oxen to provide the force necessary to cut through the heavy sodded soil. Very few farmers of the day owned eight oxen. They had to combine their land, share their oxen and cooperate in plowing and planting. The farmers merging their land holdings eventually led to a manorial society.

It may be difficult to imagine that the invention of a new plow made such a major change in the way of life. Today farmers are able to produce all of the food needed in the U.S., while exporting excesses in significant quantities, with less than 2% of our population. However, before the advent of engines and motors (steam, gasoline and electric), men and women struggled to grow enough food to keep from starving. In the Middle Ages, 90% of the population worked in agriculture to produce much less to eat than we do today.

Oxen were used as a source of power for plowing and transport on the farms. However, oxen are very slow and they consume large quantities of grass and grain. At that time, horses were of limited use because of inadequate harnesses and saddles and the frequent problem of splitting hoofs. (Early harnesses fitted about the horse's neck tended to strangle the animal.) These problems were alleviated when a newly designed horse collar, which transferred the load to the shoulders of the horse, was introduced and when iron horse shoes were fitted to their feet. Horses gradually replaced the oxen because they were 50% faster, worked longer and required less food.

In early warfare, the horse was not very effective in frontal attacks because stirrups for the rider did not exist. Consequently, horse mounted warriors were of limited value, because they could not adequately thrust their lance while staying mounted on the horse. With the advent of the stirrup, the horsemen could stand-up, lean forward and thrust the lance through the defenders' shields without losing their seat. The simple addition of stirrups to saddles had a profound influence on society. Feudalism evolved with an aristocracy based on warriors conquering and defending landholdings. A new battle strategy with armored knights mounted on large strong horses was dominant for several centuries [5].

In early history, small technological improvements produced major changes in the form of government, the type of society and the way people of all classes lived. It is remarkable that for about 5,000 years, humans were the prime source of power for most activity. Progress in agriculture and engineering proceeded together to provide more food, better housing, roads and bridges and eventually horse- and oxen-powered plows and wagons.

1.3   <u>ENGINEERING AND THE INDUSTRIAL REVOLUTION</u>

Civilization has always been power limited; even today, space travel is severely constrained by the inadequacies of rocket power. Until the last three centuries, animals, windmills and waterwheels provided the only power available for agriculture, construction, milling, mining and transportation. Progress was often slow because power was not available when and where it was needed. Life, in the power starved 17th century, was rather bleak for the general population. The people worked from dawn to dark as farmers or craftsmen. Only those endowed with landed estates and/or money lived what you would consider a modestly comfortable life.

The Development of Power

The first of several breakthroughs in developing new sources of power occurred in the 18th century. In Great Britain, **Savery**, **Newcomen** and **Watt** developed stationary steam engines. While these early steam engines were woefully inefficient (only 0.5%), they were produced in relatively large numbers (500 existed in 1800) and were employed to pump water and to power textile and flour mills.

Because the steam engines were replacing horses as power sources, James Watt conducted experiments to measure the power a **brewery horse** could provide. His measurements indicated the brewery horse generated 32,400 (foot-pounds/minute) of power. The horse's output was rounded to 33,000 (foot-pounds/minute), which is still used today as the conversion factor for one horsepower (HP).

Most of the 18th century was devoted to innovations directed toward improving several different models of steam engines. Applications were largely limited to stationary power sources because steam engines, of that day, were large, heavy and inefficient. However, with improved metallurgy and machining methods, it was possible in the 19th century to reduce the size and the weight of the engines while maintaining their output. These improvements permitted the steam engine to be adapted to power riverboats, ships, locomotives and later even a few steam-powered automobiles.

The development of the steam engine and its adaptation to several modes of transportation formed the foundations of the industrial revolution. The life style of almost everyone was markedly changed with the emergence of steam-powered factories and much more effective transportation systems. Fewer people were employed in agriculture, but many more were working under appalling conditions in factories and mines. Twelve-hour days, seven days a week, with only Christmas as a holiday during an entire year was the norm. Manufactured products, such as clothing and housewares, were available to a larger share of the population at reduced prices, but the margin between earnings and the money necessary for a factory worker, miner or farmer to live remained very narrow. Steam power had relieved many millions of men from brutally hard labor, but they were released from one dull job for another. On a more positive note, the progress in agriculture and industry enabled the population to triple during this period (1660 to 1820).

1.4   <u>ENGINEERING IN THE 19TH AND 20TH CENTURIES</u>

For the eight thousand years of recorded history prior to about 1800, the advance of technology was extremely slow. Agriculture dominated society with most of the population growing food that was consumed locally. Factories were being developed and simple products needed on the farm, in the home or by the military were produced. Long distance transportation of goods was usually limited to sailing ships on the seas or horse drawn barges on rivers or canals.

During the 19th and 20th centuries, technology literally exploded. The many scientific discoveries of the preceding millennium expanded knowledge sufficiently to allow countless engineering applications. To illustrate, let's define modern technology to consist of the following five elements:

1. Abundant available power
2. Available food in sufficient quality and quantity to insure a nutritious diet for everyone
3. Transportation by air, land and water with associated infrastructure to provide safe, rapid, convenient and inexpensive access
4. Communication between everyone, anywhere and anytime at a reasonable cost
5. Information (knowledge) storage, retrieval, and processing in seconds at any location, anytime and anywhere at negligible cost

While the advances in technology have been astonishing, particularly in the last 50 years, there are still significant deficiencies. Rocket power for launching space vehicles currently constrains space travel. Traffic jams in large cities extend the working day for many millions of people commuting to their jobs. Today about 40,000 people are killed in accidents on U. S. highways every year. Travel on airlines is becoming unreliable, difficult and often unpleasant. Communication in many parts of the world is still not available to the public. However, the recent introduction of cell phones has greatly improved communication in underdeveloped countries. The cost of cell phone towers is much less than the expense of installing land lines required for conventional phone service. Many people are just beginning to explore the different ways to use the massive amount of information that has recently become available to those fortunate enough to own a personal computer with a broadband Internet connection.

You may debate the exact year of some invention. However, the exact date is rarely important because an invention usually evolves into a commercially successful product after several iterations with improvements or modifications by several designers. You should not worry about the exact dates. Instead, recognize the continuous progress and observe the shift in emphasis from one period to another. In power, technology evolved from animals, to steam, to internal combustion (gasoline) and to distributed electrical energy. Power, from gasoline fueled internal combustion engines became mobile and electricity was widely distributed. In transportation, technology evolved from the horse and buggy and sailing ships to automobiles, airliners, nuclear powered submarines and space shuttles. In communications, technology progressed from the pony express to the telegraph, telephone, facsimile machines, cellular telephones and the Internet. In the information business, technology is currently evolving from a hard copy library system to a digital multi-media information source available at any time, to everyone, at nearly zero cost.

Have these advances in technology changed society? It is believed that they have had a profound beneficial effect. Farmers produce an overabundance of food with a small number of personnel working in agriculture. Factory workers manufacture most of our consumable products with less than 20% of our work force. The workweek is shorter, and we have many more affordable products. Most families own at least two cars and home ownership is high. Concerns are expressed regarding the improvements in the standard of living over the past decade, particularly for those people at the lower end of the income spectrum. Technology has probably not helped this group as much as others. Global trade has permitted many of the semi-skilled tasks to be moved to third world countries, where the cost of labor is a small fraction of the labor costs in developed countries. Consequently, many of the previously well-paid factory jobs in manufacturing in the U.S. have been moved offshore and this process has negatively affected those with limited marketable skills.

1.5   GREATEST ENGINEERING ACHIEVEMENTS—20TH CENTURY

The National Academy of Engineering (NAE) working with 27 professional engineering societies complied a listing of the 20 greatest achievements in engineering during the 20th century. What made these achievements great? They are notable because of the very significant benefits of each accomplishment to all members of society. An excellent publication outlining the engineering achievements that transformed our lives is presented in Reference [7].

1. Electrification—in the 20th century, widespread electrification gave us power for cities, factories, farms, and homes and forever changed our lives. Thousands of engineers made it happen, with innovative work in fuel sources, power generating techniques and transmission grids. From streetlights to supercomputers, electric power makes our lives safer, healthier and more livable.
2. Automobile—perhaps the ultimate symbol of personal freedom, the automobile is a showcase of engineering ingenuity and the world's major transporter of people and goods.
3. Airplane—with its speed and efficiency, air travel makes personal and cultural exchange possible on a global scale.
4. Water Supply and Distribution—water is our most precious resource, and improvements in its treatment, supply and distribution have changed our lives profoundly, virtually eliminating waterborne diseases in developed countries and providing clean and abundant water for communities, farms and industries.
5. Electronics—electronics provide the basis for countless innovations including digital cameras, iPods, iPhones, TVs, and computers to name only a few. Electronic products improve the quality and convenience of modern life.
6. Radio and Television—radio and television were major agents of social change in the twentieth century, entertaining millions, expanding the horizons of those living in remote areas of the world and recording history making events.
7. Agricultural Mechanization—tractors, combines and hundreds of other agricultural machines increase the productivity of farms, reduce the need for manual labor and improve our ability to feed our citizens and others in the world.

8. Computers—perhaps the defining symbol of 20th century technology, the computer has transformed businesses and lives around the world, increased our productivity and opened up access to vast amounts of knowledge.

9. Telephone and Cell Phone—a cornerstone of modern life, the telephone brings the human family together and enables the communications that enhance our lives, industries and economies.

10. Air Conditioning and Refrigeration—air conditioning and refrigeration make it possible to transport and store fresh foods and adapt the environment to human needs. Once luxuries, they are now common necessities which greatly enhance our quality of life.

11. Highways—vast networks of roads, bridges and tunnels connect our communities, enable goods and services to reach remote areas and encourage economic growth.

12. Spacecraft—the human expansion into space is perhaps the most awe-inspiring engineering achievement of the 20th century. The development of spacecraft has thrilled the world, broadened our knowledge of the universe and led to thousands of new inventions and products.

13. Internet—in a few short years, the Internet has become a vital instrument of social change, transforming business practices, educational pursuits and personal communications. By providing global access to news, commerce and vast stores of information, the Internet brings people together and adds convenience and efficiency to our lives.

14. Imaging—from tiny atoms to distant galaxies, modern imaging technologies have expanded the reach of human vision with tools such as X-rays, telescopes, radar and ultrasound.

15. Household Appliances—household appliances have greatly reduced the labor involved in everyday tasks, giving us more free time and enabling more people to work outside the home.

16. Health Technologies—medical imaging devices, artificial organs, replacement joints and biomaterials are a few of the advanced health technologies that improve the quality of life for millions.

17. Petroleum and Petrochemical Technologies—petroleum and petrochemicals provide fuel for cars, homes and industries and the basic ingredients for products ranging from aspirin to zippers.

18. Laser and Fiber Optics—lasers are used in supermarket scanners, surgical devices, satellites and other products. Today fiber optics provides the infrastructure to carry huge amounts of information, spurring a growing revolution in telecommunications.

19. Nuclear Technologies—the harnessing of the atom led to dramatic changes in the 20th century by providing a new source of electrical power, improving techniques for medical diagnosis and treatment, and transforming the nature of war forever.

20. High-performance materials—from the building blocks of iron and steel to the latest advances in polymers, ceramics and composites, engineering materials are used in thousands of important applications.

1.6   <u>NEW UNDERSTANDINGS</u>

In this discussion of the historical development of technology, we have been optimistic about the very positive effects of technology on society. In the past the public accepted, without serious questions, all forms of technology and the risks that were involved. However, the situation has changed to a remarkable extent in the past 40 years. Our leadership (engineers included) has, to a large degree, lost the public trust. An excellent example of this loss pertains to the technology employed in the generation of electricity using nuclear power. Since the Three Mile Island incident that occurred in Pennsylvania in 1979, the general public in the U.S. has opposed nuclear power until very recently. There is a valid public perception that the technological risks associated with nuclear power were significantly understated. John O'Leary, a deputy director for licensing at the Atomic Energy Commission (AEC), stated that "the frequency of serious and potentially catastrophic nuclear incidents supports the conclusion that sooner or later a major disaster will occur at a major generating facility" [8]. This statement was made several years prior to the accident at Three Mile Island. In spite of this warning of the likelihood of a serious accident, federal, state or local governments have not planned for orderly emergency evacuation of nearby residents or for containment measures to limit the possible contamination from an accident.

The Chernobyl accident, which occurred in the former Soviet Union in 1986, proved that O'Leary was correct. The No. 4 reactor at the Chernobyl facility exploded sending a radioactive plume in the air so high that an alert nuclear plant operator in Sweden detected it. Many died in this radioactive explosion and many more have died since then from radiation poisoning.

In recent years many environmental organizations who once opposed nuclear energy have recognized that generating base load electricity with nuclear energy is preferable to its generation by coal. Nuclear power plants do not emit CO_2 that is believed to contribute to global warming. Also new designs of nuclear reactors have incorporated fail safe features to prevent a loss of cooling accident.

The nuclear industry and the government agencies regulating it, world wide, have not been candid with the public. Nor has the public been very interested in an intelligent debate—resorting instead to large, ugly demonstrations to express their displeasure. The public needs an accurate and honest assessment of the risks of any nuclear power project by the industry, the regulators and experts. The risks must be weighed against the benefits to society.

The purpose here is not to argue the case for or against nuclear power. Nuclear energy is used to illustrate the fact that understating or poorly assessing risk brings a severe penalty—loss of public trust followed by a ban on technical development in the subject area. As engineers, we must significantly broaden our perception regarding technological risk. Today, many engineers dismiss failure due to human, operator or pilot error because they are not machine errors. This attitude is **absolutely wrong**. The **human operator and the machine** constitute a system and affect public safety accordingly. The machine and its operator must be included in system design since they both constitute an interacting set of weaknesses and capabilities [9].

Another new understanding pertains to geography. Prior to the Second World War (WW–II), international trade was very limited. As a country we produced what we consumed, mined our own minerals and pumped our own oil. Geography was very important because it constrained trade by law and by distance. After the war, laws were changed and trade agreements (GATT, NAFTA, etc.) that lowered or eliminate tariffs were arranged. International trade was encouraged so much so that the U.S. now consistently imports more goods and materials than it exports. The annual deficit in the trade balance for the U.S. goods and services was $540.4 billion in 2012, down from $559.9 billion in 2011. This deficit represents the equivalent to more than 12 million high-income jobs paying $45,000 a year.

Previously, the cost of shipping limited trade among countries separated by large distances. However, following WWII many technological improvements were implemented in the shipping business: the super tanker was developed—greatly improving the efficiency of shipping huge quantities of oil. Also very large containerized ocean vessels were developed reducing the cost and time of loading and unloading cargo. Large efficient diesel engines were installed on the ships permitting increased speed and significant reductions in shipping time and costs.

With the time and costs of shipping greatly reduced and the legal barriers to trade removed, the country's commerce is conducted in a global marketplace. Engineers compete worldwide to develop world-class products or services. Products designed in the U.S. may be produced anywhere in the world, whereas products designed in Europe or Asia may be produced in the U.S. Today, design and production activities are located to minimize costs and to maximize benefits to the customers and/or to the corporate entities.

The third new understanding is in the role of the government (federal, state, county and city). Prior to WWII, governments were small and taxes were relatively low. The primary role of the federal government was to insure national security. State governments worried mostly about transportation infrastructure, and the local governments concerned themselves with education. Today, the situation is markedly different. Governmental agencies, at all levels and in response to their interpretation of the law (either old or new), issue regulations which pertain to topics of concern to society including the protection of the environment, safety, health and energy conservation.

You can debate the wisdom of many of the regulations, but that is not the issue. The regulations currently exist and new ones will continue to be issued at an alarming frequency. As an engineer, you must be prepared to serve society-at-large as reflected by these government regulations. If you question the wisdom of some of the regulations, the best approach is to become involved in the political process so that you may have the opportunity to influence them. Engineers tend to resist constraints because they limit the freedom of their designs and add to the cost of the product. You may think regulations are government imposed constraints, but they are really barriers imposed by society-at-large. Society controls the regulatory process because it has the voting power to change the politicians responsible for the laws and of the bureaucrats formulating the regulations.

1.7 BUSINESS, CONSUMERS AND SOCIETY

Engineers serve three constituencies—business, the customer and society-at-large. Business leaders (management) look to engineers to develop products, services and processes that meet global competitive demands. Consumers seek more convenient, reliable, enhanced and value-laden products at reduced prices. Society-at-large, through elected politicians, trial lawyers and public interest groups demand action leading to solutions of problems involving safety, health, energy conservation and improving the environment. It is an overlapping set of constituencies, because society encompasses business, governments and consumers. To discuss the issues raised by serving three constituencies, consider the U.S. automotive industry as an example. Marina Whitman [10] has described many aspects regarding the three-way demands.

Business demands are simple although they are often difficult to meet. Any business must be profitable to continue to exist. Losses can occur over the short term, but if the red ink is prolonged the company is forced by its creditors into bankruptcy. Making a well-defined profit every quarter is the usual goal of management. For many years following WWII, it was relatively easy for U.S. businesses to make money. Bombing and shelling during the war had destroyed most of the manufacturing capability and a significant part of the infrastructure in every developed country in the world except in the U.S. It was easy to be a leader, when the competition was without factories and adequate infrastructure. For the automobile industry, the golden years began to erode in the late 1960s, and survival became a life or death battle in the 1980s. Problems for the auto industry came from global competition, which continues to this day.

To show the extent of the market penetration by imported automobiles, recall that the imports accounted for only 4% of the market in 1962, but they controlled more than 50% of the market by 2008[2]. This devastating loss of market share by US car makers was not limited to the auto industry; several other industries including optics, consumer electronics, ship building, steel, etc. suffered even more serious losses.

Engineers and management in U.S. firms had lost the competitive edge after decades of modest competition from the Europeans. The Japanese automobile companies (Toyota, Honda, Nissan, etc.) provided more affordable, reliable, appealing, fuel efficient and value-laden cars than the American or European companies. In June of 2009 both General Motors and Chrysler Corporations declared bankruptcy. With loans from the federal government the two companies emerged from bankruptcy, but with fewer plants and employees. In recent years, both companies have improved their product lines and the quality of their products; however, regaining the market share of the 1960s will be a very long and difficult process due in part to high legacy costs, which still exist.

---

[2] The impact of the imported automobiles on the economy was mitigated by establishing assembly plants in the U. S. by foreign firms such as Honda, Toyota, Nissan, BMW, Volkswagen, etc.

The second constituency, the consumers, has been well recognized in the past decade. The importance of keeping the customer pleased is foremost in the minds of both management and engineering. Systematic methods for establishing the consumers' needs have been developed and placed into practice [11]. Many companies in the U.S. currently follow a product development process that begins with the customer interviews and proceeds through all of the phases of product development. The newer methods [11, 12] tend to insure that the customers' requirements are more accurately defined by the product specification before beginning a development.

The third constituency, society-at-large, generates the most challenging requirements. Society-at-large is represented by government agencies and a variety of public action groups. Officials of the government agencies (bureaucrats) produce regulations that define requirements for automobiles that may affect safety, fuel conservation or the environment. The public action groups create pressure and influence politicians in setting public policy that eventually generates new requirements.

Engineers often have some difficulties with regulations and/or proposals from the public action groups. Lawyers write the laws and public servants write the regulations. Because lawyers and public servants are often technically illiterate, they sometimes seek technical support prior to drafting binding regulations. However, government bureaucrats sometimes place regulations into effect prior to a thorough study by knowledgeable representatives of the public and engineering communities. The difficulty with this procedure is that the demands of the customer, including style, comfort, affordability, reliability, performance, convenience, etc., may be in serious conflict with societal imposed regulations on safety, fuel efficiency, emissions, etc.

The interaction of the government agencies representing the public-at-large, business and the consumer is woefully inadequate. The public and business needs solutions often in a relatively short time. Yet, the regulatory system is too cumbersome to provide rapid solutions. The awkwardness comes from inflexibility on the part of the government and inadequate cooperation between the automakers. Competition on the part of business to increase market share is understandable, but safety features should be excluded from the normal competitive secrecy. The government agencies must become much more flexible and recognize that they cannot decree away all danger.

The difficulties between the government and business are not new. An interesting summary of the regulatory relationship between government and business was recently published by Jasanoff [14]. She is quoted below:

> Studies of public health, safety and environmental regulation published in the 1980s reveal striking differences between American and European practices for managing technological risks. These studies show that U.S. regulators on the whole were quicker to respond to new risks, more aggressive in pursuing old ones and more concerned with producing technical justifications for their actions than their European counterparts. Regulatory styles, too, diverged sharply somewhere over the Atlantic Ocean. The U.S. processes for making risk decisions impressed all observers as costly, confrontational, litigious, formal and unusually open to participation. European decision making, despite important differences within and among countries, seemed by comparison almost uniformly cooperative and consensual, informal, cost conscious and for the most part closed to the public.

The assessment by Sheila Jasanoff gives us clear direction. It is important to reduce the adversarial relationship between business and government. With engineers serving both government and business, we must work to produce a regulatory system with more flexibility, cooperation, cost consciousness and consensus.

1.8   CONCLUSIONS

The standard of living in the U.S. will be determined by the interplay of three powerful influences:

- New and rapid technological advances.
- Business (management) response to global consumer demands.
- Social demands as evidenced by legislative and regulatory requirements.

Engineers have always provided the leadership in technological advances, and they are expected to continue to provide business and the public-at-large with cutting-edge, world-class technology.

Business requires capital, technology and astute management to remain competitive. In addition to providing the technological base for a company, engineers frequently serve in management. If a management career path appeals to you, plan on extending your education in a business school. The combination of a Bachelor of Science degree in engineering and a Master of Science in business provides a very solid foundation for a career path in a technically oriented business.

In the past, engineers usually have not been actively engaged in societal issues. They lack patience in dealing with the public and become irritated by the public's lack of technical literacy. They fume at the inefficiencies of government agencies and their lack of flexibility. They become outraged at the arrogance of bureaucrats and at their lack of concern for time and costs. They withdraw from public forums and focus on new developments and products. In the meantime, the public-at-large grows wary of many important sociotechnical systems and the government resorts to litigious processes that impede real progress.

It is unfortunate that engineers have not been a significant force in dealing with societal issues. When society, business and the customer create conflicting demands, engineering expertise is essential in crafting well-balanced solutions. The conflicting demands provide new opportunities and complex challenges for engineers. To take advantage of these opportunities, engineers of tomorrow will require enhanced communication skill, greater disciplinary flexibility, a better understanding of the mechanisms of regulatory agencies and a much broader perspective relative to societal demands.

REFERENCES

1. Friedman, T., The World is Flat, Farrar, Straus and Giroux, New York, NY, 2005.
2. Singer, C. E. J. Holmyard, A. Hall, and T. Williams, Eds. A History of Technology, Oxford University Press, New York, NY, five volumes, 1954-1958.
3. Hughes, T. P., "From Deterministic Dynamos to Seamless-Web Systems," Engineering as a Social Enterprise, ed. Sladovich, H. E., National Academy Press, Washington, D. C. 1991, pp. 7-25.
4. Sladovich, H. E., ed. Engineering as a Social Enterprise, National Academy Press, Washington, D. C. 1991, pp. 7-25.
5. White, L., Jr., Medieval Technology and Social Change, Oxford: Clarendon Press, NY, 1962
6. Kirby, R. S., S. Withington, A. B. Darling, and F. G. Kilgour, Engineering in History, Dover Publications, New York, NY, 1990.
7. Constable, G. and B. Somerville, A Century of Innovation: Twenty Engineering Achievements that Transformed Our Lives, Joseph Henry Press, Washington, D. C., 2003.
8. Ford, D. F., Three Mile Island: Thirty Minutes to Meltdown, Viking, New York, NY, 1982
9. Adams, R., "Cultural and Sociotechnical Values," Engineering as a Social Enterprise, ed. Sladovich, H. E., National Academy Press, Washington, D. C. 1991, pp. 26-38.

10. Whitman, M. v. N., "Business, Consumers, and Society-at-Large: New Demands and Expectations," Engineering as a Social Enterprise ed. Sladovich, H. E., National Academy Press, Washington, D. C. 1991, pp. 41-57.

11. Clausing, D. Total Quality Development: World-Class Concurrent Engineering, ASME Press, New York, NY, 1994.

12. Schmidt, L., et al, Product Engineering and Manufacturing, 2nd Edition, College House Enterprises, Knoxville, TN, 2002

13. Payne, H. "Misguided Mandate," Scripps Howard News Service, Knoxville News-Sentinel, January 12, 1997, p. F-1.

14. Jasanoff, S. "American Exceptionalism and the Political Acknowledgment of Risk," Daedalus, Vol. 119, No. 4, 1990, pp. 61-81.

EXERCISES

1.1   Consider a product that you or your family has purchased in the past month or so. Write a brief paper describing both the positive and negative impacts of that product on society.

1.2   The Federal Aviation Administration enacted a regulation affecting the mailing of packages weighing more than 12 ounces. The regulation requires you to present the package to a postal clerk for mailing and prohibits the mailing of the package from a postbox. Write a paper covering the following issues:

- Describe the regulation in more detail.
- Why is the FAA writing a regulation affecting the U.S. Post Office?
- Do you believe that the regulation will be effective for its intended purpose? Please give arguments supporting your viewpoint.
- What actions will the post offices (40,000 of them) have to take to make the regulation effective? What actions do the post offices actually take with regard to the regulation?
- Will these actions be costly? Estimate the costs to both the public and the individual. Assume that a postal clerk is paid $20/h and that an individual considers his or her free time worth $20/h.
- Give your assessment of the cost to benefit ratio for this regulation?

1.3   Horses were not used extensively to relieve man from brutal work or to provide a significant advantage in military actions until nearly 1,000 AD. Write a brief paper describing both the social and technical reasons for the very long time required to effectively implement the use of horses in either military or commercial endeavors.

1.4   Write a paper comparing the lifestyles, as you imagine them, for a man or woman living today and living in the year 1,200 AD. Did technology make a difference in the quality of life?

1.5   Find the Website for the National Academy of Engineering and click on Greatest Achievements for the 20th Century. Select one of the 20 achievements and prepare a paper describing the history of that particular achievement.

1.6   Find the Website for the National Academy of Engineering and click on Greatest Achievements for the 20th Century. Select one of the 20 achievements and prepare a paper describing the time line for that particular achievement.

1.7   Explain why the public-at-large was opposed to power generation with nuclear energy prior to 2003? Was the public-at-large correct in their collective assessment? Explain why you formed this opinion.

1.8   In recent years the public's perception of nuclear power is slowly changing because of the publicity associated with global warming. Explain why nuclear power is slowly gaining acceptance because of the fear of global warming?

1.9   Why does a global marketplace exist today? Does global trade improve our standard of living? What does our current trade deficit have to do with wages for factory workers? Does technology help or hinder the balance of trade deficit?

1.10   Describe our immigration laws as they exist today. Are they being enforced? If not why not? What is your opinion about illegal immigration? Does it affect our unemployment rate? What do you believe the country should do about the 12 million illegal immigrants already residing in the U. S.?

1.11   What social science courses should you take during your undergraduate program to broaden your prospective and enhance your understanding of socio-technical issues?

CHAPTER 2

THE ENGINEERING PROFESSION

2.1   WHY ENGINEERING IS A PROFESSION

You may consider engineering from two different viewpoints—as a course of study pursued in an accredited college of engineering and as an occupation. Let's discuss these two different viewpoints, because they are both important in establishing engineering as a profession.

As a student in an engineering college, you are presented with a structured curriculum. You are required to take many credit hours of mathematics, chemistry, physics and perhaps biology in your first two years of study. Mixed with mathematics and science courses are several additional courses in engineering science. These are essentially applied science courses taught by engineering instructors. In the last two years of the curriculum, many discipline oriented engineering courses are required. These provide the basic analytical methods necessary for success after graduation, when you begin to practice engineering. Also included in the final two years are courses intended to provide realistic design experiences related to product development. This schedule of courses is designed to prepare you to practice engineering immediately upon graduation. Near the conclusion of your studies, you will be presented with an opportunity to take the first of two examinations leading to a professional license to practice engineering—the Engineering Fundamental Examination. It is advisable to take this examination before graduation because it provides you with an opportunity to test your analytical skills. It is also an important attribute that you can add to your resume. Passing this examination is a certification of your fundamental analytical skills.

Your ability to practice engineering immediately upon graduation is the primary reason for such a highly structured curriculum containing so many technical courses. The on-the-job training opportunities available to you in your first position depend upon the policies of the company that is paying your salary. Some large firms provide a transition period where you serve as an engineer trainee with close supervision and rotate from one division to another to provide an overview of the company's operations. However, if you take a position with a smaller corporation, usually there is little formal training, and you are expected to earn your salary as soon as you begin working.

In addition to at least three years of mathematics, science, engineering science and engineering courses, you will be required to take several courses in other colleges on campus. The university has general education requirements, which dictate that all students become familiar, if not proficient, with basic premises in the social sciences, fine arts and the humanities. The engineering profession endorses this out-of-college exposure, because it broadens one's perspective and encourages important assessments of contemporary issues.

Let's now consider the second viewpoint—engineering assignments undertaken after graduation. Whether you will be working as an engineering professional depends to a large degree upon the position you accept. Some engineering graduates are offered positions with investment banking and brokerage houses at very attractive salaries. Managers at brokerage houses like the analytical skills developed in engineering programs and find that engineering graduates perform well in assessing investment opportunities. These graduates pursue an interesting and lucrative career, but they are not practicing engineering.

At the other extreme, a graduate of a civil engineering program takes a position with a construction firm that designs and builds highways, bridges and large public buildings. In this position, public safety is an important consideration and licensing is essential because State laws require it. The graduate is expected to pass the Engineering Fundamental examination, and then—after several years to gain practical experience—he or she will take the Principles and Practice of Engineering examination to become licensed. During the period between the two examinations, the individual works with licensed engineers, who supervise his or her work and certifies it to be accurate and correct.

For those graduates working in industry, licensing is usually less of an issue. Most corporations have a small staff of licensed engineers who certify work when certification is required. However, most of the engineers are working on products where public safety is not an issue. In these cases, professionalism is not a matter of licensing, but it is more a matter of attitude. Professionals are usually salaried and their remuneration is independent of the hours worked. They do not punch a time clock and they are expected to devote the time necessary to finish a task on schedule. Engineers assume responsibility and are committed to the project and to a development team. They cooperate with management, and freely share ideas and concepts to advance the welfare of the corporation and their division. They may decide to move from one corporation to another, but loyalty and respect are important considerations even after they switch employers.

2.2   CHARACTERISTICS OF ENGINEERING STUDENTS

Students entering colleges of engineering today are bright, with average Scholastic Achievement Test (SAT) scores ranging from 1,800 to 2,100 in most colleges in the U. S. In fact, the higher ranked engineering colleges attract students with average scores of 2,000 or above. In addition, engineering students are usually in the top 5 or 10% of their high school graduating class with grade point averages ranging from 3.8 to 4.0+. Student applications to the admissions office of the university often include strong letters of recommendation from their teachers, counselors and principals certifying the student's stellar abilities.

It is well recognized that the abilities of entering engineering students in mathematics is exceptional. SAT scores for the mathematics portion of the examination approach or exceed 700 for most students. What is less well recognized is that engineering students also have excellent verbal skills. With average SAT scores for verbal skills of exceeding 600, engineering students usually rank above their peers in the colleges of arts, humanities and social sciences.

Most engineering students are men; however, 19.9% of the bachelor degrees in engineering were awarded to women in the 2013-2014 academic year. The percent of master and doctoral degrees award to women in 2013 were 24.2% and 22.2%, respectively. The interest of women in engineering depends on the discipline, as shown in Table 2.1. Statistics show that women are attracted to chemical, biomedical, environmental and industrial engineering disciplines, but not to petroleum, electrical, computer or mechanical engineering programs.

The geographic representation of students in a class depends strongly on whether the university is state supported or privately funded. State universities often restrict enrollment of out-of-state students to a small proportion of the class. A tuition differential is also imposed making it more expensive for out-of-state students to attend. Consequently, undergraduate classes in state universities tend to be much more homogeneous with students representing the backgrounds and cultures of the region. Privately funded universities usually do not have such restrictions and often vigorously seek students with diverse backgrounds from different states and countries.

Table 2.1
Percentage of BS Degrees awarded to women by discipline (2014)

Discipline	Percent	Discipline	Percent
Biomedical	40.6	Nuclear	16.0
Chemical	36.3	Computer Science Outside Engineering	13.2
Agriculture & Biological	33.7	Aerospace	13.8
Industrial or Manufacturing	31.6	General Engineering	26.3
Engineering Management	21.1	Computer Science Within Engineering	13.7
Metallurgical and Materials	29.8	Petroleum	15.9
Mining	11.7	Electrical/Computer	13.6
Architectural	26.0	Mechanical	13.5
Engineering Science Engineering Physics	17.4	Computer Engineering	12.0
Civil	19.4	Environmental	48.0
Civil/Environmental	27.3	Electrical	13.7

Clearly, engineering students are well prepared academically to begin their studies. They have proven themselves in high school as scholars and class leaders. They have gained the respect of their teachers, peers and family. However, the retention rate for engineering students is appalling. Nearly one half of the entering class of well-qualified students either fail-out or withdraw from the program with most leaving in the first two years. While the exact retention rate varies from college to college across the U. S., the problem is widespread. Why then do so many students leave the engineering program without completing the requirements for a B. S. degree in one of the engineering disciplines? Answers to this perplexing question will be explored in Chapter 3.

2.3 SKILL DEVELOPMENT

Pursuing a successful career in engineering is strongly dependent on your ability to develop a number of important skills. Of course, you enter the program with many skills, but it is necessary to enhance them and to add new ones to your repertoire. The engineering curriculum is designed to develop these additional skills and to markedly improve your current skills. As you proceed through the program, you will find that your ability to perform challenging engineering tasks improves and expands each year.

The Accrediting Board for Engineering and Technology[1] (ABET) has developed a list of requirements for all of the disciplines offered in the accredited colleges of engineering in the U. S. Each department in the college must demonstrate to a team of ABET visitors (every six years) that their graduates have the skills and knowledge listed in Table 2.2 to maintain the accreditation of each engineering program for which a B. S. degree is offered.

---

[1] The Accreditation Board for Engineering and Technology, Inc., 111 Market Place, Suite 1050, Baltimore, MD 21202, certifies all programs (not departments) leading to accredited engineering degrees in the U. S.

Table 2.2
Listing of knowledge and skills that engineering graduates must demonstrate.
ABET Criteria 2000

(a) Ability to apply knowledge of math, engineering, and science
(b1) Ability to design and conduct experiments
(b2) Ability to analyze and interpret data
(c) Ability to design a system, component or process to meet needs
(d) Ability to function on multi-disciplinary teams
(e) Ability to identify, formulate and solve engineering problems
(f) Understanding of professional and ethical responsibility
(g) Ability to communicate effectively
(h) Broad education necessary to understand the impact of engineering solutions in a global and societal context
(i) Recognition of the need for, and the ability to engage in life-long learning
(j) Knowledge of contemporary issues
(k) Ability to use techniques, skills, and tools necessary for engineering practice

Before examining this list, let's consider a somewhat different set of skills and abilities for engineering graduates prepared by a study committee for the American Society of Mechanical Engineers (ASME). The ASME listing [1], in priority order, is presented in Table 2.3.

Table 2.3
Skills considered important for new Mechanical Engineers
With B. S. degrees, Priority ranking

1. Teams and Teamwork	11. Sketching and Drawing
2. Communication	12. Design for Cost
3. Design for Manufacture	13. Application of Statistics
4. CAD Systems	14. Reliability
5. Professional Ethics	15. Geometric Tolerancing
6. Creative Thinking	16. Value engineering
7. Design for Performance	17. Design Reviews
8. Design for Reliability	18. Manufacturing Processes
9. Design for Safety	19. Systems Perspective
10. Concurrent engineering	20. Design for Assembly

As you explore both of these lists, the emphasis on skill development in colleges of engineering should become evident. The importance of these skills, as determined by two major engineering societies, should guide your efforts in achieving the proficiencies necessary to insure a successful career.

Design

Most engineering graduates join corporations and participate in some way related to either product or process development. If it is a manufacturing company, the product may be a vacuum cleaner, a lawn mower, an automobile or some other item. The point is that you will design, manufacture, assemble, ship and service an array of products. Design is essential to the success of product development. If the design is flawed, the product or service offered will fail. The company and your career may suffer. If the design is optimal, the product will be profitable, the company will prosper and your career will advance. This is an over simplified viewpoint, but one that has considerable merit.

Design is perhaps the most difficult of the engineering skills to acquire because it is as much an art as a science. While there are well-defined design procedures that you will learn (some of them introduced in this book), one often intuitively senses the good and bad elements of a design. Also, experience is important, because design is a complex activity that incorporates many different tasks that cross discipline lines. Examine Table 2.3 and note the various descriptors ASME used with the word design. For example, engineers design for reliability, for performance, for safety, for cost, for manufacture, and for assembly and one could add "ease of maintenance" and others to the list.

As you proceed through the engineering program, the opportunities you have to practice design will be severely limited. The emphasis in engineering education is on developing analysis methods used to predict reliability and performance of products and systems. Some engineering programs are limited to a single capstone design experience in your final year of study. Other programs have two or three design courses where you will have the opportunity to design a new product or to modify the design of an existing product or process. Engineering programs with more than three design courses (a total of nine credit hours in a program of about 128 credit hours[2]) are rare in the U. S.

Many colleges of engineering participate in design competitions that are sponsored by the professional societies. These competitions involve the design of robots, large model airplanes, automobiles, steel bridges, and human and engine powered vehicles, etc. You are encouraged to become involved in one or more of these projects, because they provide you with an opportunity to participate on an interdisciplinary design team and to gain valuable experience in the design process. Usually you can obtain credit towards your graduation requirements for this design competition by registering for a technical elective, which has been established by the faculty member in charge of the design project.

Another method for gaining design experience is by participating in the co-op engineering option or working off campus. Many students work to partially fund their expenses while pursuing their education. If you are able to find a position that involves design, manufacturing or assembly, the experience gained is probably worth more than the salary you earn. Try to obtain a position in a company that produces products rather than with a retail organization. While experience in either type of organization is valuable, the production experience is more applicable to engineering.

Teamwork

In this course, you will be required to participate on a development team. There are three reasons for this requirement. First, the project is too ambitious for an individual to complete in the time available. You need the collective efforts of the entire team to develop a vehicle during a single semester. The development teams will be pressed time-wise to complete this project on schedule.

Second, you must begin to learn teamwork skills. Experience has shown educators that most students entering the engineering program are seriously deficient in team skills. From elementary through high school, the educational process has focused on teaching you to work as an individual, often in a setting where you competed against others in your class for the best grades. However in engineering, you are expected to function as a team member where cooperation, following and listening are as important as individual effort. Leadership is important in a team setting, but cooperation and following the lead of others are critical elements for successful team performance.

The final reason for participating on a development team is to better prepare you for the real world that you will enter upon graduation. You will probably be assigned to a development team very early in your career, if you take a position in industry. A recent study by ASME, the results of which are shown in Table 2.3, ranked teamwork as the most important skill to develop in an engineering program. Teamwork was also the first skill, in a list of 20, considered important by managers from industry.

---

[2] The number of credit hours required varies from engineering program to program and college to college. The 128 credit hours cited here is the typical requirement for many programs.

Hopefully, this course will be instrumental in exposing you to team working skills so necessary for a successful career. A detailed discussion of teamwork skills is presented in Chapter 6.

Communication

Communication skills are vitally important in every aspect of your life. In your personal life you must be able to accurately convey your thoughts to your family, friends and peers. Professionally, it is just as important to communicate effectively. A great idea is of little value if you cannot express it with sufficient clarity for it to be accepted by management or your associates. With family and friends, most of your communication is by informal conversation. Occasionally, you write letters but more frequently you will text, e-mail or use Facebook or Twitter to communicate. You can be casual in communicating with friends and family because they are accommodating and overlook your shortcomings. You must be much more careful in your professional communications. Messages must be clear and unambiguous. The information conveyed must be accurate and timely. The presentation must be a concise and to the point; long rambling prose is not appreciated.

Three different modes of communication are used in engineering—**writing, speaking and graphics**. All three are important and engineering versions of all three modes of communication different to some degree from commonly accepted methods.

It is important that you enjoy writing, because engineers often have to prepare several hundred pages of reports, theoretical analyses, memos, technical briefs and letters during a typical year on the job. Advancement in your career will depend on your ability to write well. You will be taking several courses offered by the English Department and by departments in Social Sciences, Arts and Humanities that require writing assignments. These courses will help you with the structure of your composition and the development of good writing skills. Most of the assignments in these courses will be to write essays, term papers, or to study selected works of literature. However, there are several differences between writing for an engineering company and writing to satisfy the requirements of courses such as English 101 or History 102. These differences will be described in a Chapter 8 titled **Technical Reports**.

The design briefing is important to both the product development process and to your career. Information about the product must be effectively transmitted to your peers, management and others involved with the project. Clear messages that accurately define problems, which the development team will address, are imperative. On the other hand, ambiguous messages are often misunderstood, hinder the definition of the problem and lead to delays in implementing solutions. The design briefing provides an opportunity to review the status of a specific product development. It also permits peers to share their ideas with you, and affords management an open forum for assessing the quality of your work and the progress made by your development team. Because the professional presentation is critically important, Chapter 7 titled **Design Briefings** has been included in this textbook. This chapter describes some valuable techniques for properly delivering your message to different audiences—strangers, peers, team members and management representatives.

Engineering graphics is the most important method of communication for presenting design concepts and details. When attempting to communicate design ideas, you will find writing and speaking insufficient to express your thoughts and concepts. A more visual technique for communicate is imperative. It is much more effective to present your ideas by means of drawings, sketches, pictures, and different types of graphs. Visuals aids, such as drawings and photographs, convey your ideas quickly and with remarkable accuracy.

There are two general approaches used in preparing drawings and graphs. The first is a manual approach where drawings and graphs are prepared by hand using a few simple drawing instruments. The second utilizes a computer and suitable software programs that greatly facilitate the preparation of drawings or graphs. A discussion of engineering graphics is provided in Chapter 9. CAD tutorials are also available that deal with Creo Parametric, SolidWorks and Autodesk Inventor.

Design Analysis

In engineering, you will attempt to predict the performance of a product prior to its final design and fabrication. Obviously, you would not want to design and construct a bridge to have it fail after a few months or years in service. You must be able to predict with confidence that the bridge will not fail over its entire design life (perhaps a hundred years or more). Constructing analytical models and performing analysis is essential in making accurate predictions regarding performance. In modeling, you reduce your structure into a number of different components or subsystems that are amenable to analysis. You then apply analytical methods that enable you to accurately determine performance parameters for each model. The analytical results permit verification of the adequacy of the design. As such, analysis and design are coupled. You usually begin with a design concept; then you subject this design to an analysis. The results of the analysis are then used to improve this design enhancing its performance or its reliability.

A significant portion of the engineering curriculum is devoted to developing your analytical skills. About one year, of the four-year curriculum, is devoted to courses in mathematics and science that provide the foundation for engineering courses. In most colleges of engineering, at least another year is devoted to engineering and engineering science courses where analytical and computational methods are taught. In a typical engineering program, your analytical and computational skills are honed more than any other skill listed in either Table 2.2 or 2.3.

Your performance (read this as grade point average, GPA) will depend on your ability to solve problems posed on hourly or final examinations. These problems are crafted to test your understanding of a few engineering principles[3]. The mathematics involved is usually not overly complex and the physics entailed is even more straightforward. You will be provided with several useful tips on problem solving methods in Chapter 3.

Experimental Analysis

In predicting the performance of a structure or machine component, you will often conduct experiments and measure the performance parameters. While it is usually more expensive to perform experimental studies than analytical ones, the costs are warranted when you are not certain that the analytical modeling accurately reflects reality. Sometimes converting the physical reality of a product to an analytical model requires many assumptions. You will often perform carefully designed experiments to verify these assumptions and to insure the successful performance of the product. Extensive tests on almost all products are performed before they are released to the market. The purpose of these tests is to insure safety of the product, and to provide management assurance that the product will meet the guarantees specified in the warranty pertaining to performance, life, durability and safety. In this class, you will be required to test a prototype of an autonomous vehicle that your team has developed. The test is to determine if the prototype's performance meets its specification.

There are four different phases of an experimental program. The first phase is to design an experiment so that it provides the answers required to verify the modeling or to insure the adequacy of a design. The second phase is to conduct the experiment using appropriate instrumentation that will provide accurate and appropriately timed measurements of the controlling performance parameters. The third phase is to analyze the data obtained using suitable statistical methods that predict and bound experimental errors. The final phase is to correctly interpret the data to verify the adequacy of a design or to provide information that may be used to enhance the design.

---

[3] In practicing engineering, you will find that analyses are based on only a dozen or so engineering principles. The methods developed in engineering courses may appear complex, but they are based on a small number of fundamental concepts.

The laboratory classes offered in conjunction with chemistry and physics courses are not designed to provide you with skills for designing experiments or for making measurements. These are highly structured laboratory experiences intended to reinforce theories and principles introduced during lecture. The curriculum of many of the engineering disciplines provides a course on electrical circuits and instrumentation. This course introduces linear circuit theory and gives some information about sensors, transducers and both analog and digital instrumentation. A course of this type is strongly recommended because it provides some of the basic information needed to plan, design and execute experiments.

You are encouraged to enroll in other courses that provide additional exposure to experimental methods. This may be possible with a careful selection of technical electives that are available in the third and fourth year of your program. You might also consider an assignment working in a laboratory directed by one of the professors in your department. Many professors lead significant research programs, and in conducting these studies, they often perform new and novel experiments. The experience gained in such a setting is often more valuable than that obtained in a traditional engineering measurements course.

Understanding Professional and Ethical Responsibilities

When you graduate and begin practicing engineering, you will be expected to follow a professional code of ethics. An example of the ABET code of ethics is presented in Chapter 4. While there are many other codes, each endorsed by a professional society, they are similar. Basically the codes require you to uphold and advance the integrity, honor and dignity of the engineering profession. The codes then list fundamental principles and canons that should govern professional behavior. The fundamental principles advanced by ABET are:

1. **To be selective in the use of our knowledge and skills so as to ensure that our work is of benefit to society.**
2. **To be honest, impartial and serve our constituents with fidelity.**
3. **To work hard to improve the profession.**
4. **To support the professional organizations in our engineering discipline.**

Personal behavior out-of-class or away from the work place is also important. The results of several polls indicate that Americans are less ethical today than in previous generations. Sixty-four percent of individuals in a sample of 5,000 admit to lying if it does not cause real damage. An even larger percentage of those sampled (74%) will steal providing the person or business that is being ripped off does not miss the item pilfered [3]. Many students (75% high school and 50% college) admit to cheating on an important exam [4]. Many students have not developed a moral code to use as a guide for their behavior. Relatively minor transgressions of a generation ago have been replaced with more serious problems involving mass murder, drug and alcohol abuse, pregnancy, suicide, rape, robbery and assault. Some of the reasons for these changes are explored later in Chapter 4.

Ethical behavior is important in both your professional and personal lives. Recall the golden rule—do unto others as you would have them do unto you—to guide your behavior. It is a very simple rule, and it is effective.

Committing to Life-Long Learning

Most engineering programs require successful completion of about 128 credit hours for graduation with a B. S. degree. This number is higher by about eight credit hours than the requirements for a B. S. degree from any other college on campus. Engineering educators have always required an extra effort measured in credit hours for their students prior to graduation. This belief is based on the concept that the student should be capable of practicing engineering immediately after graduation.

During the past twenty years the number of credit hours required for graduation has decreased by about 10%, while the amount of important material necessary to practice has increased significantly. Moreover, the scope of the fields in all of the engineering disciplines continues to evolve and to expand. Engineering educators recognize that they cannot increase the number of credit hours to accommodate the evolution and expansion of the knowledge base for each discipline. Instead, it becomes essential that all engineering graduates recognize the need to continue their studies after graduation.

Upon graduation many of you will take a position in industry and be challenged with a new environment and new assignments. You will be able to measure your strengths and weaknesses against this new setting. Perhaps you will conclude that you need advanced courses in your discipline, or basic courses in some other engineering discipline. After more time with a company, you will probably be given an opportunity to lead a small engineering group and become a first level engineering supervisor. In this case you may decide to enroll in a few courses in the business school or to pursue a masters program in business administration. No one really understands with certainty what knowledge will be important in the future. The idea is to stay flexible and be prepared to devote time each week to study and professional development.

You are fortunate that many opportunities exist for continuing education. Some companies bring instructors into its facilities to focus on topics of immediate concern to the organization. Professional societies offer short courses on timely topics in their discipline. Well-known universities offer short courses each summer with outstanding professors teaching in their specialty. Local universities offer a wide array of courses on a regular basis that lead to M. S. and Ph. D. degrees in a number of engineering disciplines. Professional societies offer many well-written books that cover advances within their discipline. Online educational offerings are relatively new and provide an effective means to gain knowledge in a wide variety of courses.

> ## It is essential that you recognize graduation with a B. S. degree as the beginning of your studies and not the end.

2.4 THE ENGINEERING DISCIPLINES

2.4.1 Enrollment and Salaries

The U. S. Military Academy at West Point, established by an act of Congress in 1802, was the first college of engineering in the U. S. Military engineering was the first discipline, but in the 19th century several other programs were introduced, and the field of engineering began its division into many different disciplines each with its special emphasis [5]. The number of different disciplines grew slowly in the 19th century and included:

1. Civil engineering
2. Mechanical engineering
3. Electrical engineering
4. Chemical engineering
5. Industrial engineering

In more recent years, the number of engineering disciplines has continued to grow with more and more technical specialties converted into degree programs. Recently the American Society of Engineering Education (ASEE) [6] published the undergraduate enrollment by engineering discipline during the 2013-2014 academic year. These results are presented in Table 2.4. The total full-time undergraduate enrollment increased from 2010 to 2014 by 26.3% to 569, 274.

The total number of B. S. degrees awarded in engineering in 2014 was 99,173 that represents a 26% increase over the decade since 2005. A large number of degrees were awarded to graduates from mechanical, electrical and computer engineering programs; however, in recent years many departments of electrical engineering have divided their program and they are now offering two different degrees—an electrical engineering degree and a computer engineering degree. Current trends show students tending to favor the mechanical engineering program when they enroll. In addition to the B. S. degrees in engineering, 51,690 M. S. and 11,309 Ph. D. degrees were also awarded in engineering in 2014.

Table 2.4
Full-time undergraduate enrollment by engineering discipline, (2014)

Discipline	Enrollment	Discipline	Enrollment
Mechanical Engineering	127,094	Architectural Engineering	3,117
Electrical/Computer Engineering	93,995	Metallurgical & Materials Engineering	6,426
Civil Engineering	51,6590	Engineering Science Engineering Physics	3,166
Other	53,4774	Biological and Agricultural Engineering	5,233
Computer Science within Engineering	55,482	Petroleum Engineering	8,465
Chemical Engineering	45,813	Environmental Engineering	5,121
General Engineering	33,220	Civil/Environmental Engineering	4,2625
Aerospace Engineering	19,973	Nuclear Engineering	2,026
Biomedical Engineering	28,434	Engineering Management	1,727
Industrial & Manufacturing Engineering	19,517	Mining Engineering	1,137

Mechanical and civil engineering programs also have large enrollments. The curriculum in both of these programs is relatively stable in comparison with electrical and computer engineering. However, some civil engineering departments are establishing a separate degree program in environmental engineering, and this separation necessitates significant changes in the curriculum.

A large number of engineering programs are classified as **other** in Table 2.4. This category contains many programs that are offered by only a few accredited colleges of engineering in the U. S. and include:

- Ocean engineering
- Marine engineering
- Fire protection and safety engineering
- Ceramic engineering
- Systems engineering
- Geological engineering

The enrollment in each discipline reflects perceived demand for the graduates and, to a small degree, the starting salaries for graduates from each program. Average starting salaries for University of Maryland graduates from several of the engineering disciplines with B. S. degrees are listed in Table 2.5.

The difference in starting salaries for engineers in different disciplines is not large and the data differs to some degree depending on its source. These differences vary somewhat from year to year with the petroleum, chemical and computer engineers usually leading the other disciplines by 5 to 10% and civil and environmental engineers usually trailing the other disciplines by 5 to 10%.

Table 2.5
Annual starting salaries for B. S., M. S. and Ph. D graduates in engineering
Source: Survey by the A. James Clark School of Engineering (2014)

Discipline	Mean B. S.	Mean M.S.	Mean Ph. D
Aerospace Engineering	$65,500	$72,000	$91,100
Bioengineering	$69,600	NA	$86,500
Chemical Engineering	$70,300	$78,000	$92,000
Civil & Environmental Engineering	$59,600	$71,600	$90,800
Computer Engineering	$79,300	$85,400	$98,000
Electrical Engineering	$68,800		
Fire Protection Engineering	$62,500	$72,800	NA
Materials Engineering	$66,400	$	$88,900
Mechanical Engineering	$64,200	$73,700	$89,600
Telecommunications	NA	$79,900	-
Mean Salary for Majors	$66,300	$80,300	$92,800

Salaries for engineering graduates are significantly greater than comparable salaries for other majors such as business, education, health sciences, humanities and social sciences as shown in Table 2.6. The difference is significant (52.2%) with the median annual salary for architecture and engineering majors at $83,000.00 compared to $61,000.00 for a B. S. graduate in all majors. Employment opportunities in engineering continue to be strong in spite of an economy that remains relatively week and high unemployment in most fields.

Table 2.6
Annual starting salaries by discipline
Source: Forbes Article November 19, 2014

Broad Category	Salary	Category	Salary
All Majors	$36,045	Electrical Engineering	$57,030
Accounting	$44,525	Computer Engineering	$56,576
Economics	$41,118	Mechanical Engineering	$56,055
Construction	$45,591	Software Design	$54,183
Psychology	$36,973	Chemical Engineering	$53,622
Nursing	$43,481	Computer Science	$52,237
Humanities and Liberal Arts	$39,162	Civil Engineering	$51,622
Mathematics and Sciences	$47,952	Chemistry	$43,344
Social Work	$36,639	Biology	$38,806

Median income of engineers differs considerably from the starting salaries as shown in Table 2.7. The median income is higher than the starting salaries because the engineers' compensation increases time on the job and the experience gained. The higher salaries after 5 or 10 years also reflect the fact that some engineers have earned M. S. and Ph. D. degrees, which lead to higher salaries. The median salaries depend on the engineering discipline with higher salaries commanded by those working in computer, general and aerospace engineering. Those working in civil and environmental engineering tend to earn lower salaries. However, after about 15 to 20 years of service, salaries tend to level out for most engineers with smaller incremental gains from year to year in the later stage of their careers. The demand for all disciplines of engineering is expected to grow in the coming decade as the baby boomers begin to retire.

Industries that paid the highest salaries to B. S. graduates in 2013 were energy related involving extraction of oil and natural gas. The next highest starting salary was paid by the computer industry for electrical and computer engineers. Mechanical engineers followed with slightly lower starting salaries.

Table 2.7 Median annual salaries for engineers by discipline, Ages 25-59, 2013
Source: Georgetown University Center on Education and the Workforce

Discipline	Salary
Aerospace	$90,000
Architectural	$80,000
Biomedical	$70,000
Chemical	$96,000
Civil	$83,000
Electrical	$93,000
Engr. Technology	$66,000
Environmental	$76,000
General	$81,000
Industrial	$81,000
Metallurgical	$98,000
Mechanical	$87,000
Petroleum	$136,000

2.4.2 Descriptions of Engineering Programs

Electrical Engineering

The **electrical engineering** program provides the basic background needed to work in broad fields of electronics, communications, automatic control, computers, materials processing, electro-magnetics, signal processing, and power utilization. The program places emphasis on the fundamentals of mathematics and science leading to careers in research, development, design or operations in such diversified areas as electronics, control systems, computers, communications systems and equipment, biomedical instrumentation, radar and navigation, power generation and distribution, consumer electronics and industrial devices.

The Institute of Electrical and Electronics Engineers (IEEE), the professional society representing electrical engineering, describes the purpose of electrical engineering as:

> Helping advance global prosperity by promoting the engineering process of creating, developing, integrating, sharing, and applying knowledge about electrical and information technologies and sciences for the benefit of humanity and the profession.

According to the U. S. Labor Department about 154,250 electrical engineers worked in 2011 to develop a wide array of electronic products ranging from simple household appliances to sophisticated missile intercept systems. The field is so large that the IEEE has divided its organization into 10 regions, 36 technical societies, 4 technical councils, approximately 1,200 individual and joint society chapters, and 300 sections. If this discipline is of interest to you, visit the IEEE web site at http://www.ieee.org/ for a much more complete description[4].

Electrical and electronics engineers have a median salary of around $95,780; the highest 10% of those earn approximately $143,200 per year, and the lowest 10% of electrical and electronics engineers earn an average of $65,700 per year. Electrical and electronics engineers with a BS earn an average starting salary of approximately $64,138 per year; those with a MS have an average starting salary of around 20% higher; and those with a Ph D can expect a starting salary of approximately 40 to 50% higher.

Mechanical Engineering

Mechanical engineers apply the fundamental principles of mechanics, thermo-sciences and design to fulfill the needs of society. These principles are covered in detail in the subjects of solid and fluid mechanics, thermodynamics, heat transfer, design, systems analysis, control theory, vibrations and machine design. Mechanical engineers work on a wide variety of different assignments. One of the primary areas is the energy field where mechanical engineers are responsible for generating and distributing electricity, designing heating and air conditioning systems and developing refrigeration systems. A second significant area of activity is in the design of a wide range of product such as vehicles, appliances, office equipment and machinery. Finally, many mechanical engineers work in production facilities developing manufacturing processes and overseeing production.

The American Society for Mechanical Engineering (ASME International), the professional society representing mechanical engineering, describes its mission as:

> To promote and enhance the technical competency and professional well being of our members, and through quality programs and activities in mechanical engineering, better enable its practitioners to contribute to the well being of humankind.

Mechanical engineering is one of the oldest disciplines with a broad range of industrial activities. Mechanical engineering and electrical engineering are the largest engineering disciplines. While membership in the ASME is about 125,000, employment by mechanical engineers in industry probably approaches one million. The organization of the ASME is divided into 39 technical divisions that publish technical papers and sponsor technical conferences.

Mechanical engineers earn a median salary of $87,140 per year in the USA. The highest 10% earn approximately $126,430, and the lowest 10% earn an average of $52,000. Mechanical engineers with a BS degree can expect a starting salary of about $61,523; those with a MS degree can expect a starting salary about 20% higher; and those with a Ph. D can expect a starting salary of approximately 40 to 50% higher.

Civil Engineering

Civil engineers are concerned with many societal problems including environmental quality, building and maintaining the nation's infrastructure and developing safe and rapid transportation systems. Topics of study include structures, soil mechanics, construction materials, environmental engineering, water resources, pavements and transportation. This program emphasizes fundamentals of mathematics, sciences, engineering principles and engineering methods leading to careers in engineering practice

---

[4] The Web sites for all of the professional societies are listed in Table 2.7.

and/or graduate education. An option in **environmental engineering** or a separate degree in this discipline is often available in civil engineering departments.

The American Society for Civil Engineering (ASCE), the professional society representing civil engineering, describes its function as:

> When new developments occur in research and practice, ASCE's technical divisions and councils and their respective technical committees bring information to the membership through conferences and workshops and by publications such as manuals of practice, pre-standards, journal articles and policy statements.

The technical activities of the ASCE include more than 14 annual conferences and workshops and the publication of 10 journals. More than 5,000 ASCE members participate by serving on technical committees. Their work addresses programmatic thrust areas such as sustainable environment, sustainable transportation, extreme environment and information technology and construction and materials.

Civil engineers have an average salary of $87,130 per year, with the highest 10% of them earning a salary of around $128,110 and the lowest 10% of them earning a salary of around $54,000. Civil engineers with a BS can expect an average starting salary of around $55,200; civil engineers with a MS have an average starting salary about 15% higher; and those with a Ph. D can expect a starting salary of approximately 35 to 45% higher.

Environmental engineers have an average salary of $86,340 per year, with the highest 10% of them earning a salary of around $125,380 and the lowest 10% of them earning a salary of around $54,000. Environmental engineers with a BS can expect an average starting salary of around $54,751; civil engineers with a MS

Computer Engineering

Computer engineering is a relatively new discipline that is still emerging in the many colleges of engineering in the U. S. Many electrical engineering faculties are dividing the traditional electrical engineering curriculum into two separate programs—one dealing with power, electronic systems and communications and the other concerned with computers and information technology. Computer engineers work in the computer industry designing hardware and writing software. The discipline combines technical knowledge from digital electronics, signal transmission and processing with a programming emphasis of computer science.

The IEEE Computer Society, an element within the IEEE organization, has emerged as the professional society representing computer engineers. The mission statement for the IEEE Computer Society is given below:

> The IEEE Computer Society is dedicated to advancing the theory, practice, and application of computer and information processing technology. Through its conferences and tutorials, applications and research journals, local and student branch chapters, technical committees and standards working groups, the society promotes an active exchange of information, ideas, and technological innovation among its members. In addition, it accredits collegiate programs of computer science and engineering in the United States.

With more than 100,000 members, the IEEE Computer Society is the world's leading organization of computer professionals. Founded in 1946, it is the largest of the 36 societies organized under the umbrella of the Institute of Electrical and Electronics Engineers (IEEE).

Computer hardware engineers earn a median salary of approximately $110,650. The highest 10% of computer engineers earn an average of $160,610, and the lowest 10% have an average salary of $65,000. Computer engineers with a BS degree can expect a starting salary of approximately $66,945; those with a MS degree have an average starting salary about 20% higher; and those with a Ph. D can expect a starting salary of approximately 40 to 50% higher.

Chemical Engineering

Chemical engineering is concerned with the processing of materials and the production or utilization of energy through molecular or sub-molecular changes. Chemical or atomic reactions and physical changes are included in this field. Chemical engineers begin their process designs with well-known chemical reactions previously discovered by chemists. They use this information to design large-scale chemical plants to produce sizeable quantities of raw materials, petroleum products, pharmaceuticals, plastics, etc.

The American Institute of Chemical Engineers (AIChE) is a nonprofit organization providing leadership to the chemical engineering profession. Representing 60,000 members in industry, academia, and government, AIChE provides forums to advance the theory and practice of the profession, upholds high professional standards and ethics and supports excellence in education. Institute members range from undergraduate students, to entry-level engineers, to chief executive officers of major corporations. The AIChE mission statement is given in the bullet listing below:

- Promote excellence in chemical engineering education and global practice;
- Advance the development and exchange of relevant knowledge;
- Uphold and advance the profession's standards, ethics and diversity;
- Enhance the lifelong career development and financial security of chemical engineers through products, services, networking, and advocacy;
- Stimulate collaborative efforts among industry, universities, government and professional societies;
- Encourage other engineering and scientific professionals to participate in AIChE activities;
- Advocate public policy that embraces sound technical and economic information and that represents the interest of chemical engineers;
- Facilitate public understanding of technical issues;
- Achieve excellence in operations of the Institute.

Chemical engineers earn an average salary of approximately $103,590 per year. The highest 10% of chemical engineers earn an average of $156,980 per year, and the lowest 10% receive an average salary of $64,000. Chemical engineers with a BS have an average starting salary of $66,281; those with a MS start with a salary about 20% higher; and those with a Ph. D can expect a starting salary of approximately 40 to 50% higher.

Fire Protection Engineering

Fire protection engineering is a unique profession that builds upon the basic tools of several other engineering disciplines. The fire protection engineer may be responsible for analyzing the level of fire safety in modern or historic buildings, nuclear power plants and aerospace vehicles; designing fire protection systems for high-rise buildings and industrial complexes; conducting post-fire investigations; and researching new technologies for fire detection and suppression.

Fire protection engineers apply engineering principles to protect people and their environment from the unwanted consequences of fire. These principles are applied to understand the nature and characteristics of fire, fire growth and products of combustion, as well as to consider the response of structures, people, processes, materials and systems to fire.

The professional society in the field, the Society of Fire Protection Engineers (SFPE), has 51 chapters in 8 countries. According to the SFPE, fire protection engineers have always been in great demand by corporations, educational institutions consulting firms and government bodies around the world. Graduating from a fire protection engineering program usually guarantees an immediate place in the workforce, most likely at a high starting salary. The median salary for fire protection engineers is significantly above that of other engineering disciplines. Data is not available from the U. S. Bureau of Labor Statistics for Fire Protection Engineers.

Industrial Engineering

Industrial engineers design the systems that organizations use to produce goods and services. In addition to working in manufacturing industries, industrial engineers serve to insure quality and productivity in places such as medical centers, communication companies, food service, education systems, government, transportation companies, banks, urban planning departments and an array of consulting firms. Industrial engineers educate and direct these groups in the implementation of Total Quality Management (TQM) principles. Today, especially popular functions are manufacturing, health care, occupational safety and environmental management. Industrial engineering is the most people-focused discipline in engineering. Those who pursue careers in this discipline usually have strong leadership skills and a commitment to working with teams of managers, scientists and other personnel to solve important problems. They enjoy helping organizations serve human needs and accommodate many different concerns

The Institute of Industrial Engineers (IIE) is the society that serves the professional needs of industrial engineers and others involved with improving quality and productivity. Its 25,000 members stay current with new developments in their profession through Institute's life-long-learning approach, as reflected in the educational opportunities, publications, and networking opportunities offered. Members also gain valuable leadership experience and enjoy peer recognition through numerous volunteer opportunities.

Industrial and manufacturing engineers earn an average salary of approximately $79,000 per year. The highest 10% of earners earn an average salary of around $115,000 and the lowest 10% earn an average of $51,000. Industrial and manufacturing engineers with a BS have an average starting salary of about $55,000. Those with a MS can expect to start with a salary of approximately 15% higher; and those with a Ph. D can expect a starting salary of approximately 35 to 40% higher.

Aerospace Engineering

Aerospace engineers are concerned with the physical understanding, analyses, and creative processes required to design aerospace vehicles operating within and beyond planetary atmospheres. Such vehicles range from helicopters and other vertical takeoff aircraft at the low-speed end of the flight spectrum to spacecraft operating at thousands of miles per hour during entry into the atmospheres of the earth and other planets. In between these extremes are general aviation and commercial transports flying at speeds below and close to the speed of sound, and supersonic transports, fighters, and missiles which cruise at speeds greater than the speed of sound. Among the subjects studied are aerodynamics, flight dynamics, flight structures, flight propulsion, and the synthesis of all these principles into one system with a specific application such as a complete transport aircraft, a missile, or a space vehicle.

The nonprofit American Institute of Aeronautics and Astronautics (AIAA) is the principal society serving the aerospace profession. Its primary purpose is to advance the arts, sciences and technology of aeronautics and astronautics and to foster and promote the professionalism of those engaged in these pursuits. Although founded and based in the United States, AIAA is a global organization with nearly

30,000 individual professional members, over 50 corporate members, thousands of customers worldwide and an active international membership.

Aerospace, aeronautical and astronautical engineers earn an average salary of $100,000 per year, with the highest 10% of them earning $125,000 and the lowest 10% on a salary of $65,000 per year. The average starting salary for aerospace engineers with a BS is $58,940 per year; those with a MS can expect to start with an average salary of about $66,670 per year; and those with a Ph D start at an average salary of approximately $68,060.

Materials Engineering

Materials engineering involves the study of mechanical, physical, and chemical properties of engineering materials, such as metals, ceramics, polymers and composites. The objective of a materials engineer is to predict and control material properties through an understanding of atomic, molecular, crystalline, and microscopic structures of materials. A materials engineer is an essential member of a team responsible for synthesis and processing of advanced materials for manufacturing. A graduate's work may vary from automobile or aerospace applications to microelectronics manufacturing. Opportunities are available through these industries in areas of research, quality control, product development, design, synthesis and processing operations.

The American Society for Materials (ASM International) serves as one of the technical societies representing materials engineering. For many years it has provided a means for exchanging information and professional interaction. Benefits to members include:

- Access to information for extending and maintaining professional skills.
- Opportunities for professional development through continuing education.
- Network locally and worldwide.
- Achieving professional recognition.

Material engineers earn an average salary of approximately $91,150 per year. The highest 10% of earners earn an average salary of around $138,450 and the lowest 10% earn an average of $51,000. Industrial and manufacturing engineers with a BS have an average starting salary of about $65,979. Those with a MS can expect to start with a salary of approximately 20% higher; and those with a Ph. D can expect a starting salary of approximately 40 to 45% higher.

Biomedical Engineering

Biomedical engineering is the newest engineering discipline, integrating the basic principles of biology with the fundamentals of engineering. With the rapid advances in biomedical research and the economic pressures to reduce the cost of health care, biomedical engineering will play an important role in the medical environment of the 21st century. Over the last decade, biomedical engineering has evolved into a separate discipline bringing the quantitative concepts of design and optimization to problems in biomedicine. The opportunities for biomedical engineers are wide ranging. The medical device and drug industries are increasingly investing in biomedical engineers. As gene therapies become more sophisticated, biomedical engineers will play an important role in bringing these ideas into real clinical practice. Finally, as technology plays an ever-increasing role in medicine, there will be a larger need for physicians with a solid engineering background. From biotechnology to tissue engineering, from medical imaging to microelectronic prosthesis, from biopolymers to rehabilitation engineering, biomedical engineers are in demand.

The Biomedical Engineering Society (BMES) serves as the professional society representing the interests of biomedical engineers. BMES is a relatively new society having incorporated in 1968. Its purpose is to promote the increase of biomedical engineering knowledge and its utilization.

Biomedical engineers earn an average salary of $91,760 per year. The lowest 10% earn approximately $53,000 per year, and the highest 10% earn around $139,350 each year. The starting salary for biomedical engineers with a BS is an average of $60,698, and those with a MS average about $72,070 per year to start.

Engineering Science and Engineering Mechanics

Engineering mechanics is a field of study that permeates all major engineering disciplines. Solid mechanics deals with analytical formulation of direct and reaction forces on structures and the resulting response of the structures to these forces. The discipline of engineering mechanics is based on applied mathematics and large scale computing. Fail-safe design is another dimension in this discipline. Because of the general applicability of this subject to many aspects of engineering design and manufacturing, opportunities are available in a broad spectrum of industries and government laboratories. The American Society for Mechanical Engineering (ASME International), the professional society representing mechanical engineering, also represents engineering mechanics.

Agricultural Engineering

Agricultural engineering is in transition with more emphasis currently being placed on biological aspects rather than traditional agriculture topics. As such many programs carry some reference to biology in their titles—for example, **BioResource and Agricultural Engineering**. Agricultural engineers are trained to creatively apply scientific principles in the design and development of new products, systems, and processes for the conversion of raw materials and power sources into food, feed and fiber. They are also concerned with protecting the environment and worker health and safety. The diversity of knowledge and skills that agricultural engineers possess is valuable in the agricultural and agribusiness industries. The agricultural engineer develops skills in design and problem solving that are based on fundamental principles in the engineering sciences including mathematics and physics, computer applications, communication, teamwork, instrumentation and biology. Agricultural engineers are distinguished from other engineering disciplines by their commitment to meeting human and animal needs for food, feed, fiber and ensuring a sustainable, safe living and working environment. Biological and economic constraints will continue to make this a challenging career opportunity.

The American Society of Agricultural Engineers (ASAE) is a professional and technical organization dedicated to the advancement of engineering applicable to agricultural, food, and biological systems. Its 10,000 members, representing more than 90 countries, serve in industry, academia and the public.

Petroleum Engineering

Petroleum engineering is primarily concerned with producing oil, gas, and other natural resources from the earth through the design, drilling, and use of wells and well systems. The petroleum industry develops methods for conveying fluids into, out-of, or through the earth's subsurface for scientific, industrial, and other purposes while remaining mindful of the ecological needs for safety. The curriculum in petroleum engineering provides a proper balance between fundamentals and practice. Graduate engineers are prepared for life-long learning but are capable of being productive contributors immediately. Petroleum engineers are currently in high demand in the industry, and their starting salaries are consistently among the top in the nation. The curriculum includes study of design and analysis of well systems and procedures for drilling and completing wells; characterization and evaluation of subsurface geological formations and their resources; design and analysis of systems for producing, injecting, and handling fluids; application of reservoir engineering principles and practices for optimizing resource development and management; and use of project economics and resource valuation methods for design.

The Society of Petroleum Engineers (SPE), with more than 50,000 professionals from oil and gas-producing regions around the world, represents the technical interests of this discipline. Its mission is to provide the means to collect, disseminate and exchange technical information concerning the development of oil and gas resources, subsurface fluid flow and production of other materials through well bores for the public benefit. The Society also provides opportunities through its programs for interested individuals to maintain and upgrade individual technical competence in these areas. The SPE accomplishes this mission through an international schedule of meetings and exhibitions, periodicals, short courses, books, electronic publications and section programs.

Mining Engineering

Mining engineering involves prospecting for mineral deposits; planning, designing, and operating profitable mines; processing and marketing the extracted minerals; insuring safe and healthy working conditions; and protecting and restoring the land during and after mining projects. Mining engineers use technologically advanced equipment, machines, robotics and computers. Those in charge of mine design, plan and test mines using computer simulators before breaking ground. Mining engineers in management use mine scheduling software to plan mining activity when operation is underway. Surface mining operations use larger mobile equipment than any other industry in the world. Mining methods and equipment are also applied to the removal of earth and rock outside the mining industry. Each year, an individual requires an equivalent of 40,000 pounds of new minerals and energy equal to that produced by burning 15 tons of coal. With each different mineral comes a different mine site and a different location, giving the mining engineer a very diverse work environment, and opportunities to work anywhere in the world.

The Society for Mining, Metallurgy, and Exploration (SME) is an international society of 18,000 professionals with members in nearly 100 countries. Services available to SME members include: publications, professional registration, peer-review of technical papers, future leader and college accreditation programs, meetings and exhibits, public education and SME short courses.

Mining engineers earn an average salary of approximately $100,970 per year. The highest 10% of earners earn an average salary of around $159,010 and the lowest 10% earn an average of $51,000. Mining engineers with a BS have an average starting salary of about $65,000. Those with a MS can expect to start with a salary of approximately 15% higher; and those with a Ph. D can expect a starting salary of approximately 35 to 40% higher.

Nuclear Engineering

Nuclear engineering is concerned with the science of nuclear processes and their application to the development of various technologies. Nuclear processes are fundamental in the medical diagnosis and treatment fields, and in basic and applied research concerning accelerator, laser and super conducting magnetic systems. Utilization of nuclear fission energy for the production of electricity is the current major commercial application, and radioactive thermal generators power a number of spacecraft. For the longer term, electricity production based on nuclear fusion is expected to become an increasingly important segment of the field. Nuclear engineers are concerned with maintaining expertise in the design and development of advanced fission reactors, performing basic and applied research in the development and ultimate commercialization of fusion energy, developing both institutional and technical options for radioactive waste and nuclear materials management and in fostering research in nuclear science and applications, with emphasis on bioengineering, detection and instrumentation and environmental science.

The discipline is represented by the American Nuclear Society (ANS). The ANS has a diverse membership composed of approximately 12,000 engineers, scientists, administrators and educators representing corporations, educational institutions, and government agencies.

Manufacturing Engineering

Manufacturing is a prime generator of wealth and is critical in establishing a sound basis for economic growth. Manufacturing education is important in achieving and maintaining a long-term competitive position for U. S. industry in the current global economy. The program covers concepts such as design for manufacture, flexible manufacturing, new communication and information networks and the impact of automation on human experience. Achieving and maintaining a long-term competitive economic position requires students to be able to analyze manufacturing systems for overall technical, economic and environmental performance. Manufacturing engineers should also be able to quantify trade-offs between these performance characteristics in relationship to the technology deployed in the manufacturing system. Many mechanical and electrical engineers find their careers in manufacturing.

The Society of Manufacturing Engineers (SME) is the world's leading professional society serving the manufacturing industries. Through its publications, expositions, professional development resources and member programs, SME influences more than 570,000 manufacturing executives, managers and engineers. The SME has some 75,000 members in 70 countries and supports a network of hundreds of chapters worldwide

Table 2.8
Web site addresses for the professional societies

Professional Society	Web Site Address
Electrical and Electronics Engineers (IEEE),	http://www.ieee.org/
American Society for Mechanical Engineering (ASME International)	http://www.asme.org/
American Society for Civil Engineering (ASCE)	http://www.asce.org/
IEEE Computer Society	http://www.computer.org/
American Institute of Chemical Engineers (AIChE)	http://www.aiche.org/
Institute of Industrial Engineers (IIE)	http://www.iienet.org/
American Institute of Aeronautics and Astronautics (AIAA)	http://www.aiaa.org/
American Society for Materials (ASM International)	http://www.asm-intl.org/
Biomedical Engineering Society (BMES)	http://mecca.org/BME/BMES/
American Society of Agricultural Engineers (ASAE)	http://asae.org
Society of Petroleum Engineers (SPE),	http://spe.org/
Society for Mining, Metallurgy, and Exploration (SME)	http://www.smenet.org/
American Nuclear Society (ANS).	http://www.ans.org/
Society of Manufacturing Engineers (SME)	http://www.sme.org/
Society of Fire Protection Engineers (SFPE)	http://www.sfpe.org

2.5 ENGINEERING FUNCTIONS

Engineers from all disciplines are involved in a wide variety of functions ranging from sales to maintenance. The more common engineering functions will be briefly described in the following subsections.

Design and Development

Companies continuously work on their product lines to maintain or increase their sales. Their ability to release a stream of high-quality, high-performance, reliable products that are competitively priced dictates their success and the profits they earn. **Design and development engineers** work to create new and profitable products. These "new" products are often redesigned models of an older product. Important in these redesigns are improvements in performance and reliability at lower costs. On rare occasions design and development engineers create an entirely new product that establishes a new market. Apple's Iphone and Ipad are recent examples of new products. Check the price of a share of Apple's common stock by Googling AAPL and note the effect of designing, manufacturing and marketing successful new products.

Design and development engineers interact with marketing personnel to establish the customer's needs. They convert the customer's needs to a product specification and then consider a large number of different design concepts that fulfill this specification. The best design concept is selected from among these options, and then the product is developed and manufactured. During this process, the design and development engineers work with manufacturing engineers, production engineers and test engineers to insure a timely and cost effective approach for producing a safe and reliable product.

Testing

Society today is increasingly litigious. It is a common occurrence for trial lawyers to seek remedies for damages due to what may be perceived as unsafe products. Product liability is a very serious concern. When a company releases a product for the market, it is **imperative that it be safe**. The only questions are—how safe and under what conditions? **Test engineers**, as the name implies, test products to ascertain if their performance specifications have been met. They conduct tests that enable them to predict the life of the product and the number of failures that will occur with time in service. They provide management with the data needed to determine warranty costs. They insure management that the product is safe to release. During the design and development phase, they may test subsystems to provide data helpful in designing the product. Basically test engineers verify analytical methods for predicting performance. When the analytical methods are not sufficient, the test results are essential to prove the adequacy of the product.

Design Analysis

Almost every design and development team includes one or more members with excellent skills in design analysis. **Design analysis engineers** are responsible for modeling, where components from the product are converted to simplified models that may be analyzed. The analysis conducted may be in closed form where well-known mathematical formulas are used to produce results that predict performance or insure safety. In many instances, closed form solutions are not available and numerical techniques, such as finite element models, are used to generate solutions. In recent years, a large amount of specialized software has become available that enables design analysis engineers to solve increasingly complex problems, rapidly and accurately.

The close interaction of the design analysis engineer with the design team during the development cycle is vitally important. If a problem is identified by analysis early in the design cycle, it can be corrected quickly and the costs of the required modification are minimal. However, if the problem goes undetected until hardware is produced and tested, the costs to correct the problem are dramatically higher and the time required correcting the problem may delay introducing the product to market.

Manufacturing

Manufacturing engineers serve on the design and development teams to provide expertise on tooling and manufacturing methods. As the components of a product are designed, many decisions are made that affect the cost of manufacturing and assembling the product. It has been established that a major fraction of the total life-cycle cost of a product is committed in the early stages of design [8]. Quality cannot be manufactured or tested into a product; it must be designed into the product. Manufacturing engineers provide the design and development team with the knowledge of manufacturing equipment and processes available within their facilities and available from qualified suppliers. As the design proceeds, they identify manufacturing requirements and begin the design of any specialized tooling required for production. They design the manufacturing cells for producing component parts. They certify vendors who will supply externally purchased components. They design the lines, cells and the tooling used in assembly.

Manufacturing engineers also work with the quality control department to establish procedures to insure the quality of each component manufactured. They select machine tools and instrumentation to insure timely adjustments for manufacturing processes to remain within the control limits established by quality assurance personnel.

Production

Production engineers also participate on the design and development team. Their role is to ensure the flow of material and components to the manufacturing facilities and to the assembly lines. Often there is a need to purchase components or materials that require long lead times for delivery. In some cases, the lead-time is comparable to the design time. The production engineer must identify these items, anticipate the number required and place the order before the design is complete. After the design is complete, they work with purchasing personnel to insure delivery of components just-in-time to be used in production. They attempt to minimize the work-in-progress inventory. They work with personnel from the sales department to match production to anticipated demand for the product. They work with personnel from shipping to ensure that the product moves from the production facilities to shipping where it is delivered to customers in a timely manner. Their goal is to meet the demand of the customers with a minimum of work in progress and finished product in inventory.

Maintenance

Most students own an automobile or have one available to them. Have you ever tried to travel from point A to point B and found the car would not start? This problem improves your understanding of the need for maintenance. Manufacturing facilities often operate on a 24/7 schedule (this schedule implies the equipment operates 24 hours a day seven days a week). **Maintenance engineers** work to keep the equipment in operation. They seek to avoid breakdowns with scheduled maintenance. They plan for periodic lubrication of bearings, replacement of parts subjected to high rates of wear, replacement of belts and scheduled inspections. They develop a recording system that provides a database for predicting the time between failures and identifies the equipment that exhibit high failure rates. They design monitoring instrumentation that enables the facilities to be shut down in a controlled manner prior to failure of any critical parts. They supervise the maintenance crews that perform the repair work. If you enjoy high-stress associated with inevitable breakdowns, consider this engineering function for a career.

Sales

Sales engineers provide an interface between the customer buying technical products and the company selling them. They work with the engineers and purchasing agents representing the customer and provide technical information necessary for the customer to intelligently purchase a product or a system. Many engineering products and/or systems are complex with the availability of several different options. Sale engineers must know the details of the product and/or the systems, provide an overview of the options, and answer questions from the customer's representatives. People skills are as important as analytical skills in this position because the effectiveness of a sales call is often affected by personalities.

The sales engineers also seek information regarding the need for new products to better match the customer's requirements. In some cases, the product line offered by the sales engineer's company does not meet the needs of the customer. When this occurs, it is important to ascertain the customer's requirements and the price he or she would be willing to pay for the product if it were to become available. This information is passed on the marketing representatives and is used in planning for new product developments.

Management

After demonstrating the ability to perform engineering functions, many relatively young engineers are promoted into **management** positions. If your company is organized along functional lines, the first management position is usually "section head." A section is a small group of engineers (8 to 12) that specializes in a certain area such as motors, controls, transmissions, power, safety, etc. The section manager assigns tasks to the engineers in the group and provides supervision and support to aid them in performing the work. A department manager (2nd level of management) supervises several section managers. The section manager's role is a mix of both technical and administrative skills and the department manager's role is mostly administrative.

In companies that are organized with product development teams, program managers are responsible for the design and production of a certain product. Often an engineer becomes the program manager. He or she oversees the team, leads team meetings, makes assignments, arbitrates decisions and is responsible for the development schedule and the budget. The program manager usually reports to a vice president for development. For larger development teams, the program manager's administrative skills are usually more important than his or her technical skills.

Education

Many engineering students continue their studies after completing the BS degree and pursue advanced engineering degrees. For example, 46,940 MS degrees in engineering were awarded in 2011. Those graduating with a MS degree may go into industry to serve in assignments demanding a higher level of analytical skills. Others take faculty positions in educational institutions (usually a community college or a four-year college). In these educational institutions, they usually teach three or four undergraduate classes each semester. They also perform many service functions for the college and community.

Some students complete the Ph. D. and take a faculty position with a college of engineering in what are known as research universities. They are responsible for teaching, research and service in these positions. The teaching is limited to one or two courses per semester usually divided between graduate and undergraduate courses. They engage in active research programs and seek funding from external sources (federal and state governments, industry and foundations) to finance their research. They guide the studies of graduate students working in their research topic. Publication in technical journals and writing textbooks is an important part of their scholarly activities. Participation in technical societies, by presenting papers, organizing sessions and serving as officers of the society, are ways for performing community service. Many faculty members are also active consultants to both industry and government.

Consulting

Consulting engineers work on a retainer from government agencies or industry. They are independent businessmen and women who basically offer engineering services at a price. In years past, it was unusual for an engineer to select this career path. Most companies and government agencies hired their own specialists. The downsizing of companies and government has changed this practice. Many companies find that outsourcing engineering work is more cost effective. They call on engineering consultants with a given specialty as required and pay fixed prices for well-defined engineering tasks. The advances in communications—fax, cell phones, e-mail and the Internet—have enhanced the flow of information essential to conducting a successful consulting engineering business. Being in an office at a company's location, eight hours a day five days a week, is not required. A consultant can solve a problem or create an engineering document from a distance while maintaining communication with any client in any country for as long as is necessary to complete an assignment. The availability of high-speed, low-cost computers has enabled the consultant to compete effectively with an in-house analyst if the consultant has access to the same software and the data needed to perform the assignment.

2.6 REWARDS

For many years, engineering graduates have received relatively high starting salaries and most have usually found interesting positions available upon graduation. However, salaries later in a person's career depend upon many different factors including:

- Years of experience
- Level of technical and supervisory responsibility
- Type of employer
- Discipline and/or function
- Degree level
- Strength of the economy

Salaries also depend on current salary-wage policies, individual mobility, working conditions, and local and regional salary scales. Many significant factors, such as challenge, productivity, creativity, technical and supervisory responsibility and job commitment are dependent on the individual's initiative. However, other factors affect salaries that the engineer cannot control. These include company size, company policies, fringe benefits and promotion opportunities. Industry and company growth, labor-management relations, and research and development funding also affect earnings. There is considerable variation, even for engineers with similar position, responsibility and experience.

It is not possible to consider all of the parameters that affect salaries in this discussion. However, the three most important factors affecting engineering salaries are: degree level, years of experience and management responsibilities. Engineers with advanced degrees command a higher average salary than those with only a BS degree. Also, those with a Ph D earn more than engineers with a MS degree for the same class of work. Finally, for all degree levels, management responsibilities are rewarded with higher salaries. The differences are significant as indicated in Table 2.9.

Table 2.9 Median Annual Salaries for Engineers based on experience, supervisory responsibility and level of education: 2012 AAES Engineering Salaries

Employer	Number of years after B. S.					
	0	**5**	**9-10**	**13-16**	**21-25**	**35+**
Non-supervisory						
B.S.	$63,858	$73,100	$78,227	$85,470	$94,508	$94,719
M. S.	-	$75,937	$87,218	$94,749	$101,378	$102,408
Ph.D.	-	-	$95,705	$102,451	$112,088	$143,857
Supervisory						
B.S.	-	-	$85,765	$92,308	$103,754	$116,321
M. S.	-	$119,998	$111,714	$120,281	$128,075	$127,795
Ph.D.	-	-	-	$126,758	$139,927	$162,764

All engineers gain in salary with years of experience. The gains are more rapid in the early years, with increases of about twice the increase in the cost of living regardless of the degree level. However, after about 20 to 25 years of experience, the rate of increase in annual salary decreases. In the last decade of a typical engineering career, an individual's compensation may not keep up with the increases in the cost of living. These data in Table 2.9 are for median salaries, and you may prove to be an exception with either higher or lower rewards for your services. But the message in this graph is clear. Make your mark early in your career. After you reach the age of 45 to 50, significant salary gains are difficult for many engineers to achieve.

Another factor not considered in a discussion of salaries is the satisfaction derived from completing an engineering assignment. With sufficient experience, many assignments typically those given to younger engineers lose their challenge. (Been there, done that.) With an advanced degree, an employer is paying you more and has higher expectations of your capabilities. Accordingly, the assignments tend to be more challenging, more visible and more rewarding. Success in one tough assignment leads to another and continued success leads to promotion and recognition.

2.7 PROFESSIONAL REGISTRATION

Today about 30% of graduate engineers are registered and licensed as "professional engineers." Professional registration is optional for many engineers, unlike medicine or law where registration is mandatory before one may practice. When public safety is an issue, registration is required for engineers. Civil engineers often design buildings, highways, bridges and other structures that would endanger the public if they failed in service. Consequently, civil engineers participating in these activities must register and become licensed to practice.

You are encouraged to begin the four-step registration process for the simple reason that you do not know when in the future you will be required to demonstrate your qualifications. Sometimes evidence of your BS degree is sufficient for qualification, but in other instances it is not. Also as you approach graduation you are very well prepared to begin the registration process.

State boards of registration control the procedure for registration. There are minor variations from state to state, but the guide shown below is typical of the process required:

- Graduation from a four year engineering program that has been accredited by ABET.
- Achieving a passing grade on the "engineering fundamentals" examination.
- Practicing engineering for a specified number of years to gain engineering experience. The state board of registration specifies the number of years required.
- Achieving a passing grade on the "principles and practice of engineering" examination.

The "engineering fundamentals" examination may be taken before graduation. In fact, it is recommended that you take it in either April or October of your senior year[5]. The examination questions are designed to test your skills in fundamental subjects. Because engineers of all disciplines take the same examination, it is evident that the coverage must be on subjects common to all disciplines—mathematics, physics, chemistry, and engineering science. The examination is scheduled for eight hours. After passing the examination and graduating, you receive a certificate that designates you as an Engineer-in-Training.

The "principle and practice of engineering" examination is discipline specific. The civil, mechanical, electrical, etc. engineers must pass a discipline-oriented examination. While the topics covered are more focused, the questions probe the depth of your knowledge in the discipline. After passing this examination, you are eligible to become a licensed professional engineer and to use a seal when certifying your work.

2.8 <u>SUMMARY</u>

Two reasons are cited for classifying engineering as a profession. The first is based on a highly structured curriculum that prepares you for a position as an engineer upon graduation. You are expected to be productive almost immediately after assuming a beginning position with an industrial firm. The second reason is based on the type of work performed after graduation. When practicing in industry or for the government, one is engaged in engineering activities. A professional engineer is almost always salaried and is not compensated directly for overtime. Nor are you docked for time off. Engineers assume responsibility and are committed to the project, development team and the company or agency. They devote the time and effort necessary to complete a project on schedule and on budget.

Statistics are cited for students entering engineering colleges. The students admitted to colleges of engineering are top notch: student leaders and scholars ranking in the top 5 or 10% of their high school graduating class. Average scores for both mathematics and verbal in the SAT examinations are outstanding. About 18.4% of those graduating from engineering in 2011 were women, but the percent of women in a typical class depends strongly on the discipline (see Table 2.1).

Pursuing a successful career in engineering depends on your ability to develop a number of different skills and understandings. These include:

- The ability to design.
- The ability to perform well on a development team.
- The ability to communicate in many ways and in different settings.
  - o Writing letters, memos, reports, specifications, etc.
  - o Participating in discussions, and presenting design briefings.
  - o Preparing two and three-dimensional drawings.
  - o Preparing slide presentations.
- Performing design analysis and interpreting the effect of the results on the design.
- Designing and conducting experiments and interpreting the data acquired.
- Understanding professional and ethical responsibilities.
- Understanding the need for life-long learning and actively pursuing this goal.

The most popular of the engineering disciplines are described and a listing of the undergraduate enrollment by discipline is shown in Table 2.4. Starting salaries for beginning engineers by discipline shows some differences with chemical and computer engineers leading civil engineers. The technical

---

[5] The engineering fundamentals examination is offered twice a year in April and October. It is usually offered at several of the colleges of engineering in each State.

emphasis of each of the popular disciplines is described. Professional societies representing each discipline are discussed.

Engineers of all disciplines perform many different functions in their professional activities. The tasks commonly conducted in each of these functions include:

- Design and development
- Testing
- Design analysis
- Manufacturing
- Production
- Maintenance
- Sales
- Management
- Education
- Consulting

Rewards (salaries) for engineers depend on many different factors. However, three factors: degree level, years of experience and management responsibilities are the most important. Advanced degrees and management responsibilities are rewarded with significantly higher salaries. Experience is also important, but the salary gains due to experience are only significant for about the first 20 to 25 years. After that period, salary gains are modest and barely equal to inflation.

Finally, the advantages of professional registration are discussed. A guide to the four-step process involved in becoming a registered professional engineer is provided. All engineering students are encouraged to take the eight-hour "engineering fundamentals" examination late in their junior year or early in their senior year.

REFERENCES

1. Valenti, M. "Teaching Tomorrow's Engineers," Special Report, Mechanical Engineering, Vol. 118, No. 7, July 1996.
2. Zhang, G. Engineering Design and Pro/ENGINEER, Wildfire Version 5.0, College House Enterprises, Knoxville, TN, 2010.
3. Patterson, J. and P. Kim, The Day America Told the Truth: What People Really Believe about Everything that Really Matters, Prentice Hall, New York, NY, 1991.
4. Sommers, C. H., "Teaching the Virtues," Public Interest, No. 111, Spring 1993, pp. 3-13.
5. Grayson, L. P., The Making of an Engineer: An Illustrated History of Engineering Education in the United States and Canada, John Wiley & Sons, New York, NY, 1993.
6. Yoder, Brian L. "Engineering by the Numbers," Engineering Statistics, ASEE Web Site at http://www.asee.org. Brian L. Yoder is the director of data research and programs for the American Society for Engineering Education. Contact at b.yoder@asee.org.
7. N. Fogg, P. Harrington, and T. Harrington, College Majors Handbook with Real Career Paths and Payoffs: The Actual Jobs, Earnings, and Trends for Graduates of 60 College Majors, 2nd Edition, JIST Publishing, Indianapolis, IN, 2004.
8. Anon, Improving Engineering Design: Design for Competitive Advantage, National Research Council, National Academy Press, Washington, D. C., 1991.
9. Anon, "American Consulting Engineers Council Directory", American Consulting Engineers Council, 10015 15th Street, N. W., Washington, D. C. 2005.

10. Maureen Byko, "2012 AAES Engineering Salaries: Update Presents the State of the Profession", Journal of Metals, Vol. 65, no. 3, 2013.
11. U. S. Bureau of Labor Website, http://www.bls.gov/oes/current/oes_stru.htm#17-0000.

EXERCISES

2.1   Write a two-page paper on why engineers should be considered professionals.

2.2   Write a two-page paper comparing the engineering profession to the medical profession.

2.3   Write a two-page paper comparing the engineering profession to the legal profession.

2.4   Write a two-page paper comparing the engineering profession to the accounting profession.

2.5   Write a one page paper citing your reasons for selecting engineering as a course of study.

2.6   Write a brief paper giving the reasons why you believe bright, well-qualified students leave the College of Engineering after only one semester.

2.7   Why do you believe that bright, well-qualified students often fail courses in their first year of study?

2.8   Prepare a list of the skills that you currently have and provide a grade (from 1 to 10 with 10 being exceptional and 1 for totally lacking this skill) for your achievement level in each skill.

2.9   Add to the list in Exercise 2.8 your goal for an achievement level in each skill at the conclusion of this course.

2.10   Do you believe that you will learn new skills in this Introduction to Engineering Design course? Why?

2.11   List the skills you have in priority order. Do you believe your best skill is the most important skill? If so, why? If not, why?

2.12   What is the purpose of the ethical codes endorsed by the professional societies?

2.13   Write a two-page paper describing your current ethical standards. Do you believe these standards will change during your tenure at college? Do you cheat when the opportunity arises?

2.14   Are you committed to life-long learning? If so, why? If not, why? How do you intend to continue your studies after graduation with a BS degree?

2.15   Have you selected an engineering discipline? If so, state the discipline and give the reasons why you have selected it. If not, state the reasons for your delay?

2.16   If you are concerned with selecting an engineering discipline, prepare an action plan to gain the information necessary for your decision?

2.17   If you are concerned that your first choice of an engineering discipline was not correct, prepare an action plan to evaluate your choice and to change disciplines if necessary.

2.18   For the engineering discipline of your choice, describe the mission and the activities of the professional society that represents their interests. Visit their web site. Do they have a student chapter at this university? Do you plan to join?

2.19   Most colleges of engineering offer only a limited number of degree programs. What are the degree programs in the Clark School of Engineering? Does one of these programs coincide with your career objectives?

2.20   After reviewing the engineering functions, select the two that best suit your interests. Write a paper discussing why these functions interest you.

2.21   After reviewing the engineering functions, select the two that least suit your interests. Write a paper discussing why these functions do not interest you.

2.22   Write a paper discussing why engineering salaries begin to plateau after about 20 to 25 years of experience. Include in your paper how you plan to avoid this problem.

2.23   Write a paper explaining why salaries for women engineers are slightly lower on the average than men engineers.

2.24   Salary data is a moving target; consequently, the information provided here is at best an approximation. It is possible to obtain more up to date salary information that takes into account the city where the position is located by visiting the Wall Street Journal Website at http://www.careerjournal.com/salary. Click on the Salary Expert and perform a salary search that shows the differences in salaries for different disciplines and different management responsibilities. Also show the differences in salaries to account for the different cost of living in different locations across the United States.

2.25   Today the retirement age for most engineers is about 67. At what age do you plan to retire? Do you believe the age requirement to receive social security benefits will change prior to your retirement? Why?

2.26   Write a paper arguing for professional registration for engineers. Do you plan to take the "engineering fundamentals" examination when you achieve academic status as a senior?

CHAPTER 3

A STUDENT SURVIVAL GUIDE

3.1 OVERVIEW FOR SURVIVAL

As a new student in the College of Engineering, you are encountering many new challenges and meeting new friends. You have to arrange for housing, transportation and meals (the basics) in an entirely new setting. You also have to cope with a schedule of classes and find the locations of classrooms, lecture halls and laboratories in a maze of buildings scattered all over campus. Finally, you must learn to study and pass courses in a much more competitive environment than you found in high school.

This chapter will discuss some strategies for coping with the new environment and techniques that will help ensure your successful completion of your Bachelor of Science degree in engineering. In writing this survival guide, it is recognized that you are bright and have demonstrated your scholastic and leadership abilities—otherwise you would not have been admitted to the College. Success should be insured; however, many of each year's incoming class do not complete the engineering program. Students leave in the first two years because of failing grades or lack of interest. Some techniques to insure your success in completing your degree will be discussed in this section, and then several important tools for improving your performance will be described later in this chapter.

Goals and Priorities

Many students fail to establish realistic goals that enable them to control their time and direct their energy in a manner needed to succeed in a competitive environment. Long-term, intermediate-term and short-term goals must be established to focus your activities. Otherwise your efforts are diffused, sufficient progress is not made and one or more goals are not achieved. The goals should be written in priority order and they should be dated. Your goals should be reviewed and revised periodically with changes made to reflect both progress and realism.

Confidence

Success cannot be achieved without confidence. It is essential that you maintain a positive attitude with a can-do philosophy. If you start to doubt your ability to pass a test, fear will become a significant factor that blocks your ability to think clearly during examinations. It is a well-know fact that fear and/or anger limits a person's ability to think clearly and act promptly. One way to gain confidence is to prepare until you are certain that you understand the course material.

Stay Current

The clock is as important in studying engineering as it is in a football or basketball game. The game begins on the first day of class and ends at the last minute allowed for the final examination. Your instructor will try to use the time available to cover as much material as possible. There is little or no float time in an engineering course. Reading and homework will be assigned for each and every class period. Moreover, the material is cumulative—to understand the second topic it is necessary to master the first topic, etc. If you fall behind in completing assignments, it is extremely difficult, if not impossible, for most students to catch up and become current.

Attend Class

As an instructor, I was amazed at the number of students cutting classes. An hour of class time costs the student, or his or her parents, more than a ticket to a rock concert. What student would throw away a ticket to a rock concert? Yet, some students throw away many opportunities to attend class. Attendance gives you the opportunity to hear the instructor's interpretation of the material and to stay up to date on assignment and exam schedules. Attending class also gives you the opportunity to ask the instructor questions or hear other student's questions, which can often be extremely helpful. Attending class reduces your dependence on the textbook or information from other students to learn the material. You will probably use your textbook or work with other students to study, but doing these activities in conjunction with attending class will maximize your opportunities for learning, and allow you to gain a deeper understanding of the course material.

Managing Time

You are accustomed to schedules because you moved from class to class following a prescribed schedule while attending high school. The schedule in college is similar except that a class period for a specific course is only scheduled for two or three hours per week instead of an hour every day. You have the **illusion** of more free time. A typical load of 16 credit hours for a semester entails only 12 or 13 hours of classroom participation and 6 to 8 hours of laboratory involvement. Unfortunately, this is misleading because reading and other assignments require significant amounts of out-of-class time. A rule followed by many instructors is to adjust the material and class assignments to require a **time multiplier of 3 to 4 on the credit hours for the class**. This adjustment means that if you are taking a three-credit hour class, you should expect to spend 9 to 12 hours each week to attend class, complete the assignments and study the material. Your 16 credit hour schedule really involves a time commitment of **48 to 64** hours per week. Time management becomes important when considering the actual time needed to succeed in engineering courses.

Developing Good Habits

It is self evident that good study habits are essential to successfully complete an engineering degree program. Yet many successful high school students have never developed good study habits. Competition in high school was not intense and you could meet or exceed your teachers' expectations without much out-of-class work. The academic competition is much more intense in a college of engineering and your instructors' expectations are higher than you might imagine. It is essential that you establish and follow a personalized schedule for study periods for each of your classes. The time allocation must be sufficient for you to complete the class assignments. Your schedule must insure that you do not fall behind in any subject.

Managing People

While you are not yet an engineering manager, you will still interact with many people, and it will be necessary to manage your relationships with them. A short list of these people includes your instructors, roommate, classmates, teammates and friends on and off campus. Maintaining relationships requires time, and as you will soon determine, time is a precious commodity. Your instructors should command priority; fortunately they are easy to manage. All you have to do is go to class, and perform well on their examinations. Your instructor will respond with good grades. You may also meet many of your friends in classes. This social bonding is an important part of the educational process. Education should be fun and classmates help make it so.

Teammates differ to some degree from classmates, because a team is a more formal group that has been organized to perform a well-defined task such as the design of a prototype of some product, process or system. Teammates also bond socially and often remain friends long after the team has been disbanded. Friends on and off campus are important to your social life. Pursuing an engineering degree will take most of your time and energy, but it is essential that you allow some time for socializing.

3.2 TIME MANAGEMENT

Weekly Schedules

A college schedule is a weekly affair. You may have classes every day of the week, for about 15 weeks, before the final examination period. It makes sense to prepare a weekly schedule showing the time and location of your classes similar to the one shown in Fig. 3.1. There are many free software tools available for generating this type of schedule. The authors are somewhat fond of Google Calendar, and have utilized it to create the sample schedules shown within this chapter.

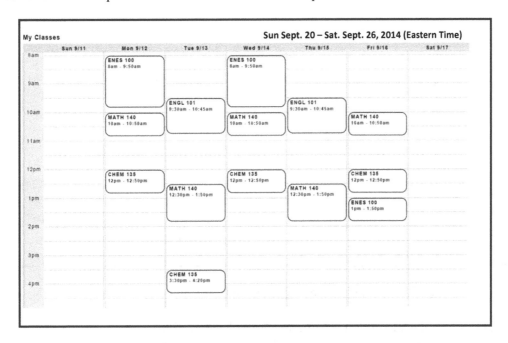

Fig. 3.1 A typical class schedule for a first semester engineering students.

The recommended load of 13 credit hours for the fall semester for first-year students is less than the typical load of 15 to 18 credit hours for the second and subsequent semesters. The lighter load is to provide you with the opportunity to adjust to a new academic and social environment. While the meeting times for your classes will differ, the number of credit hours and contact hours will probably remain about the same. The number of contact hours (18) exceeds the number of credit hours because laboratories and recitation periods are not counted with the same weight as lecture hours.

The next step in time management is to schedule study periods. How much time should you schedule? The general rule most instructors use in designing assignments is that the **average** student spends two hours studying for each hour of lecture and another hour studying for each hour of laboratory or recitation. This rule implies that a student should schedule, for example, $3 \times 2 + 1 \times 1 = 7$ hours of out-of-class study time for the CHEM 135 course. Adding the study time for all four courses required this semester gives a total of 26 hours that should be devoted to study for the **average** student. The word

average is stressed. You may be able to perform well with less than the recommended 26 hours. Or you may be having trouble with one or more courses and find 26 hours is not sufficient. The weekly schedule shown in Fig. 3.2 provides for 26 hours of study time.

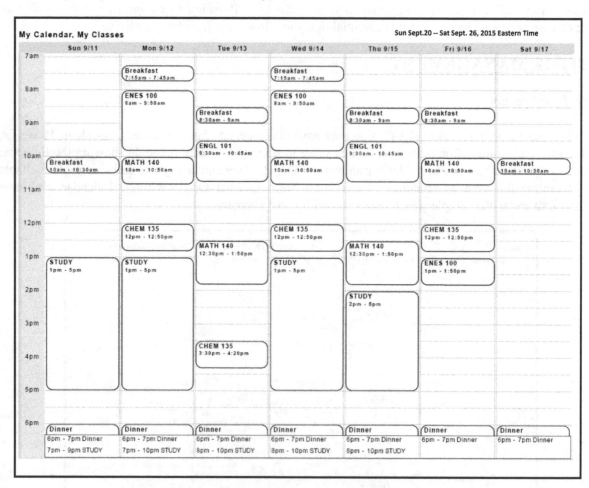

Fig. 3.2 A complete weekly schedule for a first semester student.

The open periods (unscheduled time) provide much needed flexibility in the schedule. On some weeks you will find the assignments much more demanding than other weeks. When this situation occurs, you will be able to use the open periods to study or to pursue other interests that you may have. It is also important to dedicate some time to social events – so long as this remains within moderation. Attending class and studying is sedentary. You will lose your body tone without interrupting this pattern of class, study and sleep. Try going to the gym to exercise regularly. The point is that there is a **significant amount of flexibility** within your schedule so long as you **effectively manage your time**.

You may find that the schedule shown in Fig. 3.2 is not suitable. That is okay. It is not necessary to impose this schedule on you. However, it is imperative that you prepare a weekly schedule that is more suitable, and one that allows time for:

- Attending all your classes, laboratories and recitations
- Sufficient study time at suitable times of the day for you to be effective in learning
- Adequate uncommitted (open) time to provide needed flexibility in your schedule
- A measure of social time to keep your spirits high
- Enough time for meals, sleep, exercise, and mental preparation for the day ahead

Daily Schedules

You will prepare a weekly schedule each semester. It is a guide to your daily activity, but it does not contain the detail needed to use your time efficiently. A daily schedule provides the detailed guide for allocating sufficient time to tasks essential to success in each course. The daily schedule assigns tasks to each block of time for each day of the week. For example, you may allocate your Monday study time from 7:00—10:00pm to complete a paper that is due in your ENGL 101 class the next morning. You are encouraged to complete a daily schedule each night before you retire.

Many calendar tools also have built in task lists which can be utilized very effectively to reduce the likelihood that you miss an assignment due date. If you accurately log each assignment within your task list, you can quite easily (with some practice) create a daily schedule to meet your highest priority tasks. Consider the task list shown in Fig. 3.3, which was created using a Google Calendar tool.

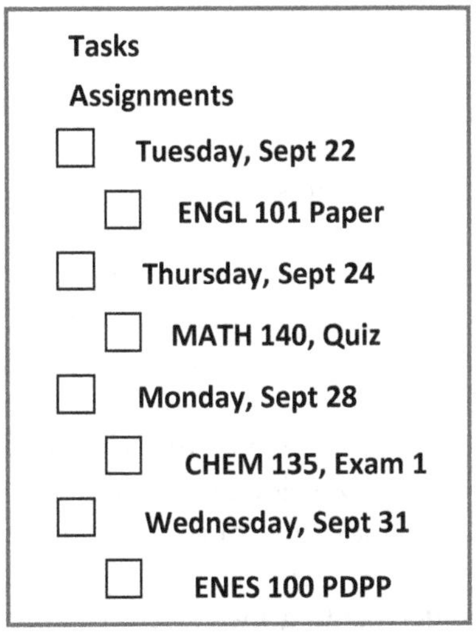

Fig. 3.3 A sample task list using Google Calendar.

Tasks

Assignments

☐ Tuesday, Sept 22

   ☐ ENGL 101 Paper

☐ Thursday, Sept 24

   ☐ MATH 140, Quiz

☐ Monday, Sept 28

   ☐ CHEM 135, Exam 1

☐ Wednesday, Sept 31

   ☐ ENES 100 PDPP

While the class structure is consistent from week to week, the assignments vary from day to day. Sometimes the assignments can be completed quickly with little effort. Other times term papers are due and research and writing requires much more time than usual. A few times each semester an hourly examination will be scheduled in each course. The time to prepare for an examination is often significant, particularly for those students who are falling behind. Because of the changing demands for time from one course to another, it is vital that several open periods be available in the daily schedule to provide you with an opportunity to engage in the extra effort needed for success.

3.3 PEER MANAGEMENT

As you move through the engineering program, you will meet hundreds of fellow students. Some of these students will be no more than faces in a common classroom, while others will become close friends. These close friends may mold your life style and behavior. Peer pressure is a recognized factor in affecting behavior. You will have an established set of behavioral and moral patterns before you arrive on campus. As you develop friendships in the classroom, dormitories, fraternities, sororities, gymnasium, etc., your behavioral patterns will change. In fact, you will become associated with several groups of friends each having different lifestyles, interests and personalities. For example, you may have a group of friends from a design team in an engineering class. You may have another group of friends from the dorms and still another from a recreational basketball team.

These groups markedly affect your social life, which is an important element of your life as you while pursue an education. It is healthy to be involved with friends, but there are two dangers of which you should be aware. First, the time involved in pursuing social activities with your friends should not interfere with any of your scheduled study time. You must remember that while pursuing a college degree, educational activities have priority.

Second, the behavior of some social groups may cause concern. Before you become deeply involved with a social group you should compare the group members' moral standards with your own. If members of the group break the law or do illegal drugs it would be in your best interest to find a different social group. It is easy to find trouble if you associate with the wrong people.

3.4 STUDY HABITS

Good study habits are important in successfully completing an engineering degree. Study habits are routines that you establish to prepare for class. There are several important aspects of study habits which affect your performance. First is timing. Some students prefer to study in the morning when they are rested and their minds are fresh. Others prefer to study very late at night when it is quiet. You should choose the time of day that is the most effective for you.

Organization is another important feature for effective study. You will be enrolled in four to six courses each semester with different requirements and different types of assignments. In addition to your time management schedule, it is important to list all assignments. Organizing the assignments, assigning priorities and making time estimates and commitments for each course are critical tasks. Time management, as discussed previously, is an important element to organization; however, priorities must be set so that assignments are completed and handed in on time. List all assignments, tasks and due dates at the conclusion of each class. Review your notes at the end of each day to establish priorities for your time as you prepare the schedule for the next day.

Developing smart study methods is also important. Most engineering students spend much of their study time working problem assignments. A typical assignment for a class may include three or four problems. The instructor probably intends the students to spend 15 to 20 minutes per problem. Students understanding the material will be able to complete the assignment in about an hour. However, if you have difficulty with one or more of the problems you will need to devote additional time to complete the work. How much time should you spend before asking for help? You probably should not devote more than about 30 minutes trying different approaches for the solution to a difficult problem. If you have not solved the problem by then, seek help from your TA, instructor or a study group.

Prepare neat homework solutions. If your homework is illegible or has many pencil/pen scratch marks on it, it may not be accepted. If you know what you are doing, most solutions can be completed with about a dozen lines and a diagram or two. Plan your approach and execute the necessary steps showing all work to arrive at the answer. If you do not show all of your work, you may not receive full credit. If you make a mistake, either erase completely or rewrite the solution. Always use a pencil because it is easy to erase after you have make a mistake. Ballpoint pens are not appropriate writing instruments for most students.

3.5 PROBLEM SOLVING METHODS

It is exceedingly important as you begin your engineering studies to develop good engineering problem solving skills. Each homework problem solution should have the following components:

1) **Basic Format:** Each problem should start on a clean side of paper (double sided is OK). You are not being graded on how little paper you can use or how small you can write! All pages should be **stapled**, with name, homework number, and section number listed. Work should be neat and legible. Most instructors reserve the option to return work ungraded if it does not meet these basic requirements.

2) **Problem Statement:** A concise (re)statement of the problem should be given. It is usually not necessary to restate verbatim the problem question from the text, but enough detail is needed to define the problem, make clear what quantities are given, and what result is being sought for the solution.

3) **Assumptions:** If any additional assumptions (not stated in the problem) are needed, they should be clearly stated.

4) **Diagrams:** A diagram indicating the coordinate axes should be drawn, and equations should be applied consistently using the convention indicated in your diagram. All symbols need to be defined, including those given as initial conditions in the problem statement or new ones that are needed for the problem solution. A free body diagram (FBD) must be included when appropriate. You may need more than a single FBD for more complicated problems.

5) **Algebraic Solutions:** Start by stating the general equations you plan to use for the solution, which should also clearly relate to the stated known and unknown variables listed from (2) and (4). An algebraic solution (e.g. in symbolic format, no numbers plugged in) should be given whenever possible, with enough steps provided for someone to logically follow your work. Numbers should be substituted into the algebraic equation to provide the numerical result with appropriate units.

6) **Numerical Results:** After substituting the numerical values into the algebraic equation, you should calculate and state the numerical result. Appropriate significant figures should be used (i.e., if your given variables only have four significant digits, your final solution should not have more than four). Your calculator can probably give you 14 digits when dividing irrational numbers, but you do not have the appropriate precision to report more than four. The final answers should be boxed so that they can be easily identified.

7) **Final Units:** You must include correct units on the final answer. This should follow consistently from the algebraic solution in step (5) above, and not simply appear with the final answer because you know what the units "should" be.

8) **Check:** You should mentally perform an arithmetic estimate of your answer to check if it is reasonable. Also, substitute the numbers into your calculator a second time to verify the result. Many careless errors are found with this simple procedure. You will also find that checking your results for accuracy, units and reasonableness is important, because errors of these types are very common.

Your instructors will expect to see all of the above components in every solution. At this stage in your development as an engineer, the problem-solving procedure is much more important than the numerical answer and problems will be graded accordingly.

3.6 <u>COLLABORATIVE STUDY OPPORTUNITIES</u>

Collaborative study is encouraged in college. However, you must recognize the difference between collaborative study and academic dishonesty. Academic dishonesty would be if you blindly copy the work of a fellow student or any other non-authorized source[1] (e.g., solution manuals, etc.). Collaborative study is when two or more people work together sharing ideas and discussing approaches for solving problems. Collaboration allows you to find errors in solutions and then to discuss these errors. Everyone involved benefits from the discussion. The individual with the correct solution increases the depth of his or her understanding by explaining the proper approach. The individual with the incorrect results

---

[1] The use of the internet to find solutions to homework problems is apparently a widespread practice among students. You should recognize that this practice is not ethical. Homework solutions should represent your independent work.

discovers the correct approach from a fellow student. This student works the problem again and has another chance to learn and to succeed. By students teaching each other, everyone's depth of knowledge is increased.

Regular meetings of small study groups are beneficial and reinforce study habits. The group bonds with time and learning becomes a social experience. Respect, confidence and esteem grow as group members become comfortable with one another.

Forming small (two to four members) study groups for each of your classes is suggested. Meet at a scheduled time after all members of the group have had an opportunity to work on an assignment independently. Compare results, share ideas and discuss difficulties. Seek knowledge from one another to determine the correct approach for solving every type of problem. If all members of the study group are encountering difficulty, seek help from your teaching assistant or instructor.

3.7 PREPARING FOR EXAMINATIONS

If you attend class, understand the material presented, and successfully complete all of the homework assignments, preparing for an examination is not difficult or time consuming. Exam preparation becomes time consuming if you have fallen behind in your daily preparation for the class and find it necessary to cram. Stress caused by cramming can be avoided if you stay current. Let's assume that you have studied regularly and are reasonably current with the material. What should you do to prepare for the examination?

A confident student with a good set of notes should try to write the examination by anticipating the instructor's questions. Anticipating the type of questions on the examination is not only possible; it is usually easy to do. In the instructor's class, he or she will emphasize a few topics and will indicate their importance. The instructor will make statements like:

- "This is really very important."
- "If you remember anything remember this."
- "I really like this derivation."
- "This is the most important principle in the course."

The instructor will usually dwell on the important topics and move quickly through the ones considered of lesser importance. Review your notes and place an asterisk beside the passages where the instructor signaled either importance or his or her personal interest. In the typical five or six week period for a preliminary examination, you may find about a half a dozen asterisks. Write a question about the topics you have identified as important and ask one of the members from your study team to solve the problems on your exam. Discuss this process with your study team and arrive at a dozen questions that the team solves. After a few examination experiences, you will gain skill at anticipating the instructor's exam questions. Sample exams and past examinations for the course should also be reviewed if they are available.

3.8 STRATEGIES FOR TAKING EXAMINATIONS

There are several strategies that you should use in taking an examination. They should enhance your performance. Let's discuss some of them:

Confidence: Study and learn until you become confident that you understand the material and that you will perform well on the examination. If you believe you will screw up the examination, you may not be comfortable enough with the material. Confident does not mean cocky, but rather comfortable with the subject and the instructor.

Calm: To be calm, you are at peace with yourself. You know that you have prepared for the examination. You are not nervous because you have demonstrated the ability to perform well on the homework assignments. You look forward with anticipation to the opportunity to demonstrate your problem solving skills.

Plan and Select: Before beginning the examination, read all of the problems. Be certain that you understand each problem. If not, ask the instructor for clarification. Then arrange them in an order from the easiest to the most difficult. Solve the problems in that order. Many instructors will arrange their examinations in this fashion with the easy problems first to give you confidence and the more difficult problems later to give range to the grades. Other instructors will arrange problems chronologically with the first problem covering material studied early in the examination period and the last problem covering more recent material.

Most hourly examinations are limited to a small number of problems. The instructor has timed the examination so that well-prepared students will have enough time to complete it. In many cases, students with experience and a deep knowledge in the subject can complete the examination in 15 to 20 minutes. You will typically have sufficient time to complete your solutions, **but not sufficient time to be studying or learning during the examination**.

Checks: Use the final ten minutes of the examination for checking your results and making sure that all of the numerical answers have appropriate units. There are three different mental checks you should go through to detect errors. The first is a simple judgment check. Does the number look reasonable or is it obviously too large or too small? Common sense often is a good guide to finding obvious errors.

The second is a numerical check. Examine the equation and make a mental estimate of the result. With practice you will soon be able to estimate the answer to within ± 10 to 20%. Next, run the numbers in the equation through the calculator to confirm the answer. In many instances, a button may have been pressed by mistake and errors result. Recalculating eliminates these careless errors.

The third check is for dimensional consistency. Sometimes errors occur because you fail to recognize the need for unit conversions or you do not make the conversions correctly. By checking an equation for consistent units, it is possible to locate and correct these errors. Equations used in the solution of engineering problems are independent of the units of measurement.

Consider an example from a physics problem, where you are to determine the distance S that a baseball falls after it is dropped from a very tall building. You seek the distance traveled by the ball after it has fallen for 6 seconds. If you assume[2] that the drag due to air resistance is negligible, the equation governing the distance the ball falls is given by:

$$S = \frac{1}{2}gt^2 \qquad\qquad (3.1)$$

where g is the gravitational constant and t is time of the fall measured in seconds.

When applying this equation, you must insure that it is dimensionally homogenous by using units that are consistent on both sides of the equation. To illustrate the various options you have in dealing with Eq. (3.1), consider the two following cases:

Case 1: Use U. S. Customary units of feet for length and seconds for time. The gravitational constant in this system of units is 32.17 ft/s2. Accordingly, you write Eq. (3.2) as:

---

[2] The drag is not negligible. It is assumed to be negligible to simplify the mathematical solution leading to Eq. (3.2). As your skills develop, drag forces will be considered and you will have the opportunity to solve the more complex mathematical expressions that result when drag forces are introduced.

$$S = gt^2/2 = (32.17 \text{ ft/s}^2)(6 \text{ s})^2/2 = (32.17 \text{ ft/s}^2)(36 \text{ s}^2)/2 = 579.1 \text{ ft} \qquad (a)$$

The time unit (s) cancels out on the right hand side of Eq. (a) and the remaining unit represents a length, expressed in feet.

Case 2: Use SI units of meter for length and seconds for time.

The gravitational constant in the SI system of units is 9.807 m/s2. Accordingly, you write Eq. (3.2) as:

$$S = gt2 /2 = (9.807 \text{ m/s2})(6 \text{ s})2/2 = (9.807 \text{ m/s2})(36 \text{ s2})/2 = 176.5 \text{ m} \qquad (b)$$

The time unit cancels out on the right hand side of Eq. (b) and the remaining unit represents a length expressed in meters.

A careful check of units on each side of the equation will allow you to locate errors caused by using inconsistent units when evaluating equations.

3.9 ASSESSMENT FOLLOWING THE EXAMINATION

Errors on an examination can be grouped into three classes.

1. Careless mistakes
2. Execution errors
3. Concept errors

Careless mistakes are when you clearly understood the material but you lacked focus. You may have had the correct numbers substituted into the correct equation, but the wrong answer for the final result. You may have accidentally reversed digits in a number or forgot to specify the units. It is recommended that you slow down and increase your concentration on the problem. The more time you spend checking your results, the more careless errors you will find and correct before the end of the examination.

Execution errors occur when you understand the principles covered on the examination, but lack skill in applying these principles to solve one or more specific problems. Examples include leaving out a force in a physics problem or using inappropriate compound combinations in an equation representing a chemical reaction. Execution errors indicate a need to solve more and different homework problems. The more problems you independently solve correctly before the examination, the better prepared you will become for problem solving during the examination.

Concept errors are much more serious. They occur when you do not use the correct principle in your attempt to solve a problem. Concept errors indicate that you do not understand the material. These errors signal that you may have difficulties in earning a passing grade in the course. Hopefully, the instructor will identify this type of error with a comment on the graded examination. If an explicit comment is not made, the instructor may signal his or her concern by giving you zero credit for your unsuccessful efforts toward a solution. If you have concept errors it is important that you seek help from the instructor as soon as possible. It will be necessary to devote more time to the course until you have mastered the material.

It is important to carefully review your performance following an examination. The instructor will usually devote class time to show the correct procedure for solving every problem on the examination. You will have the information necessary to study your mistakes. Classify them and adapt your study habits to improve your performance. If you find an execution error, ask questions to your study group or your instructor until you clearly understand the correct solution to the problem.

Unfortunately, many students do not attempt to learn from their errors. It is essential that you understand why you made the error because it would be senseless to lose points again for the same error. Engineering is a discipline that builds from a basic foundation to more advanced subjects. If you do not have a strong understanding in the foundation, your performance in the more advanced subjects will be severely impaired. You will repeat errors and fall further behind. Make it your business to identify your errors on an examination, correct your procedures and **never repeat the same error**.

3.10 <u>SUMMARY</u>

An overview for your success as a student in a College of Engineering has been presented. Success as a student of engineering depends on setting goals and priorities, developing confidence in your ability to perform and being dedicated to staying current in every course. Also important is attending class, managing time, developing good study habits, managing your relationships with many new people and seeking help when the need arises.

Time management requires generating schedules in which your time is carefully designated to activities such as attending class, studying social meetings, and maintaining your physical well being. Weekly and daily schedules have been described in considerable detail. These schedules are very important tools that will help you to effectively manage your time commitments.

A brief discussion was given on peer management. You will be interacting with several new groups of friends in the next few years. Your relations with some of these groups will be very helpful. Study groups and design teams provide opportunities for collaborative learning and support. Groups formed to workout in the gym or to play club sports help maintain your physical well-being. Most social groups provide you with enjoyment, which is very important during your education. However, it is important to compare the ethical standards of a group with your own set of standards. Participate in a group in which you are comfortable.

Study habits important to your success have been discussed. Advice has been given on how to study smart and how to prepare homework assignments that are acceptable and respected by your instructor. Study groups have been encouraged because of the many benefits of collaborative study.

An eight-step approach to problem solving was reviewed. While eight steps may seem tedious when you are trying to hurry through an examination, experience shows that the process actually saves time and reduces errors.

Collaborative study opportunities were described. Study teams or design teams are encouraged for all of your classes. Meet at scheduled times after each team member has had an opportunity to work independently on an assignment, and then compare and discuss approaches and results. Student to student learning is effective. There will be occasions when you cannot determine how to solve a problem on your own but your study partner does know how (or vice versa). Respect, confidence and esteem grow as the group members bond and become comfortable with one another.

There are resources and guidelines in the College of Engineering to help you succeed in your studies. The counselors, instructors and teaching assistants are ready to help you. However, they usually will not initiate the process. It is your responsibility to seek help. In the end, the key to being successful in a College of Engineering requires that you:

- Prepare in advance for every class
- Arrive on time and stay awake and alert
- Ask questions when you become confused
- Complete homework assignments correctly, completely and on time
- Stay current with all assignments

REFERENCES

1. Combs, P. and J. Canfield, <u>Major in Success: Make College Easier, Fire Up Your Dreams, and Get a Very Cool Job</u>, 3rd Ed., Ten Speed Press, Berkeley, CA, 2000.
2. O'Brien, P. S., <u>Making College Count: A Real World Look at How to Succeed In and After College</u>, Graphic Management Corporation, Green Bay, WI, 1996.
3. Hanson, J., <u>The Real Freshman Handbook: An Irreverent and Totally Honest Guide to Living on Campus</u>, Houghton Mifflin, Boston, MA, 1996.
4. Carter, C., <u>Majoring in the Rest of Your Life: Career Secrets for College Students</u>, 3rd Ed., Farrar Straus and Giroux, New York, NY, 1999.
5. Davis, L., <u>Study Strategies Made Easy</u>, Specialty Press, New York, NY, 1996.
6. Dodge, J., <u>The Study Skills Handbook: More Than 75 Strategies for Better Learning</u>, Scholastic Trade, New York, NY, 1995.
7. Luckie, W. R., et al, <u>Study Power Workbook: Exercises in Study Skills to Improve Your Learning and Grades</u>, Brookline Books, Cambridge, MA, 1999.
8. Tufariello, A., <u>Up Your Grades: Proven Strategies for Academic Success</u>, vgm Career Horizons, Lincolnwood, IL, 1996.
9. Hjorth, L., <u>Claiming Your Victories: A Concise Guide to College Success</u>, Houghton Mifflin, Boston, MA, 2000.
10. Jensen, E. and T. Kerr, <u>Student Success Secrets</u>, 4th Ed., Barron's Educational Series, Hanppauge, New York, NY, 1996.

EXERCISES

3.1   Prepare three different lists of goals—short term, intermediate and long term. Define the time for accomplishment of each set of goals and provide your motivation for setting each goal.

3.2   Describe your plan for staying current in each and every subject for which you have registered this semester.

3.3   Prepare a weekly schedule for your activities this semester similar to one shown in Fig. 3.2. Also list time reserved for class, study and social activities. How many open hours were you able schedule?

3.4   Describe the provisions in your schedule for taking care of your physical and mental health.

3.5   Prepare a daily schedule for each day of the week. Identify your open periods and discuss a strategy for using them when an examination is scheduled.

3.6   Do you know the procedure to apply for scholarships, grants or financial aid?

3.7   Have you joined a study group (team) for this class? If so, when and where do you study?

3.8   Have you ever visited your instructor's office during his or her office hours? Is he or she ready to receive you? Did you find the help provided useful? Do you plan on visiting again when you encounter difficulties with the course material?

3.9   Which is the most difficult course you are taking this semester? How much time do you allocate for study in this course? Is that sufficient? What other strategies do you employ to improve your performance other than study time?

3.10 Explain the methods you use to check answers during an examination?

CHAPTER 4

ENGINEERING ETHICS AND DESIGN

4.1   INTRODUCTION TO ETHICS

4.1.1 Professional, Organizational and Personal Ethics

Technology represents the potential for both benefit and harm to society. Engineered products can kill or injure large numbers of people, either "all at once" in catastrophic injury or one at a time in "routine" injuries. Technological development is often "asymmetrical" in that those who benefit from the technology may not be those at risk of injury. Because the developers and users of technology may not internalize the full costs of injuries, they may have little incentive to avoid potentially harmful innovations. The **public** safety problem of engineered products has therefore always been a matter of social, not merely private concern. Before society fully accepts a technology, the public typically demands guarantees that public safety will be protected. Society (governments) often creates safety regulations to help protect the public.

But while regulation may be one method of protecting public safety, there are many reasons why it is inadequate. Technology develops with time and there is a need for flexibility in the regulations. If engineers can be trusted to design for safety, society can delegate responsibility for product design. This delegation requires an assurance that public safety will be protected by the engineers without governmental supervision, acting as **professionals** serving the public interest. This concept is the foundation for professional ethics in engineering design. **The Code of Ethics approach does not fully substitute for the legal system, but it provides an important complement to an often overworked, rigid or otherwise inadequate regulatory system.**

Public safety is of course only one of the ethical issues in engineering, but it can be used to explore many of the most difficult problems of ethics and the practice of engineering. Ethics is a very complex and culturally driven subject. On a **personal** level ethics deals with issues of character, loyalty, honesty, decency and all the other virtues that make interpersonal dealings more comfortable and efficient. On an **organizational** level it describes the informal and formal rules that govern the internal functioning of an organization, which in turn create the **public face** for the organization. Finally on a **professional** level it describes the mutual expectations of those who practice a given profession and the acceptable public expectations, which control the profession.

This chapter will primarily focus on **professional ethics** in the design process. This in no way detracts from the importance of other ethical issues for engineers. Many aspects of personal ethics are inherent in the professional ethics of engineers, so the importance of those traits is recognized. The engineer's responsibility within an organization is also part of professional ethics.

4.1.2 Professions and Ethics

Professions can be defined as groups of trained individuals who mutually agree to enforce standards designed to serve a **moral** ideal.

> **Moral—relating to the practice, manners, or conduct of men as social beings in relation to each other, as respects right and wrong, so far as they are properly subject to rules.**

The origin of the concept of professional ethics is generally attributed to physicians. The **oath of Hippocrates** in its many forms is thought to establish the beginnings of a concept of professional Ethics. For example the oath states

> **I will prescribe regimens for the good of my patients according to my ability and my judgment and never do harm to anyone.**

This obligation, to keep the **patient's** well being at the center of medical practice, establishes a far different relationship between physician and patient than between seller and customer. Attorneys also developed canons of ethics that governed their actions. The clear goal of such canons was to ensure that their public responsibility outweighed their private interests. As early as the 13th century attorneys in Britain took the Lawyer's oath.

> *You shall doe noe Falshood nor consent to anie to be done in the Office of Pleas of this Courte wherein you are admitted an Attorney. And if you shall knowe of anie to be done you shall give Knowledge thereof to the Lord Chiefe Baron or other his Brethren that it may be reformed. You shall Delay noe Man for Lucre Gaine or Malice*
> *You shall increase no Fee but you shall be contented with the Old Fee accustomed. And further you shall use your selfe in the Office of Attorney in the said office of Pleas in this Courte according to your best Learninge and Discrecion.*

This oath, which is the ancestor to all those used by attorneys in the USA today, establishes the key components of professional ethics:

1. A professional will be personally honest, and insist that all they deal with be honest.
2. When unethical conduct is found, the professional will **blow the whistle** on the offender.
3. The professional will give those matters, entrusted to his or her care, his or her best professional effort.
4. The professional will deal fairly with the public in carrying out professional duties.

It is very important to understand that a professional often has defined duties owed to persons other than those who are paying the person's salary or fee. Professionals can owe duties to other specific persons or even the public as a whole. These may be described as the **beneficiary** of the professional's actions. In law and medicine for example ethics fundamentally defines the **beneficiary** who is owed the highest duty. The military lawyers who are defending the **unlawful combatants** in Guantanamo are upholding the highest ethical responsibility a lawyer can have—the obligation to protect your client no matter who is the employer. Similarly corporate physicians who protect the safety of research subjects against the financial interests of their employers are acting in accordance with an ethical code that is over 2,000 years old. Lawyers and physicians take enormous pride in these professional standards and feel betrayed when a member of their profession **lets down the side**, as President Clinton did when he misled a court.

A professional cannot do something which violates ethical obligation to the beneficiary, even if the employer wants it. A professional is given special privileges in return for a commitment to the public interest, not merely the person's private interest. When a profession is accepted by society, the public is invited to depend on the commitment of professionals to their ethical norms, usually expressed in codes of ethics. Under this **Professional Codes of Ethics** approach, the commitment to the public interest is enforced by professionals themselves.

4.2 ENGINEERING AS A PROFESSION

Doctors and lawyers are clearly professionals, but what about engineers? Do engineers actually accept this same understanding of their role? Do they see any obligation to protect the public from the decisions of their employers or their clients? Or do they merely consider themselves highly trained technical employees, who owe the public no special duty? Do they think that if they comply with all rules and regulations they are acting ethically even if the product is dangerous?

The answer is not easy. Engineering is clearly an emerging profession, and the concept of serving the public as opposed to private interests is by no means accepted around the world. However, at least in the U. S., the claim of engineers to professional status and commitment to public safety has been made for many years. As one court stated:

> **An obligation to follow good engineering practice (as distinct from the FDA's Good Manufacturing Practices) may arise for the unlicensed engineer, as a professional obligation. The Court views such professional obligations—as the FDA undoubtedly does—as inherent in all medical manufacturing. [1]**

4.2.1 The History of Engineering Ethics in the U.S.

Engineering codes of ethics in the U.S. date to just before the First World War. In the beginning, many such professional codes were anti-competitive and primarily designed to enhance the dignity and income of the profession. However, by the 1970s engineers were clearly making the claim to true professional status directly connected with service and obligations to the public. The bedrock public interest principles of professional engineering ethics were found in statements such as the Canons of Ethics of the Accreditation Board for Engineering and Technology (ABET). Among the most important are the Fundamental Canons:

Engineers, in the fulfillment of their professional duties, shall:

> 1. Hold paramount the safety, health and welfare of the public.
> 2. Perform services only in areas of their competence.

These Canons acting together require engineers to both fully understand the products and systems that they are designing and insure those products meet a high standard of public safety. They have become widely accepted around the world. As one author describes Engineer's obligations:

> **Protect the health and safety of citizens in the host country**. The requirement to hold paramount the health and safety of the public is explicitly recognized in most engineering codes, and so it must have the status of a universal norm for engineers.[1]

The profoundly difficult task engineers set for themselves in these Canons should be recognized. Arguably neither attorneys nor physicians make such a sweeping commitment to **hold paramount** the general public interest. Attorneys and physicians are taught that they are normally serving the public interest by devoting their honest and ethical skills to their client or patient. The assumption is that the benefit to the public is derivative of their professional treatment of the patient or client.

Engineers have a much more complex problem, because the beneficiary of their actions is supposed to be the public but the **paying client** is often the party pressing them to design some product that may

---

[1]Charles E. Harris, Jr., Internationalizing Professional Codes in Engineering Science and Engineering Ethics (2004) 10, 503-521.

potentially put the public at risk. This requirement unquestionably puts the engineer in a difficult position, because unlike the doctor or lawyer, they do not have a beneficiary sitting across the room watching their actions. They also may lack strong peer support in protecting the public.

Engineered products routinely function correctly within their intended design envelopes. This is a tribute to the skill and ethics of the designers. Catastrophic failures due to failures in the process of engineering decision making are fairly rare. But when such failures do occur, they force a reexamination of the assumptions and processes that produce the decisions as well as the ethical issues involved.

One peculiarity of Engineering in the United States is that some disciplines are primarily represented by state licensed professional engineers who are permitted to put the terms PE after their names. These tend to be fields in which engineers work as **consultants** for **clients** or those where regulatory requirements for PE approval are necessary. In other fields, engineers tend to work as employees for large firms and PE status is rare or relatively unimportant. The question might be raised whether this changes the ethical responsibility of the engineer. This is an ongoing debate. Many but not all non licensed engineering societies have also adopted codes of ethics.

Engineers have a vested interest in preserving and enhancing the confidence of the public. To gain public confidence, it is essential that they all behave in an ethical manner. All three types of ethics are important. **Personal** ethics set a foundation. Because engineering developments are sponsored, financed and controlled by businesses, **organizational** ethics are critical. Finally, compliance with **professional** norms is the last and most important step.

4.3 <u>RIGHT, WRONG OR GRAY</u>

Christina Sommers [2] develops an interesting viewpoint regarding what is considered right, wrong and controversial by society-at-large. She argues that there are some ethics related issues that are not clearly defined, and that society has not reached a consensus regarding the correctness of these issues. You can develop cogent arguments for and against a specified viewpoint for controversial issues such as abortion, affirmative action, handguns, gambling, capital punishment, nuclear weapons, etc. She refers to these controversial issues as **dilemma** ethics that should be distinguished from **basic** ethics. With **dilemma** ethics, society-at-large is not certain about what is right, wrong or gray. There are several attitudes prevalent in society—sometimes the issue is so important to segments of our population that demonstrations are organized to elevate one viewpoint or another. Occasionally a group can get its ethical viewpoint enacted into law, but without the full backing from the majority of society such a law will often fail.

These dilemma issues reflect the public ethical face of a society. The term **moral ecology** can describe the overall environment in which a profession functions. In some instances, society changes its moral ecology and its law. An example of a failed moral change in this country was prohibition, which was enacted by a constitutional amendment ratified in 1919. The government wanted to make it illegal to consume alcoholic beverages, but society-at-large wanted to drink. Bootlegging, racketeers and speakeasies followed bringing crime and violence to many neighborhoods. When it became evident that the law was causing more problems than it solved, it was repealed in 1933.

From an engineering standpoint, dilemmas include topics such as nuclear weapons development or response to global warming. These are essentially political decisions, which engineers have the right and duty to influence as citizens, and where they may provide specialized input. However, the profession neither has nor claims any unique insight. No attempt will be made in this textbook to resolve any of the dilemma issues. These issues have been unresolved for decades. These issues will require much more time and understanding before our society is willing to reach the consensus necessary for resolution. It is much more productive to consider **basic** ethics where right and wrong or white and black are much more clearly recognized by the majority of society.

4.4   PERSONAL ETHICS

Many people are confused about the value of personal ethical behavior. The results of several polls indicate that Americans are less ethical today than in previous generations. Sixty-four percent of a sample of 5,000 individuals admits to lying if it does not cause real damage. An even larger percentage of those sampled (74%) will steal providing the person or business that is being ripped off does not miss the item [3]. Many students (75% high school and 50% college) admit to cheating on an important exam [4]. Most students have not developed a moral code to use as a guide for their behavior.

Virtues

To understand personal ethics we can start with the four traditional cardinal virtues[2]:

- Prudence
- Justice
- Fortitude
- Temperance

Combined together these virtues have been considered the foundations of **right living** in Western societies for the past 2,300 years.

Prudence refers to careful forethought, good judgment and discretion. In other words, think before you act and exercise care and wisdom. If you act, will your action cause problems, now or later, for you or anyone else? Are you careful about your remarks concerning others? One of my rules is not to speak ill of others under any circumstances. Is the action that you are about to undertake in your best interest? Will that action be appreciated by your family, friends, coworkers, managers and peers?

Justice involves fairness, honor, giving everyone their due. It is clearly wrong to lie, cheat or steal. Justice is not the same thing as legality. Actions may be lawful but unjust, such as slavery was earlier in our history.

Fortitude is about courage and persistence. Will you stay with an idea, even if it is not popular, if you know that it is the "right" approach? Showing determination and resolution is sometimes difficult when the risks of failure are high or when peers encourage you to abandon your idea. Fortitude also discourages rashness

Temperance refers to moderation in pleasure seeking. Temperance can also mean control of human passions such as anger, lust, hostility, revenge and spite. Most individual behavior can be judged by very clear and well-understood virtues that are part of **basic** ethics. Many institutions teach and counsel compliance with these norms as the basis for human interactions.

4.5 ETHICS AND LAW

Governments (federal, state and local) also play a role in defining ethical behavior, because they generate laws and set public policy. Laws (mostly at the state level) are written on the basis of the **police power**, which is the general power to protect the health, welfare and safety of the people. Moral or even religious values routinely influence the creation of law. While notions of just and unjust laws are beyond the scope of this chapter, the role of ethics in the creation and enforcement of law is very important to engineering design.

---

[2] These virtues are of course described by their western European traditional names. However, all are represented in both Islamic and Eastern philosophies

The purpose of most laws is to prohibit a well-defined set of vices. For example, it is illegal to sell drugs, rob banks, commit burglary, kill or abuse another human, or engage in prostitution. It is also illegal to park overtime at a parking meter or start a business without a license. However these violations are not the same type of unlawful act. Legal scholars describe two different types of unlawful acts based on essentially ethical considerations.

> **Malum in se** is Latin for **evil in itself** and would normally apply to acts that are universally described as morally and ethically wrong. Such laws would derive from principles of right and wrong and violations are considered as involving **moral turpitude**.

> **Malum prohibitum** is Latin for **wrong because it is prohibited**. These are acts that are unlawful but not otherwise immoral. Violation of a copyright for example is usually considered malum prohibitum.

Personal ethical violations tend to resemble offenses under **malum in se** concepts. Rape, murder and robbery are all malum in se. Professional ethical violations, such as contempt of court by an attorney, can often be considered **Malum prohibitum**. Former President Clinton, for example, was disbarred for a professional ethical violation that did not meet the stricter standard of the criminal law.

The difference is important because an ethical or legal violation can occur without the individual feeling they did anything **wrong**. "I complied with all regulations"—"I was only following orders"—"I was told it was out of my job responsibility" are routine justifications heard after a disaster. But only acts which are **malum in se** require a guilty mind. "Knowing violations" are ethically easy, because they involve the "guilty mind" or **mens rea** that is routine in the law. But other acts without such a guilty mind may be just as unethical.

On the other hand, some acts may even involve legal consequences are not considered by the law to have any moral element at all. For example, in the law there is no **moral element** in breach of contract. You may have to pay damages, but you have done nothing morally wrong in the eyes of the law. Whether such conduct is ethical or not depends on the field.

The relationship between law and ethics is therefore complex. In a few cases, the legal system has overruled the profession, such as on limitations on professional advertising or restraints on competition. In other cases, such as insider trading, it has added the force of law to what was already considered an ethical violation. The physician who is legally allowed to perform an operation for which she is unqualified is acting in an unethical manner. An engineering ethical problem however can arise without bad intent due to the decentralized nature of engineering decision making.

4.6 ETHICAL VIOLATIONS WITHOUT BAD INTENT

A professional ethical violation can occur even if the professional does not believe they were doing **wrong**. A common case is where an engineer "doesn't know what he or she doesn't know". Recall that the code requires:

Engineers, in the fulfillment of their professional duties, shall:

> 1. Hold paramount the safety, health and welfare of the public.
> 2. Perform services only in areas of their competence.

For example consider the stairway at the University of Maryland's Comcast center, shown in Fig. 4.1. The professionals who designed the northeast exit of the Comcast Center created a "funnel" that could cause a lethal crowd crush. A pinch point is obvious halfway down the left side. The engineers who approved the design simply did not recognize the hazard, possibly because their training did not include "crowd crush". Brannigan [5] has described this ethical issue as **design process failure**, and it can occur any time critical design decisions are disaggregated. In such cases, different design concepts, knowledge or assumptions from different design groups combine to create unsafe products, even if no individual wanted to create the safety hazard. Engineers have an ethical duty to make sure that the entire system for which they are designing a component holds paramount the public safety.

Fig. 4.1 Comcast Center as constructed on the University of Maryland campus. Note the two funnels created by the stair rails.

4.7   ORGANIZATIONAL ETHICS

Corporations are entities that under law are treated as **juridical persons**. A corporation, acting like a person, has the right to conduct business and the obligation to pay taxes. Just as individuals are judged by their ethical behavior, corporations are judged by their ethical conduct [7]. The chief executive officer (CEO) for a corporation sets the moral and ethical tone for the organization. Any large business is organized with several layers of management. If the CEO operates the business with the highest ethical standards, the middle and lower level managers will normally follow the example set at the highest level in the corporation. However, if lax ethical standards are permitted, a pattern is set and questionable ethics will become the corporate style and standard.

In an excellent paper, Baker [6] has developed a list of issues that occur daily in a typical corporation. The manner in which a corporation deals with these issues establishes its character. Baker's list shown below includes:

1. How people, employees, applicants, customers, shareholders and the families who live near our facilities are treated?
2. Is everyone free of systemic or individual practices of discrimination?
3. How do we spend our shareholders' money on our expense accounts as an institution and as individuals?
4. What is the level of quality that goes into our products? Do we meet our customers' expectations for quality? Do we meet our own standards for quality?
5. What is our concern for safety, not only for our employees, but also for our customers and our neighbors in communities in which we operate?
6. How "ethically" do we compete?
7. How well do we adhere to the laws of the locales, regions and nations in which we do business?
8. What is acceptable gift giving and gift taking?
9. How honest are our communications to our employees and our public advertising?

10. What are our corporate and personal positions on public policy issues, and how do we promote those positions?

This is a long list, but we certainly could add more items to this catalog of concerns—sustainability and the environment for instance. Adding more items is not as important as recognizing that corporate decisions are made on a daily basis by its managers and employees. These day-to-day decisions determine the corporate character. The ethical judgments that serve as the basis for most of these decisions will depend on individual personal values (virtues) and experience. The quality and consistency of the ethics applied in these decisions will markedly affect the success of the corporation, its management and its employees.

4.8 PROFESIONAL ETHICS FOR ENGINEERS

A personal code of ethics is the cornerstone to achieving a meaningful sense of moral values. Professional ethics builds on a well-understood personal code of ethics. Hopefully, the preceding sections of this chapter will be helpful in any attempt you make to establish a sense of right and wrong.

There are several codes of ethics for engineers. Most of the founding societies of engineering have developed and distributed codes and guidelines for professional conduct. The codes of ethics for the different engineering societies are all similar. The fact that each society has their own code is more to insure a complete distribution of the code than to pursue unique ethical issues.

Accreditation Board for Engineering and Technology*

CODE OF ETHICS OF ENGINEERS

THE FUNDAMENTAL PRINCIPLES

Engineers uphold and advance the integrity, honor and dignity of the engineering profession by:

I. using their knowledge and skill for the enhancement of human welfare;

II. being honest and impartial, and serving with fidelity the public, their employers and clients;

III. striving to increase the competence and prestige of the engineering profession; and

IV. supporting the professional and technical societies of their disciplines.

THE FUNDAMENTAL CANONS

1. Engineers shall hold paramount the safety, health and welfare of the public in the performance of their professional duties.

2. Engineers shall perform services only in the areas of their competence.

3. Engineers shall issue public statements only in an objective and truthful manner.

4. Engineers shall act in professional matters for each employer or client as faithful agents or trustees, and shall avoid conflicts of interest.

5. Engineers shall build their professional reputation on the merit of their services and shall not compete unfairly with others.

6. Engineers shall act in such a manner as to uphold and enhance the honor, integrity and dignity of the profession.

7. Engineers shall continue their professional development throughout their careers and shall provide opportunities for the professional development of those engineers under their supervision.

111 Market Place, Suite 1050, Baltimore, MD 21202-4012

*Formerly Engineers' Council for Professional Development. (Approved by the ECPD Board of Directors, October 5, 1977)

AB-54 2/85

Fig. 4.2 ABET's code of ethics.

Let's consider the Code of Ethics of Engineers, presented in Fig. 4.2, which was sponsored by the Accreditation Board for Engineering and Technology (ABET). The code is divided into two parts. The first part deals with four fundamental principles, and the second part contains seven fundamental canons. The Engineers' Council for Professional Development first advanced this code in 1977. The fact that the code remains without modification for more than 30 years indicates that professional ethical values are as constant as personal ethical values.

The fundamental principles seek to insure that the engineer will uphold and advance the integrity, honor and dignity of the profession. These goals are common to your personal goals that were discussed previously; however, as indicated below, the approach to achieve the professional goals is different:

1. We are selective in the use of our knowledge and skills so as to ensure that our work is of benefit to society.
2. We are honest, impartial and serve our constituents with fidelity.
3. We work hard to improve the profession of our discipline.
4. We support the professional organizations in our engineering discipline.

The seven fundamental canons in the ABET Code of Ethics of Engineers, presented in Fig. 4.2, are explained in considerable detail in guidelines that are used to expand and clarify the relatively brief statements that represent the **regulations** that shape our professional behavior.

4.9   ETHICS IN LARGE ENGINEERING SYSTEMS

Engineers design and build many different products each year that are included in large and complex sociotechnical systems. Usually these products and the systems are conservatively designed with adequate safety factors, carefully tested, and perform well in service for extended periods of time. These products provide a much needed service without endangering either individual or public safety. However, from time to time mistakes are made in the initial design. These mistakes are usually detected in prototype testing and eliminated prior to releasing the product to the market place. In very rare circumstances, a mistake or several mistakes are overlooked, for a variety of reasons, and the system is released and placed in service with an unacceptably high probability for failure. An undetected mistake often leads to a catastrophic accident. The space shuttle system clearly falls into this category. If you are a space supporter, you may argue that the risks are worth the benefits. But if this is the case, an unprepared and untrained high school teacher should not be invited to ride along to enhance the image of the space program. It is one thing to order a career astronaut into peril, which they understand and accept, but a totally different proposition to invite an uninformed civilian to participate in a very dangerous mission.

In Chapter 5, the probability of failure (or an accident) will be discussed in considerable detail. You must understand that there is always some risk of failure when designing high performance systems. It is important to learn to accept a trade-off between risk, safety and performance—it is an inherent part of the process. In most complex sociotechnical systems (air transportation for example), there is a small but finite risk for failure with a subsequent loss of life and property. The public must know this risk. It is the responsibility of the engineering community, industry and the government to alert potential customers to the dangers involved. Knowing the risk, the customers can decide whether or not they want to use the system.

Some people worry more than others and place a very high value on safety. They require a probability of failure of nearly zero. Do you know someone who will not travel by flying? (John Madden, the popular professional football announcer, travels from game to game each week on a special bus). Other folks love the thrill of a risk, and they are willing to accept a much higher probability of an accident. They think hang-gliding or skiing on black diamond slopes is great sport.

4.10   THE NASA SPACE FLIGHT PROGRAM—A CASE STUDY

4.10.1 The Challenger Accident[3]

With this background on risk, safety and performance, let's begin a discussion of the Challenger accident [7]. The Challenger was one of the original four orbiters built by the National Air and Space Administration (NASA) to serve the space shuttle system.   The rocket fuel tank , containing liquid hydrogen, on the Challenger 51-L mission exploded 73 seconds after launch on Tuesday, January 28, 1986. The crew of six and a schoolteacher passenger were killed, and the space shuttle was lost. Among the dead was Dr. Judith A. Resnik, a graduate of the Clark School of Engineering at the University of Maryland

The Challenger accident has been selected as a case study, because it illustrates several ethical issues in the engineering and management of large, complex and inherently dangerous systems. Some issues that will be raised are:

1.   The design of the shuttle and the selection of the contractors involved many political considerations [7].
2.   The lack of communications between key people and organizations was a significant factor in the accident.
3.   The interface between the upper-level administrators (business managers) and the engineers was an important element in the decision to launch on that disastrous morning.
4.   Public attention and opinion, not safety, markedly affected the decision-making process.
5.   The risk potential was not known by the public and not appreciated by top administrators and politicians directly involved in the launch decision.
6.   Christa McAuliffe, a high school teacher and mother of two children, was killed in the accident. Why was she invited to fly on such a dangerous mission?

Background Information

To set the stage for the accident, you have to go back to the early 1970s. NASA had been very successful with the Apollo missions (even with the problems of Apollo 13), and looked forward to larger and more aggressive space endeavors. They proposed an integrated space system that would include a space station, space shuttle, space tug and manned bases on both Mars and the moon. This agenda sounded wonderful until the public examined the price tag. The public did a quick look and wanted no part of it. A poll indicated that the public believed that Apollo had been too costly. Politicians, always driven by the polls, took note and reduced NASA's budget. NASA recognized the need for a new, cost-effective project to follow Apollo that the public (and politicians) would buy. Responding to these political pressures, NASA proposed the space shuttle that would serve the military, the scientific community and the rapidly growing commercial business of placing satellites in orbit. The space shuttle system, illustrated in Fig. 4.3, was marketed as a relatively routine space transport system. Even the name "shuttle" connected with the airline shuttle services that routinely fly every hour from one large city to another.

To make the space shuttle system a commercial success, NASA proposed a fleet of four orbiters [7], which would eventually fly on a weekly basis (NASA initially set a goal of 160 hours for the turnaround time for an orbiter). They planned on nearly 600 flights in the period from 1980 to 1991. The early estimate of the cost of a launch was $28 million with a payload delivery cost of $100 to $270 per pound. After some

---

[3] On June 1, 2014 the New York Times released a video title "Challenger and Columbia and the Nature of Calamity," which presents a detailed description of the events leading to both accidents. We strongly suggest you Google the title and watch the video to see a demonstration of system failure and management failure at NASA and Morton Thiokol. See also the Retro Report by Clyde Haberman.

operational experience, the cost estimates proved to be completely unrealistic. Launches actually cost on the order of $280 million, and the cost to place a pound of payload in orbit on the shuttle was in excess of $5,200 [8]. Early experience with the space shuttle system indicated that NASA could not hold their schedules and their costs were running more than a factor of 10 higher than the original estimates. Because of its poor performance, NASA was struggling to improve its image as 1985 ended.

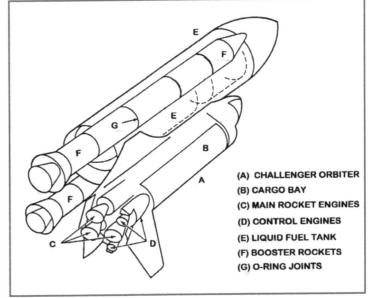

(A) CHALLENGER ORBITER
(B) CARGO BAY
(C) MAIN ROCKET ENGINES
(D) CONTROL ENGINES
(E) LIQUID FUEL TANK
(F) BOOSTER ROCKETS
(G) O-RING JOINTS

Fig. 4.3 Illustration showing the main components on the NASA space shuttle.

Prior to the launch of the Challenger on January 28, 1986, twenty-four shuttle flights had been made. These flights were all successful in that they returned to earth with all crewmembers safe. They also showed the operational capabilities of the shuttle system in placing commercial satellites into orbit, repairing satellites in space and salvaging malfunctioning satellites. However, these flights also indicated that the shuttle system had many serious problems. The three main liquid rocket engines were too fragile with many critical components. Some of the tiles in the heat shield needed repair and/or replacement after every flight. The computers and the inertial navigation systems experienced occasional failures. The brakes and landing gear were stressed to the limit when the orbiter (an 80 ton dead-stick glider) landed at speeds ranging from 195 to 240 MPH. Finally, the seals in the solid fuel booster rockets showed distress (sometimes extensive) in 12 of the 24 previous launches.

The record of the shuttle from 1981-1985 showed a consecutive series of successful launches, but with many prolonged delays to repair failing components in a very large, highly stressed system. The maintenance records showed so many problems that NASA estimated it took three man-years of work preparing for a launch for every minute of mission flight time. Clearly, the word shuttle to describe such a transportation system is a misnomer.

The Solid Propellant Booster Rockets

The explosion of the main fuel (liquid hydrogen) tank on the Challenger was due to the failure of the O-ring seals on the solid fuel boosters that were adjacent to the hydrogen fuel tank. To understand the seals and their purpose, it is essential that you appreciate the size and function of the two booster rockets used to provide much of the thrust necessary for the launch. They are enormous cylinders—12 feet in diameter and 149 feet tall. Each cylinder is filled with 500 tons of a solid propellant consisting of a rubber mixture filled with aluminum powder and an oxidizer—ammonium perchlorate. When ignited, the propellant burns to produce an internal pressure in the motor case of about 450 psi) at a temperature of about 6,000 °F. The expanding gasses exit the rocket motor case through a nozzle, to produce about 2.6 million pounds of thrust from each booster. These booster rockets are essentially very **big** bottle rockets!

Big was the nub of the problem. The solid rocket boosters were made by Morton Thiokol in Utah, but launched from the Kennedy Space Center in Florida. They were much too long to ship across the country as a single cylinder. To circumvent this problem, the motor casing was fabricated in segments—each 27 feet long. The segments were filled with propellant and shipped by rail to the launch site in Florida. They were then assembled in a special facility near the launch pad. This procedure solved the shipping problem, but it created a different problem. The rocket casing is a pressure vessel that must contain very hot gasses at a pressure of about 450 psi. The joints, where the cylindrical segments of the motor case were fitted together, must be sealed so that these hot gasses will not leak and cause damage to adjacent components of the launch vehicle.

The seal was made with a pair of rubber O-rings fitted over the internal finger of a clevis type of joint, as shown in Fig. 4.4. The O-rings, 0.280 inch in diameter, were compressed between the two surfaces affecting a seal that prevents the pressurized fluid from leaking past the joint. A putty like compound was used to prevent the hot gases from eroding the O-rings. The pins used to lock the segments together (axial constraint only) did not clamp the clevis fingers about the center finger.

The sealing of the segmented cylinders with the O-rings was not a new concept. A single O-ring had been employed previously on the Titan III rocket effectively sealing the circumferential joints on a smaller diameter motor case. Morton Thiokol, the contractor building the boosters, introduced the second O-ring to provide a margin of safety in the event the first O-ring failed. It all sounded feasible, and the initial test firings of the booster in Utah were apparently successful.

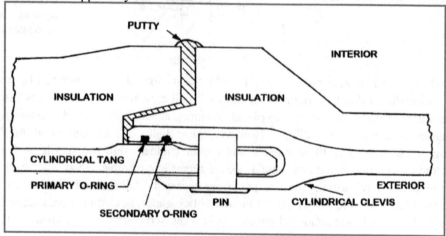

Fig. 4.4 Design of the O-ring seal used at the circumferential joints of the solid rocket motor cases.

Unfortunately, one test does not validate the safety of even a simple system. To insure high reliability of a component, many tests are required. Moreover, this test firing had little or no bearing on problems associated with the reuse of the solid rocket boosters (up to 20 times). After splash down following a launch and recovery from the Atlantic Ocean, the cylinders were shipped back to Utah to be filled and used again.

Failure of the O-ring Seal

Motion photographs taken at the final launch of the Challenger indicated that the O-ring seals failed almost immediately after ignition of the booster. The hot gasses cut through the O-rings and the joint of the booster. A hole was formed in the wall of the booster at the joint, and a flaming jet of white-hot gasses escaping from the booster rocket cut through the wall of the adjacent tank containing the liquid hydrogen. The launch was effectively over. While many things happened in the last five seconds before the devastating explosion—all bad—the penetration of the adjacent tank, which contained liquid oxygen and liquid hydrogen, spelled the disastrous end of the mission.

It was a terrible day. We as a nation were in shock, while we mourned the loss of the crew and the schoolteacher/mother. It was sometime later, when the facts were brought to the public about the space shuttle program that we learned the shuttle should not have been permitted to fly that day. The on-site engineers understood the very high probability of failure of the O-ring seals. Moreover, knowledgeable engineers at Morton Thiokol tried without success to prevent the launch.

The story containing all of the facts about the failure of the O-ring seal is too long to be covered here, but you are encouraged to read references [7 – 12], where very complete and well-written accounts are given. The essential elements leading to the catastrophic failure are listed below:

1. The circumferential joint changed shape when the motor case was pressurized, and the gap, which the O-rings filled, increased markedly in size. A schematic illustration of the new gap geometry is presented in Fig. 4.5.

2. The new gap opening was so large that the back-up O-ring probably could not seal the joint. When the booster case was pressurized, the seal depended on a single O-ring—not two.

3. The hot exhaust gases had eroded the O-rings on 12 of the previous 24 launches indicating some leakage about half of the time, and on a few occasions, significant erosion of the rings occurred. There was also clear evidence that the putty failed to keep the hot gasses from attacking the rubber O-rings.

4. The joints moved early in the launch sequence as the orbiter engines and the booster engines were ignited. This motion requires the O-ring seals to be flexible and to reseat and reseal continuously during these movements.

5. The temperature the night before the launch dropped to 22°F and had only increased to about 28°F by the time of the launch.

6. The O-ring seals were not certified to operate below a temperature of 53°F by the contractor responsible for the solid rocket boosters.

7. Tests conducted by the contractor Morton Thiokol in 1985, six months before the accident, indicated that the O-ring seal was not effective at low temperatures. (At 50°F the O-rings would not expand and follow the movement in the joint during the period of operation of the booster.) In fact, at room temperature 75°F, it required 2.4 seconds for the O-rings to reseat and seal pressurized gasses after a gap in the joint was opened.

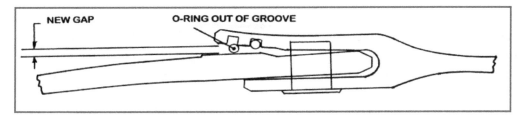

Fig. 4.5 Rotation of the joint due to cylinder pressurization opens the gap in the seal region.

The seven facts listed above show very clear evidence of two problems. First, the design of the seal in the circumferential joint was marginal and should have been fixed much earlier in the program. You do not inspect a piece of burnt and eroded O-ring and **walk away** from the problem.

Second, the launch should have been postponed due to cold weather for several different reasons. The very low temperatures were well below the limits to which the system had been certified. The rubber in the O-rings becomes very stiff at these low temperatures and cannot respond quickly enough to accommodate joint movement. The O-rings were almost guaranteed to leak. Also, there was considerable ice on the rocket motors and the fuel tank. Pieces of this ice pose a danger, because it might damaged the rocket motors, the tiles on the heat shield or the wing's leading edges of the orbiter when it separates and falls during the violent vibrations that occur just after ignition but before lift-off.

Ignoring the Problem

Problems usually do not go away when they are ignored. They persist and sooner or later failure will occur. Although the seals had failed on 12 of the previous 24 flights, the failures were not catastrophic because the leaks were small and at locations which did not endanger the adjacent components. The boosters on the shuttle operate for only a few minutes before they are cut loose to fall into the Atlantic Ocean for recovery at a later time. Those concerned with the launch schedule tolerated the small leaks for short periods of time. The trouble with the leak on the Challenger was that it was large, and the jet of hot gas issuing from the leak was pointed directly at the liquid hydrogen fuel tank. The jet of hot gas acted exactly like a cutting torch, and the huge thin-walled fuel tank was penetrated within seconds.

The O-ring seals were inadequate and the joint needed to be redesigned and modified. In fact, during the development of the boosters in 1977–79, several memos were written by engineers at the Marshall Space Flight Center indicating their concern over the design of the joint. They recommended that the joint be redesigned, because **the adequacy of the clevis joint was completely unacceptable**. The suppliers of the O-rings stated: **The O-ring was being required to perform beyond its intended design and that a different type of seal should be considered.**

Apparently there was a major breakdown in communication; none of these memos and reports from the Marshall Space Flight Center was forwarded to Morton Thiokol, the designers and builders of the boosters. A problem, properly detected by engineering at the NASA Center in charge of monitoring the technical aspects of the development of the boosters, was not pursued to its logical conclusion. Instead of insisting on a redesign of the joint, Marshall Space Flight Center approved the boosters in September of 1980. The pressures of cost overruns and repeated schedule slippage sometimes make managers (and engineers) accept flawed designs. It is very shortsighted and not in the best interest of safe design to follow this practice.

NASA accepted a booster rocket with inadequate seals. On the first launch, with Columbia, there was no reported damage to the seals. The O-ring seals failed on the second flight. One of the O-rings was burnt with about 20% of the thickness of the ring vaporized. The ring failure did not affect the mission, but it was clear that the putty was not protecting the rings from the hot gas. Marshall Space Flight Center reacted to the field experience by reclassifying the joint to a **Criticality 1 category**. This was the official recognition on the part of NASA that a seal failure could result in "loss of mission, vehicle and crew due to metal erosion, burn through and probable case burst resulting in fire and deflagration." The failure of the O-rings by erosion continued with high frequency (50% of the missions), but NASA decided to accept the situation and did not recommend remedial action.

It appears that NASA decided to live with the seal problem at least until a new lighter booster motor could be designed with improved joints. These new boosters were to be ready about six months after the Challenger exploded. A very real lesson—in too-little, too-late.

Recognizing the Influence of Temperature

From the initial development phase of the boosters, the O-ring seals had been recognized as a problem. A problem that NASA classified as very serious, but one that they must have erroneously believed did not elevate the risk to unacceptable limits. In January of 1985, a year before the accident, the Challenger was launched on a cold day at a temperature of 51 °F. An examination of the joints after the recovery of the boosters showed that a number of O-rings were severely damaged and evidence of extensive gas leakage was observed. The contractor, Morton Thiokol, and Marshall Space Flight Center reviewed the O-ring seal problems. During this review, Morton Thiokol noted that low temperatures were a contributing factor to the inability of the O-rings to seal properly. The condition was considered undesirable, but was acceptable. However, the O-ring problems persisted and NASA eventually placed a launch constraint on the space shuttle. The launch constraint would prohibit launches until the problem had been resolved or reviewed in detail prior to each launch. Unfortunately, NASA routinely provided launch waivers for each subsequent flight to avoid delays in an unrealistic launching schedule. The problem of O-ring erosion was common. Fortunately, the erosion and attendant leaks had not caused any serious difficulty until the Challenger accident. The problem was expected and accepted as routine.

A Management Decision

The very cold weather on the night before the fatal launch was no surprise. The weatherman (or woman) was on the mark in predicting the temperature and the ice that formed on structures that night. The engineers at Morton Thiokol, in Utah, were very concerned about the effects of the very low temperatures on the ability of the O-rings to function properly. Teleconferences took place between the Morton Thiokol engineers and the NASA program managers. The engineers argued to postpone the launch until the temperature increased to at least 53 °F.

The NASA program managers were very unhappy about the engineer's concerns. They did not want the extended delay required to wait for warmer weather. Senior vice presidents from Morton Thiokol sensed the displeasure from the customer and took over the decision process. Senior managers decided to keep the customer happy and reversed the decision of the Vice President of Engineering not to launch at these very low temperatures. The engineers at the end of the day sat silent.

Senior managers at Morton Thiokol signed off on the launch ignoring the fact that the temperature at launch time was expected to be 25 °F lower than the lowest temperature which engineering believed the seals would be effective (53 °F).

Senior managers at Morton Thiokol yielded to client pressure. The engineers were unanimous in their opposition to launch. Senior managers asked the engineers to prove that the O-ring seals would fail. Of course, the engineers could not state with 100% certainty that the rings would fail. Failure analysis is performed in terms of probabilities. The risk had elevated as the temperature decreased, but the engineers could not prove conclusively that the seals would fail in a manner that would detrimentally affect the mission. Senior managers at Morton Thiokol and program managers at NASA concluded erroneously that the risks were low enough to proceed with the launch.

Approvals at the Top

NASA has a four-level approval procedure to control the launch of the shuttle. This sounds good, but for a multilevel approval process to be of any value, information must flow freely from the bottom of the organization to the top of the chain of command. The technical discussions concerning the ability of the O-rings to function at the very low temperatures took place between fourth level NASA administrators and engineers and program staff at Morton Thiokol. Program managers at Marshall Space Flight Center, third level administrators, were also included in these discussions. While the discussion was extended, with clear polarization between engineering and management, not a word of these concerns was conveyed to the top two levels of management at NASA.

Approvals for the launches were given by second level administrators for the National STS Program in Houston, TX and the first level administrators at NASA's Headquarters in Washington D.C. These approvals were essentially automatic because the administrators in charge had been isolated. They were not privy to the management decision to fly with very cold O-ring seals. They were not informed of the engineering recommendation to postpone the launch and to wait until the temperature increased to at least 53 °F.

The lesson here is clear—approvals by the very high level managers are worthwhile only if they are informed decisions. If complete information is not presented to the executives for their evaluation, then their approval is meaningless. These administrators are not knowledgeable and they add no value in an informed decision making process.

4.10.2 **The Columbia Accident**

On February 1, 2003 the Columbia shuttle disintegrated as it reentered the atmosphere at the end of its 16-day mission. In Mission Control, Columbia's re-entry appeared normal until 8:54:24 AM[4] when engineers monitoring the sensors informed the flight director that four hydraulic sensors in the left wing were indicating—off-scale low—a reading that falls below the minimum capability of the sensor. At 8:55:00 AM, nearly 11 minutes after Columbia had re-entered the atmosphere, its wing leading edge temperatures reached their normal level of nearly 3,000 °F. At 8:55:32 AM Columbia crossed from Nevada into Utah, while traveling at Mach 21.8 at an altitude of 223,400 ft. At 8:58:20 AM, as Columbia crossed from New Mexico into Texas, a thermal protection tile was lost. At 8:59:15 AM the engineers monitoring sensor data informed the flight director that the sensors monitoring the pressure on both left main landing gear tires were lost. The flight director then informed the Columbia's crew that Mission Control was evaluating these sensor readings, and added that the crew's last transmission was not clear. At 8:59:32 AM a broken response from Columbia was recorded but it was cut off in mid-word. Videos made by observers on the ground less than a minute later, presented in Fig. 4.6, revealed that Columbia was disintegrating.

Fig. 4.6 A video photograph of Columbia disintegrating as it is passing over Texas while traveling at about Mach 19.5 at an altitude of about 210,000 ft.

Columbia was the first space-rated Orbiter, and it made the Space Shuttle Program's first four orbital test flights. Because it was the first of its kind, Columbia differed slightly from the other four Orbiters— Challenger, Discovery, Atlantis, and Endeavour. Built to earlier engineering specifications, Columbia was slightly heavier, and was not able to carry sufficient cargo to conduct cost effective missions to the International Space Station. For this reason, Columbia was not equipped with a Space Station docking system; hence, it had more space in its cargo bay for longer pieces of equipment. Columbia usually flew science missions and serviced the Hubble Space Telescope.

The final flight of Columbia was designated STS-107. This was the Space Shuttle program's 113th flight and Columbia's 28th. It appeared that the flight was nearly trouble-free. Unfortunately, there were no clear indications to the crew onboard Columbia or to personnel in Mission Control that the flight was in trouble as a result an impact by at least one large piece of foam insulation that occurred shortly after launch. Mission management overlooked a few clues revealed during launch that the Orbiter was in trouble and failed to take corrective action.

---

[4] Times are given as Eastern Standard Time (EST.).

Immediately following the failure of the Orbiter, NASA established a Columbia Accident Investigation Board to identify the chain of events that caused the Columbia accident [13]. Upon completion of its exhaustive study, the details of which have been omitted here, the Columbia Accident Investigation Board concluded that the cause of the accident was a breach in the thermal protection system on the leading edge of the left wing. The breach was initiated by a piece of foam that separated from the left bipod ramp of the external tank and impacted the left wing leading edge panel. The conclusion that foam separated from the external tank bipod ramp and struck the wing in the vicinity of panel 8 is clearly documented by photographic evidence, sensor data and the aerodynamic and thermodynamic analyses conducted. As was the case with the Challenger disaster, NASA engineers, scientists and managers knew of the design flaw associated with the foam bipod ramps, but collectively chose to overlook the problem instead of finding a solution.

4.10.3 Hard Lessons to Learn

Dr. Diane Vaughan, an authority on risk assessment, testified before the Columbia Accident Investigation Board and stated: "What we find from a comparison between Columbia and Challenger is that NASA is an organization that did not learn from its previous mistakes and it did not properly address all of the factors that the presidential commission identified."

Organizational failures usually occur with higher frequency in larger organizations with several layers of management. Failures happen regardless of the safeguards and systems that an organization employs. Failures are even more common in high-risk organizations like NASA. NASA uses several risk-avoidance systems that are designed to insure that the instruments and astronauts sent into space complete their missions and return safely. However, NASA has failed in three cases to achieve this objective when sending astronauts into space. The Space Shuttles Challenger and Columbia tragedies, as well as the Apollo launch pad fire in 1967, are examples of NASA failures that resulted in the deaths of 17 astronauts. Organizational failures played an important role in all three cases.

As was the case with O-ring damage in the years before the Challenger accident, the failure of Columbia did not represent the first time that foam had detached from the external tank and caused damage to an Orbiter. There had been many reports of impacts on previous flights by pieces of foam that were much smaller than the one which critically damaged Columbia. Orbiters would often return with many damage sites due to both hypervelocity and low velocity impact. Some sites were larger than 1 inch in diameter. There are numerous reports in the Problem Reporting and Corrective Action (PRACA) system that showed some of the thermal barrier damage was probably due to foam failure during launch and assent. Not only was foam impact not considered in the design of the thermal protection system, but the engineering specification for the Orbiter clearly, and erroneously, assumes that nothing should impact the shuttle during launch.

However, foam had been failing and striking the Orbiter with high frequency and a team of chemists and engineers were working on improving the foam and the techniques used for applying it. The studies of impacts with foam concluded that they did not pose a "flight safety" risk because experience showed that the small pieces of foam that had been coming off the external tank only caused small pits and craters on the Orbiters. Unfortunately, the engineers did not consider a large piece of foam traveling at high velocity[5] when they declared foam failure not to be a "flight safety" issue. The impact of such a large piece of foam on a vulnerable area of the Orbiter was unprecedented. (So was the three days of abnormal cold before the Challenger launch.)

---

[5] Photographic analysis revealed that one large piece and at least two smaller pieces of foam had separated from the bipod ramp area of the external tank. The large piece of foam was about 24 in. long and 15 inches wide. It was rotating at 18 RPS and moving with a relative velocity of about 500 MPH when it impacted the Columbia's wing.

Foam failure that regularly damaged the thermal protection system on a large fraction of the flights was not considered to be dangerous because the orbiters were returning safely with damage that could be repaired or ignored. What the engineers or management did not consider was the possibility of a larger or heavier piece of foam hitting the Shuttle in a particularly vulnerable area like the leading wing edge. This is a classic example of Vaughan's "normalization of deviance" [14] where an unpredicted anomaly becomes routine.

Normalization of Deviance or Accepting Problems Instead of Pursuing Solutions

Vaughan [13] has developed the concept of **normalization of deviance** to explain the way technical flaws escape the scrutiny of the various safety boards within a large organization such as NASA. Frequently these boards observe unanticipated problems that continue to occur on a regular basis without leading to an accident. People begin to believe the problems are benign and are lead to the pragmatic notion of **acceptable** deviance. The leadership at NASA found it very expensive and time consuming to establish the cause of some problems and to incorporate added inspections in the regular maintenance cycle of the Shuttle system without strong evidence of **flight safety** risk. Under the pressures of the Shuttle's flight schedule and limited budget, NASA did not commit significant resources to problems that its management did not classify as **flight safety** risks that could cause the loss of an Orbiter. This practice resulted in disincentives for the engineers to determine the source of problems, even though scenarios could be envisioned that would cause significant **flight safety** risks. NASA frequently cleared flights as operational based on previously successful flights that had completed their missions while exhibiting clear evidence of a design problem. This reasoning prompted Richard Feynman [11] in his analysis of the Challenger accident to remark: **When playing Russian roulette the fact that the first shot got off safely is little comfort for the next.**

The Role of NASA's Management Structure in Columbia's Accident

NASA's organizational and physical infrastructure was developed in the days of the Apollo program when it had a large budget and a sharply focused mission—to land a man on the moon before the Russians. A layered bureaucratic group of middle and upper level managers functioned well and the Apollo program accomplished its goals with only one serious mishap in 1967 and a near catastrophic mission on Apollo 13. After more than 40 years, NASA's organizational structure is nearly the same although its missions have changed markedly. There is reason to question if their organizational structure is sufficiently flexible considering its plans for the future and its ongoing projects.

For the Shuttle program, there should have been a mechanism for engineers to bypass the bureaucracy and hierarchy, especially in the pre-launch and launch processes. Suppose the engineers had succeeded in calling off the launch of the Challenger due to the effect of cold weather on the O-ring seals. The flight would have been cancelled, Challenger would have been taken off of the launch pad and the solid rocket boosters disassembled to replace the damaged O-rings. This action would have been expensive but not nearly as costly as the loss of crew and the Orbiter. Similarly, if an engineer needs a certain type of data, there should be a way to bypass the formal bureaucratic procedures to obtain the data with dispatch. Engineers have many intuitions and hunches that take time and resources to translate into analysis and data. These intuitions need to be respected, given credence, explored and welcomed by upper management [15].

Probability of Failure

Richard Feynman served on the Presidential commission that investigated the Challenger accident that occurred on January 28, 1986. It was near the end of his life, and the famous physicist devoted significant amounts of time to the investigation. After the investigation, Dr. Feynman prepared a paper describing his personal observations on the reliability of the Shuttle [16]. We suggest you read this interesting paper, because we are only reviewing a small portion of this reference. The following paragraphs are direct quotes from Feynman's personal observations.

1. It appears that there are enormous differences of opinion as to the probability of a failure with loss of vehicle and of human life. The estimates range from roughly 1 in 100 to 1 in 100,000. The higher figures come from the working engineers, and the very low figures from management. What are the causes and consequences of this lack of agreement? Since 1 part in 100,000 would imply that one could put a Shuttle up each day for 300 years expecting to lose only one, we could properly ask—"What is the cause of management's fantastic faith in the machinery?"

2. If a reasonable launch schedule is to be maintained, engineering often cannot be done fast enough to keep up with the expectations of originally conservative certification criteria designed to guarantee a very safe vehicle. In these situations, subtly, and often with apparently logical arguments, the criteria are altered so that flights may still be certified in time. They therefore fly in relatively unsafe conditions with a chance of failure of the order of one percent (it is difficult to be more accurate).

3. Official management, on the other hand, claims to believe the probability of failure is a thousand times less. One reason for this may be an attempt to assure the government of NASA perfection and success in order to ensure its annual funding from the government. The other may be that they sincerely believed it to be true, demonstrating an almost incredible lack of communication between themselves and their working engineers.

4. In any event this has had very unfortunate consequences, the most serious of which is to encourage ordinary citizens to fly in such a dangerous machine, as if it had attained the safety of an ordinary airliner. The astronauts, like test pilots, should know their risks, and we honor them for their courage. However, it is clear that McAuliffe was a person of great courage, who was closer to an awareness of the true risk than NASA management would lead us to believe.

5. Let us make recommendations to ensure that NASA officials deal in a world of reality in understanding technological weaknesses and imperfections well enough to be actively trying to eliminate them. They must live in reality in comparing the costs and utility of the Shuttle to other methods of entering space. They must be realistic in making contracts, in estimating costs and the difficulty of the projects. Only realistic flight schedules should be proposed, schedules that have a reasonable chance of being met. If in this way the government would not support them, then so be it. NASA owes it to the citizens from whom it asks support to be frank, honest and informative, so that these citizens can make the wisest decisions for the use of their limited resources.

6. For a successful technology, reality must take precedence over public relations, for nature cannot be fooled.

As one studies the Columbia's fate, we appreciate the significant value of the personal observations made by Richard Feynman about 40 years ago. It appears that his message was lost on NASA as they continued to ignore significant operational problems in their efforts to meet an unrealistic flight schedule with a constrained budget. Dr. Feynman estimated that the Shuttle flies in a relatively unsafe condition, with a chance of failure of the order of one percent. It is possible to test his estimate by reviewing the additional Shuttle flights since the Challenger accident. The Columbia mishap was the second accident in 113 flights. Thus, the probability of failure P_f of future Shuttle flights is:

$$P_f = N_f/N = 2/121 = 1.65\%$$

This estimate is the same order as Dr. Feynman predicted after the Challenger accident and 1,700 times higher than the probability of failure of 0.001% (1/100,000) believed by NASA managers.

4.11 SUMMARY

The degradation of personal ethical behavior during the past two or three generations has been described. While there are many controversial issues that can be classified as **dilemma** ethics, there are several **basic** characteristics where the difference between right and wrong is clearly defined. The basic virtues—prudence, justice, fortitude and temperance—were described. Laws are written to prohibit, by punishment, a well-defined set of actions. These laws control behavior, but only affect the specific actions on the part of individuals or corporations.

Ethics allows society to trust engineers to design products and systems with a high regard for safety. Engineers have both an individual and a collective interest in keeping that trust. Individual behavior is important, but as we learned in the Challenger explosion, it is not sufficient. Corporations (business) must operate with the highest ethical standards. The chief executive officer (CEO) establishes the standards and the lower level managers act to establish the corporate character. Design systems must also avoid ethical failures due to individuals not understanding what they don't know or information not flowing to those who have the needed knowledge.

Professional ethics were described as following patterns established first in medicine and law. Professionals get special status based on a commitment to the public. Engineering professional ethics was described by introducing the Code of Ethics of Engineers sponsored by the Accreditation Board for Engineering and Technology (ABET). The code is short with four fundamental principles and seven basic canons.

The fatal accidents of the Challenger and Columbia Space Shuttles, which failed during their missions in 1986 and 2003, were discussed. Many ethical issues dealing with individual and corporate behavior were apparent when the events leading to this fatal accident were reviewed. The bottom line is that society will only allow engineers to design and manufacture products and systems if they act at all times with the public safety as their highest duty.

REFERENCES

1. Goodman, L. S., The Historic Role of the Oath of Admission, American Journal of Legal History, Vol. 11, No. 4, (October 1967), pp. 404-411..
2. Sommers, C. H., "Teaching the Virtues," *Public Interest,* No. 111, Spring 1993, pp. 3-13.
3. Woodward, K. L. "What is Virtue," *Newsweek*, June 13, 1994.
4. George, R. P., Making Men Moral: Civil Liberties and Public Morality, Oxford University Press, New York, NY, 1994.
5. Brannigan, V. M., Teaching Ethics in the Engineering Design Process: A Legal Scholars Perspective, Boulder, CO, 33rd ASEE/IEEE Frontiers in Education Conference, S2A-19, November 5-8, 2003.

6. Baker, D. F. "Ethical Issues and Decision Making in Business," Vital Speeches of the Day, 1993

7. Jensen, C., No Down Link: A Dramatic Narrative about the Challenger Accident, Farrar, Straus and Gitoux, New York, NY, 1996.

8. *Time,* February 10, 1986

9. Lewis, R. S., The Voyages of Columbia: The First True Space Ship, Columbia University Press, New York, NY, 1984.

10. *The New York Times* April 23, 1986.

11. Feynman, R. P. What Do You Care What Other People Think? Bantam, New York, NY, 1988.

12. Report to the President by the Presidential Commission on the Space Shuttle Challenger Accident, Ayer Co., Salem, MA, 1986.

13. Anon, Report of Columbia Accident Investigation Board, Volume I, August 26, 2003, See http://www.nasa.gov/columbia/home/CAIB_Vol 1.html.

14. Vaughan, D., The Challenger Launch Decision: Risky Technology, Culture and Deviance at NASA, University of Chicago Press, Chicago, 1996.

15. Hall, J. L., "Columbia and Challenger: Organizational Failure at NASA," Space Policy Vol. 19, 2003, pp. 239-247.

16. Feynman, R. P., Personal Observations on the Reliability of the Shuttle, see Website http://www.fotuva.org/feynman/challenger-appendix.html.

17. New York Times, Video: Challenger and Columbia and the Nature of Calamity, June 1, 2014.

EXERCISES

4.1 Prepare a two page brief that distinguishes the differences among professional, organizational and personal ethics.

4.2 Why was the engineering profession much later than either the medical or legal professions in establishing a code of ethics for professional behavior?

4.3 If you have selected an engineering discipline for your studies, determine the professional society that represents this discipline. From the Internet locate the URL for its Website and then examine its code of ethics. Write a one page brief describing the code and your reaction to it.

4.4 Is there an honor code at the University where you are pursuing your studies? Have you read it? Do you abide by the rules?

4.5 Write a paragraph or two explaining your position regarding the four basic virtues.

4.6 Why does the CEO of a corporation set the standard for ethical behavior? What are some of the issues that arise on a daily basis that develop corporate character? Is it possible for a corporation with tens of thousands of employees to develop a character like an individual?

4.7 Examine the code of ethics for engineers given in Fig. 4.2, and describe your opinion of the fundamental principles or canons. Can you live with these expectations if you become a professional engineer? Do you believe ABET's code is adequate, or should it be revised to reflect a more modern viewpoint of what is right and wrong?

4.8 Suppose that you were a lead engineer working for Morton Thiokol on the evening of January 27, 1986 involved in the discussion of the O-rings. What would you have done?

- Early in the teleconference.
- At the critical stage of the teleconference.
- After management had taken over the decision process.

There is no right or wrong answer to these questions. The purpose of the exercise is to place you in a professional dilemma. Someday you may be placed in a similar situation, where there is a trade-off between safety and corporate business interests. Think about your response in advance and be prepared to deal with such a dilemma and its consequences.

4.9 Write a review of Chapter 2 "Columbia's Final Flight" from volume I of the Report of the Columbia Accident Investigation Board. Reference 12 provides the URL for the Website.

4.10 Write a review of Chapter 3 "Accident Analysis" from volume I of the Report of the Columbia Accident Investigation Board. Reference 12 provides the URL for the Website.

4.11 Write a review of Chapter 5 "From Challenger to Columbia" from volume I of the Columbia Accident Investigation Board. Reference 12 provides the URL for the Website.

4.12 Write a review of the video identified in Reference 17.

4.13 Are there government agencies that are concerned with product safety? Name them and describe the way they function.

4.14 Which government agency is responsible for the safety of your automobile. What is your assessment of its value?

4.15 Which government agency is responsible for your safety on commercial flights within the U. S.

CHAPTER 5

SAFETY AND PERFORMANCE

5.1 LEVELS OF POTENTIAL INJURY

When engineers design a new product or modify an existing one, they can create the possibility of injury. Sometimes the potential injury is minimal, but in some instances the injury may be catastrophic. In the late 1950s Lockheed introduced the L-188 Electra. Three of these aircraft crashed due to structural failure in flight before the Federal Aviation Administration (FAA) grounded the entire fleet. An investigation revealed that a torsional resonance occurred under certain operating conditions causing the engines to vibrate violently and break away from the wing. This was a very serious engineering failure, which caused the death of over a hundred innocent passengers. Clearly, this example illustrates a case where the risk far outweighs the benefits of performance of a new aircraft design. It also illustrates the seriousness of design errors and inadequate testing of prototypes of products that may pose a danger to the public.

Recently, Boeing's new fleet of Dreamliners was grounded after two incidences of smoke in the cockpit cause emergence landings. The smoke was due to a fire in the lithium-ion battery. An investigation by several groups of experts was conducted to determine the cause of the battery failures. In this case, no one was injured, but the costs of the failure to Boeing, several airlines, and the battery supplier are significant.

Even more recently General Motors has discovered a faulty ignition switch that sometime fails when the automobile encounters a severe bump. The switch mechanism slips from on to off and the engine shuts down. Consequently and the power steering and power brakes are compromised and steering and stopping become difficult. Accidents have occurred over several years, with loss of life and property. General Motors has been fine for negligence by the U. S. government and is currently engaged in a very extensive and very expensive recall to replace the faulty ignition switches.

5.1.1 Safety Concepts Important for Safe Engineering Design

Hazard and Risk

The first is the **hazard** or the type of injury which might occur (also called the **severity of injury**). The second is the **probability** of injury under various combinations of circumstances, which is also known as the **rate of injury**.

The terminology in this field is often confusing. For the purpose of this chapter, consider **risk** as the probability of injury times the specific hazard. A risk is **acceptable** if a fully informed society would consider the combination of rate and hazard to satisfy social expectations.

Intrinsic and Extrinsic Safety

We should also distinguish between safety which is **intrinsic** in the desired performance of an object or system and safety which must be **designed in** but is **extrinsic** to specified performance. In a pistol, for example, the strength of the barrel is **intrinsic** to the performance but the safety catch is **extrinsic**. **Intrinsic** safety requirements are evident in the specification for a product, but **extrinsic safety** requires an engineer to fully understand the entire working environment of the product, not merely its performance specifications. **Extrinsic** safety features require special expertise and always risk being overlooked. For example, the designers of the Titanic failed to consider the lifeboats as a system. As a result no plan was created to fill the lifeboats, and over 400 people died for whom there would have been places in the boats.

CASE STUDY — Drill Design

You are designing an electric drill that is powered from the standard 120 volt, 60 cycle single-phase power supply from a local utility company. The **performance** requirement for the drill specifies the torque applied to the drill bit. Electrical safety must be assured in the drill's design. Drill operators are at risk of electric shock, although the probability for injury is very small if the tool is properly designed. It is your responsibility as an engineer to minimize this risk, while maintaining enhanced performance of a new improved model of an electric-powered drill.

How do you protect an operator from electric shock? An operator is holding an electric motor with 120 volts across the armature coils and the field coils in his or her hands. An early answer was to "ground" the tool using the familiar "three prong plug." This solution relies on operators using the tool correctly and not subverting the grounding system, or using outlets that were improperly wired. A later concept was to build the case for the drill from a tough, durable plastic that is structurally strong and an excellent electrical insulator. The case keeps the operator from touching any part of the electrical circuit powering the motor. In fact, many companies use double insulation to provide two independent insulation barriers to prevent the operator from making contact with any part of the electrical circuit. Operators may still manage to receive an occasional shock, but it will not be easy. Accidental contact of the operator's hands with a live electrical circuit is a rare event. In fact, the probability of electrical shock is so small that a worker routinely picks up a power tool and employs it to perform some task without even a passing thought about the possibility of receiving an electrical shock. Operating such a power tool does involve risk, but it is acceptable because the **rate of injury** is sufficiently low.

5.1.2 System Safety

One of the key problems in designing for safety is the tendency to focus solely on the product being designed rather than on the entire system of which the product is a component. Let's use the drill as an example. A system safety approach might be to use battery operated tools that eliminate the hazard completely or ground fault circuit interrupters to further minimize the shock risk. System safety analysis (SSA) examines the entire design process to assess hazards, probabilities and appropriate responses. A key element is human factors engineering, which examines the human/machine interface.

5.1.3 Analyzing Risk

Consider flying in a commercial airliner on a 1,200 mile trip. How safe is it to make this trip? Most people believe that it is reasonably safe to fly commercial airliners, but they realize that there is a small probability of a fatal crash. Statistical analysis may help to refine the probability of an injury. The statistics show the probability of a fatality p_k in an airline accident is about one in a billion passenger miles flown ($p_k = 1.8/10^9$).

Hence, if you make the round trip from point A to B, your probability P_k of being killed due to an airline accident can be described as:

$$P_k = s \, p_k = (2)(1{,}200)(1.8)/10^9 = 4.32 \times 10^{-6} = 0.000432\%$$

where s is the number of miles in a round trip.

There are a number of assumptions "hidden" in this type of statistical analysis. In particular are all "passenger miles" the same? The answer is clearly no. The risk of a crash may be much more related to take off and landing rather than the miles flown. All statistical safety analysis must insure that the correct data is actually being used. It is important to note that correlation is not causation. Unfortunately, sometimes statistics are all we have with which to work.

An accident rate of 4.32 chances in a million is considered by many to be a very low probability for a fatal accident; consequently, most people do not worry much about the possibility of dying in a crash when they board an airliner. However, if you are a sales engineer flying 100,000 miles a year, every year for 20 years, your total mileage accumulates to 2×10^6 miles. Your estimated probability of dying in a crash increases to:

$$P_k = (2 \times 10^6)(1.8)/(10^9) = 3.6 \times 10^{-3} = 0.36\%$$

The probability has increased significantly if you accumulate the miles flown by a traveling professional over a 20-year period. Knowing there is one chance in about 300 that you will die in an airline crash, would you still want to be a sales engineer flying weekly to meet with customers? Could you tolerate this level of exposure? Would you be apprehensive?

Everyone has some tolerance for risk. People weigh the speed, convenience, cost and risk of flying against that of traveling by train or by car. In almost all instances, travelers select the plane for long distances and the car for short distances. Travelers select the train when it combines relatively good service (high speed with frequent trains) to convenient (downtown) locations. Most people do not contemplate the risks involved in travel, because fatal accidents are rare events, except for driving a car.

Unfortunately, fatal accidents in automobiles are not rare events. Data from the National Highway Traffic Safety Administration (NHTSA) showed that 32,719 occurred on U.S. highways in 2013. This number is significantly lower than the 40,000+ killed every year prior to 2007. In addition over 2.4 million serious injuries occur in accidents each year. Death on the highway is the leading killer of Americans 1 to 35 years old. It is clearly more dangerous to drive than to fly, when you compare the probability of fatalities and/or injuries on a per mile basis. How do you handle the higher risk associated with driving? One answer is rationalization—I am an excellent driver, and it will not happen to me. Not necessarily a true statement, but the rationalization of one's superior driving skills alleviates the worry and concern about a fatality or injury producing accident. **The real risk remains.**

When driving you should recognize that the risk of an accident is a function of your driving behavior. Driving habits (high-speed, reckless steering, tailgating, etc.) and driving skills affect, to a large degree, the level of risk involved. When flying, you are a passive participant in determining the rate of injury.

Involuntary Exposure to Risk

Clearly the exposure to risk ranges from clearly voluntary to involuntary depending on the circumstances. Whether a choice is truly voluntary is a matter of social decision. One of the major changes in safety regulation in the 20th century was to consider the exposure of workers to risk to be involuntary and therefore less acceptable. This change has been incorporated into the Office of Safety and Health Administration (OSHA) and its regulatory standards. Recreational activities such as skiing, scuba diving, horseback riding, hang-gliding, mountain climbing, dirt biking, etc. carry higher levels of risk than other activities Yet, intelligent people swarm to the ski resorts and pay large amounts of money to potentially break their bones. Why? They have voluntarily decided that the pleasure derived from the activity offsets the risk. Also, they can control the level of risk. They can choose the simpler slope and minimize the risk or the more challenging slope to maximize the thrill. However even in these cases society often sets the boundaries of acceptable risk.

Knowing the social environment in which a product will be used is critical to setting the level of acceptable risk as part of the design process. For example, airlines are expected to take very high levels of safety precautions. Some level of risk is involved in almost all products produced but it is imperative that this risk be within the social expectation of safety, while maintaining an acceptable level of performance. An appropriate balance between risk and performance must be achieved, or the product is not viable[1]. It is important that engineers, business and governmental regulatory agencies cooperate to provide a realistic assessment of the risk level and a clear statement of what is acceptable. Finally, the public should be made aware of the residual risks involved in using a product, although merely warning of a risk is not an acceptable alternative to producing a safer design, especially when the person being warned is not the person at risk.

5.2 <u>MINIMIZING THE RISK</u>

5.2.1 <u>Safety in Design: Performance Analysis</u>

The first step in safe design is to analyze the required performance and determine suitable safety factors and margins of safety. This approach is normally used with well understood systems, materials and failure modes. It is particularly useful for intrinsic safety analysis.   As noted above designing for safe performance is not exactly the same as designing for safety, although it shares many similar techniques. Designing for safety means **avoiding injury**. In a **failsafe** design the product stops performing before it can cause injury. However the simplest design case is where the performance matches the safety requirement, such as with the tension member of a bridge.

Tension Member

Let's begin a performance analysis by introducing a tension member as shown in Fig. 5.1.

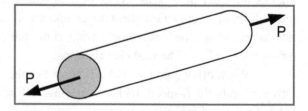

Fig. 5.1 A tension member is a long thin rod subjected to an axial load P.

When the tension rod is subjected to an axial load P, an axial stress σ develops that is uniformly distributed over the cross section of the rod. This uniformly distributed stress acting on the rod is illustrated in Fig. 5.2.

---

[1] Cost may be an issue when balancing safety, risk and performance. As a guide to cost, the Environmental Protection Agency (EPA) uses a value of $6.1 million for the "value of statistical life" in their cost benefit analyses for new regulations. However the use of such a figure in the design of product by a private company is more problematic. Indifference to a known and preventable risk is unethical and usually unlawful.

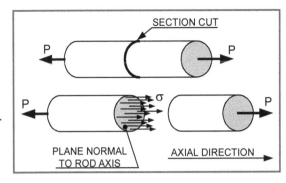

Fig. 5.2 The stress σ is distributed uniformly over the cross-sectional area of the tension rod.

The magnitude of this stress is determined from:

$$\sigma = P/A \qquad\qquad (5.1)$$

where σ is the stress given in units of pound/square inch (psi) or Newton/square meter (Pa).

P is the load expressed in units of pound (lb) or Newton (N).

A is the cross sectional area of the rod specified in square inch (in^2) or square meter (m^2).

Two examples demonstrating the technique for computing stresses in axially loaded tension members are presented below:

EXAMPLE 5.1

If the axial load P applied to a circular tension rod is 40,000 lb, determine the axial stress σ that is developed if the rod is 1.0 in. in diameter.

Solution: The cross sectional area is given by:

$$A = \pi r^2 = \pi(0.50)^2 = 0.7854 \ in^2$$

From Eq. (5.1), the axial stress σ is:

$$\sigma = P/A = (40,000)/(0.7854) = 50,930 \ psi$$

where psi is the abbreviation for lb/in^2.

At this stage of the analysis, it is not possible to interpret the significance of the result for the stress σ = 50,930 psi. More information is needed pertaining to the strength of the material from which the tension member is fabricated. When the strength of the rod's material is known, it is possible to make a comparison of the magnitude of the stress relative to this strength and remark on its significance.

Let's suppose that the tension rod used in Example 5.1 is machined from very strong alloy steel exhibiting 100,000 psi in yield strength. Do you believe the tension rod is safe? Will it retain its shape under load? Will it fail by yielding (stretching under load and not recovering completely when the load is removed)? These questions may be answered by making a simple comparison of the applied stress σ to the strength S_y of the material. In Example 5.1, the axial tensile stress σ = 50,930 psi and the strength of the material from which the rod was machined is S_y = 100,000 psi. The comparison shows that the yield **strength** of the rod is **greater** than the applied **stress**; therefore, the rod **will not fail** by yielding or breaking. But how safe is the rod? You know it will not fail; however, some measure of the safety of the rod should be established. Is the safety of the rod marginal, or is the rod too safe?

Safety Factor

To respond to the safety issue, determine the ratio of the strength to the stress and define this ratio as the safety factor SF:

$$SF = S/\sigma \tag{5.2}$$

For the stress imposed on the rod in Example 5.1, the safety factor determined from Eq. (5.2) is:

$$SF = 100,000/50,930 = 1.96$$

The safety factor is unitless. The value of SF = 1.96 indicates that the tension rod is almost twice as strong as it must be to resist yielding under the applied load. You have a reasonable safety factor and can be confident that the rod will perform safely in service. Your only concerns are the accuracy with which the load has been predicted and the quality control for the material used in manufacturing the tension rods. If the factors listed below are satisfied:

- You have extensive experience with the application
- You have a clear understanding of the failure mode
- You are certain of the magnitude of the load
- You have confidence in the manufacturing division in tracking their materials
- You have verified the strength of the materials received from suppliers
- You are assured that the environment in which the rod will operate is benign

then you usually can be satisfied that a safety factor of about two is adequate. However, if you are not certain about the load, materials and the environment, a safety factor of two is not sufficient. You should increase the safety factor to accommodate your ignorance of either the applied load or your lack of control over the material used in fabricating the rod or control of the environment.

If you do not know the failure mode, the safety factor may not be applicable. Many ships over several decades suffered a brittle failure of welded plates due to enbrittlement caused by low temperatures. That failure mode was not well understood at the time and so normal safety factors did not apply.

Safety factors ranging from two to four are commonly employed in design. Safety factors of less than two require considerable care, expense and expertise. The use of relatively low safety factors is justified only for very high-performance applications or for components that are produced in very high volume. In these special situations, the engineering analysis, prototype testing, quality control inspections, maintenance inspections and documentation must be extensive.

Margin of Safety

The margin of safety MS is sometimes used to describe the degree of safety incorporated into the design of a structural component. The margin of safety should not be confused with the safety factor as they are different quantities. The margin of safety is defined as:

$$MS = (S - \sigma)/\sigma = SF - 1 \tag{5.3}$$

Using the results from Example 5.1 and Eq. (5.3), the margin of safety is given by:

$$MS = 1.96 - 1 = 0.96 \text{ or } 96\%$$

The tension rod has a margin of safety of 96%, which implies that the strength of the material exceeds the applied tensile stress by 96%. If the load on a tension member can be determined, it is easy to size (adjust the cross sectional area) the rod to provide any margin of safety that is deemed necessary to satisfy management's concerns and to meet professional obligations to society-at-large.

System Safety Analysis

Safety factors and margins of safety only apply to well defined materials and failure modes. Numerous engineering disasters have been due to a lack of understanding of the underlying failure paths. This is especially true when the safety failure is extrinsic to performance.

CASE STUDY: Bridge Failure Due to Fire

Consider for example a tank truck fire on the bridge designed using the strength of steel as described above. Steel under tension weakens dramatically when heated.[i] On April 29, 2007, a large steel interstate highway bridge collapsed in a tank truck fire, when the steel girders supporting the bridge softened as the flames increased their temperature in excess of 1,000 °F.

The steel columns and beams in buildings are routinely encased in **fire proofing materials** to provide protection and avoid collapse. The steel in bridges is not. Only a proper system safety analysis can tell if the risk we are running by not protecting the steel is reasonable. The information to develop this case study is available on the Internet at the URLs provided below.[2]

5.3 FAILURE RATE

Component Analysis

In some systems an engineer can minimizes risk by preventing failure of each and every component in the system. This task is not easy as there are several different ways in which components can fail. Parts fail by breaking and aging, by corrosion or fatigue, by overload and burning, etc. In some instances, you may anticipate these failures and replace the parts before they malfunction. This controlled replacement of finite life parts is known as preventative maintenance. In most cases, engineers try to design each component so that failure will not occur during the anticipated life of the product.

Sometimes components fail by burning out, aging or wearing out—not necessarily by breaking or yielding. When conducting a failure analysis for wear or aging, the safety factor is normally not a relevant parameter. Instead you use some form of time to failure analysis or mean time between failures. You must cope with wear, aging or burn out by using other methods. To begin, let's recognize that all components do not wear out at the same time. Some people drive their car 25,000 miles before the brakes wear out and others may drive 60,000 miles or more. Some people can use their personal computers for several years before a component fails; yet others have problems after only a few months in service. To analyze these differences in service before failure, a mortality curve for both mechanical and electronic components is introduced in Fig. 5.3a and Fig. 5.3b, respectively.

---

[2] http://www.pbs.org/newshour/bb/science/jan-june07/overpass_05-10.html

5.3.1 **Computing a Failure Rate for Components**

To determine the failure rate FR, records of the performance with time-in-service of a large number of components are maintained. You begin your record of performance with N_0 components that are placed in service at time t = 0. After some arbitrary time t, some number N_f will have failed and the remainder N_s will have survived. At any arbitrary time, you may write:

$$N_f + N_s = N_0 \qquad\qquad (5.4)$$

With time, N_f increases and N_s decreases, but the sum remains constant and equal to N_0. The failure rate FR is defined as:

$$FR = N_f / (N_0 \times t) \qquad \text{when } t > 0 \qquad\qquad (5.5)$$

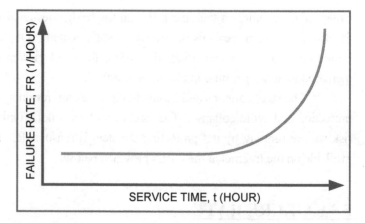

Fig. 5.3a Mortality curve for mechanical components showing failure rates with time in service.

When the results of Eq. (5.5) are plotted with respect to time, the component mortality curve presented in Fig. 5.3a is obtained. For mechanical components, where testing is performed on the assembly line, the failure rate is small when the components are relatively new. However, the failure rate increases non-linearly after some time in service because parts begin to wear, corrode or they are abused.

The failure rate for electronic components is usually characterized by the well-known bathtub mortality curve presented in Fig. 5.3b. This curve exhibits three different regions of interest—each due to a different cause. The high failure rates in the first region, $t < t_1$, are due to components manufactured with minute flaws that were not discovered during inspection and testing at the factory. In the center region of the curve, for $t_1 < t < t_2$, the failure rate FR_0 is small and nearly constant with time. Clearly, this is the best region to operate if you are to maximize the reliability of the product. Near the end of the service life, when $t > t_2$, the failure rate increases sharply with time as the components begin to fail due to the effects of aging.

For high-reliability systems, it is important to eliminate the inherently defective components associated with early failures. For mechanical systems, inspecting and testing the system's performance to eliminate flawed components produced in manufacturing accomplishes this objective. This task is not easy for the electronic component manufacturers because the feature sizes on the components are so small that inspection requires very high magnification and detailed examination of relatively large areas. Even with elaborate inspection schemes, not all of the flaws can be detected. A better procedure to eliminate the few inherently flawed components is to conduct what is known as burn-in testing prior to incorporating the components into an electronic product. With burn-in testing, all of the components that are produced are operated for a time t_1 (as defined in Fig. 5.3b). The flawed components burn out (fail) and are eliminated. The surviving components that will function with a much lower failure rate FR_0 are used in fabricating high reliability products.

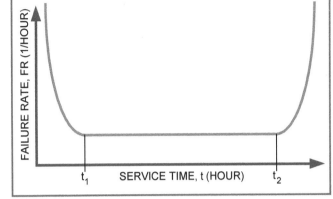

Fig. 5.3b Mortality curve showing the failure rate for electronic components with time in service.

Clearly, it is important to design components with low failure rates. If the failure rates are low, then the time between failures will be high. In fact, manufacturers sometimes cite what is known as the mean time between failures (MTBF) in their product specifications. The MTBF and the failure rate are related by:

$$MTBF = 1/FR \qquad (5.6)$$

You should understand the concept of component failure rates and how they vary over the life of the component. How is this information about the failure rate used to establish the reliability of a specific component to perform without failing over the anticipated life of a product? This topic is discussed in the next section.

5.3.2 Component Reliability

To answer the question of component and system reliability, you must introduce the concept of probability or chance of occurrence. In considering probability, let's return to the test conducted to measure the failure rate and use the data collected for N_f, N_s, and N_0. The probability of survival P_s, which varies with time t, may be determined from:

$$P_s(t) = N_s(t)/N_0 \qquad (5.7)$$

And the probability of failure P_f is given by:

$$P_f(t) = N_f(t)/N_0 \qquad (5.8)$$

Both the probability of survival and the probability of failure are functions of time. With increased time in service, the probability of survival decreases and the probability of failure increases.

From Eqs. (5.7) and (5.8), it is evident that:

$$P_s(t) + P_f(t) = 1 \qquad (5.9)$$

Substituting Eq. (5.8) into Eq. (5.9) yields:

$$P_s(t) = 1 - \frac{N_f(t)}{N_0} \qquad (5.10)$$

If Eq. (5.10) is differentiated with respect to time t and the result is rearranged, it is possible to write:

$$\left[\frac{N_0}{N_s(t)}\right]\left[\frac{dP_s(t)}{dt}\right] = \left[\frac{-1}{N_s(t)}\right]\left[\frac{dN_f(t)}{dt}\right] = -FR(t)dt \tag{5.11}$$

Consider the term $[1/N_s(t)][dN_f(t)/dt]$. It is the instantaneous failure rate FR(t) associated with a sample size N_s at time t. Next, integrate Eq. (5.11) and assume that $FR(t) = FR_0$, is a constant. The relation obtained gives the probability of survival as a function of the failure rate.

$$P_s(t) = e^{-(FR_0 t)} \tag{5.12}$$

where   e = 2.71828 is the exponential number. Let's consider an example to demonstrate the method for determining the reliability of a mechanical component.

EXAMPLE 5.2

Suppose a company maintenance records show the number of failures of a certain component over an extended period of time. Analysis of this data indicates that the failure rate is essentially constant with 1 failure per 10,000 hours of service. Does that sound good to you? Will the reliability be adequate? Let's determine the probability of survival (the reliability) of this mechanical component with time.

Solution:

Substituting the failure rate $FR_0 = 1/10,000$ h into Eq. (5.12), yields the results for $P_s(t)$ shown in Table 5.1.

Table 5.1
Reliability $P_s(t)$ of a mechanical component
with increasing service life ($FR_0 = 1/10,000$ h)

Time (1000 h)	Time* (years)	$FR_0 \times t$ (unitless)	Reliability $P_s(t)$
1	0.5	0.1	0.905
2	1	0.2	0.819
5	2.5	0.5	0.606
10	5	1.0	0.367
20	10	2.0	0.135
50	25	5.0	6.738×10^{-3}
100	50	10.0	4.540×10^{-5}
200	100	20.0	2.061×10^{-9}

*The conversion from hours to years of life is based on 2,000 hours/year.

Reliability varies markedly with service life. For a short service life, say a year, the reliability is reasonable with 81.9% chance of surviving. However, for long life, say 10 years, the reliability drops to only 13.5%. For very long life 50 to 100 years, failure is almost certain. While the failure rate of 1 in 10,000 hours appears satisfactory in an initial assessment, the reliability resulting from this failure rate is disappointing. You cannot be certain of a service life of 10,000 hours prior to failure. In fact, there is only a probability of 36.7% to survive for 10,000 hours. If a reliability of 90% is required for a service life of 10,000 hours, the failure rate FR0 must decrease to about 1 failure in 100,000 hours.

5.4 SYSTEM SAFETY AND RELIABILITY

The failure of a component may or may not cause a "safety failure" of a system. When a system is comprised of several components, its safety will depend on:

1. The probability of failure of the individual components
2. The arrangement of the components
3. Safety systems put in place to avert negative consequences

5.4.1 Series and Parallel Systems

Conceptually there are two basic system arrangements that are possible—series or parallel. In **series systems**, failure of a single component causes system failure. In **parallel systems**, multiple independent failures are required before the entire system will fail.

An automobile can illustrate both the series and parallel arrangement of components. For lighting the highway during the night, an automobile is equipped with two headlights. This is a parallel arrangement because the design incorporates two identical components (the headlights) that perform nearly the same function. If one headlight fails, you can still drive. Your visibility is impaired to some degree and the system has been compromised; however, it has not completely failed. You also have notice of the first failure before the second component fails. Similarly, automobiles have two or even three independent braking systems, because you may not notice the failure of one system. On the other hand, suppose you intend to start your engine. Consider the components involved in this action (ignition switch, battery, wiring, solenoid relay, starter motor, spark plugs, fuel, fuel pumps, etc.). If any of these components should fail, the engine will not start when you turn the key or push the start button. A complete series of successful components is required for the system to function. To start your engine, every component must properly function because the failure of a single component will result in a failure of the entire system.

In general, an automobile's inability to stop is considered a greater risk than its inability to start. Shutting the engine down rarely causes an accident. However, for an aircraft shutting down the engine is much more hazardous, so they have more redundant ignition systems.

5.4.2 Reliability of Series Connected Systems

Let's consider the reliability of a system involving three components that are in a series arrangement, as shown in Fig. 5.4.

Fig. 5.4 A system comprised of a series arrangement of three components.
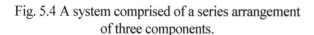

Because all components in this series arrangement must operate successfully for the system to perform satisfactorily, the probability of survival of the system (P_s^s) is a product function given by:

$$P_s^s = P_{s1} P_{s2} P_{s3} \qquad (5.13)$$

where the superscript s refers to the entire system of components.

Combining Eqs. (5.12) and (5.13) enables system reliability to be expressed in terms of the failure rate of the individual components as:

$$P_s^s = e^{-(FR_1 + FR_2 + FR_3)t}$$ (5.14)

When the number of components in a system increases in a series arrangement, the system reliability decreases markedly. Also, the system reliability continues to decrease with time in service.

EXAMPLE 5.3

Consider the case where n components are arranged in a series where n is a variable that increases from 1 to 1000. Let's also assume that the reliability of each of the n components is the same ($P_{s1} = P_{s2} = \ldots\ldots = P_{sn}$) at some time t_X during the operating life of the system. Evaluate the system reliability for three different values of the component reliability—P_s = 0.999, 0.99, and 0.90.

Solution: The results obtained from Eq. (5.13) are shown in Table 5.2.

Examination of the results presented in Table 5.2 clearly shows the detrimental effect of placing many components in a series arrangement. For components with a high reliability (P_s = 99.9%), the system reliability P_s^s drops to about 90% with 100 series arranged components. However, with components with a lower reliability (P_s = 90%), the system degrades to a reliability of less than 1% when the number of components approaches 50. The lesson here is clear—if you connect a large number of components together in a series arrangement to develop a system, then the individual component reliability must remain extremely high over the entire service life of the system.

Table 5.2
Reliability of a system P_s^s with n series arranged components as a function of component reliability P_s at time t_X

Number of Components, n	$P_s = 0.9990$	$P_s = 0.990$	$P_s = 0.900$
1	0.999	0.990	0.900
2	0.998	0.980	0.810
5	0.995	0.950	0.590
10	0.990	0.904	0.349
20	0.980	0.818	0.122
50	0.950	0.605	*
100	0.905	0.366	*
200	0.819	0.134	*
500	0.606	*	*
1000	0.368	*	*

* System reliability of less than 1%.

5.4.3 Reliability of Systems with Parallel Connected Components

Recall the description of a highway lighting system on an automobile with a pair of headlights. This system is an example of a parallel arrangement because both headlights must fail before the system fails. It is possible to drive with one light, although the State Troopers might warn the driver of the dangers of doing so. The human body has several parallel systems; you have two hands, two eyes, two legs, two feet, three lungs, two kidneys and four fingers. These parallel systems increase a person's ability to function in the event of a serious mishap involving one or more of the body's critical systems.

A system with a parallel arrangement of three identical components is represented with the model shown in Fig. 5.5.

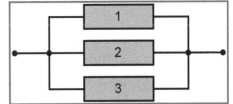

Fig. 5.5 System with a parallel arrangement of three identical components.

Parallel systems are considered to be redundant because all of the components must fail before the system fails. Accordingly, the relation for the probability of system failure is written as:

$$P_f^s = P_{f1} \, P_{f2} \, P_{f3} \tag{5.15}$$

Substituting Eq. (5.9) into Eq. (5.15) gives:

$$1 - P_s^s = (1 - P_{s1})(1 - P_{s2})(1 - P_{s3}) \tag{5.16}$$

Expanding Eq. (5.16) and reducing the resulting expression gives:

$$P_s^s = P_{s1} + P_{s2} + P_{s3} - P_{s1} \, P_{s2} - P_{s2} \, P_{s3} - P_{s1} \, P_{s3} + P_{s1} \, P_{s2} \, P_{s3} \tag{5.17}$$

Let's examine Eq. (5.17) to determine if increasing the number of parallel components improves reliability.

EXAMPLE 5.4

Suppose the probability of success P_s for all the components in a parallel system is the same. However, P_s is considered as a variable increasing from 0.1 to 1.0. Show the effect of component redundancy on system reliability P_s^s by considering a parallel system comprised of one, two and three components.

Solution:

The results obtained from Eq. (5.17) are shown on the next page in Table 5.3. An examination of the results presented in Table 5.3 shows that the degree of redundancy improves the reliability of a system with a parallel arrangement of components. If you have relatively poor component reliability, say 60%, and employ two of these components in a parallel arrangement, the system reliability improves to 84%. Add a third component to the parallel arrangement and the reliability increases to 93.6%. Adding more components in parallel will continue to improve the reliability, although it is a case of diminishing returns.

The improvement in reliability by employing parallel (redundant) components in designing a system is a distinct advantage. However, one rarely if ever, enjoys a free lunch. The corresponding disadvantages are the increased costs and the added power, weight and size of the system. These are significant disadvantages, and redundant design is used only when component reliability is too low for satisfactory system performance or when a failure produces very serious consequences. For a more complete discussion of component and system reliability, see reference [1].

Table 5.3
Effect of the degree of redundancy on system reliability P_s^s
$$P_{s1} = P_{s2} = P_{s3}$$

Component Reliability	Single Component	Two Components	Three Components
0.10	0.10	0.19	0.271
0.20	0.20	0.36	0.488
0.30	0.30	0.51	0.657
0.40	0.40	0.64	0.784
0.50	0.50	0.75	0.875
0.60	0.60	0.84	0.936
0.70	0.70	0.90	0.973
0.80	0.80	0.96	0.992
0.90	0.90	0.99	0.999
1.00	1.00	1.00	1.000

5.4.4 Redundancy and Safety Systems

Parallel design is only one example of redundancy. A wide variety of types of redundancy can be identified in biological systems that are relevant analogies to technical systems. The biological categories include:

Duplicated organs or systems: Duplication exists in the human body in many ways. Humans have multiple kidneys, genes and other components that can fully function independent of the duplicate unit. A building analogy might be duplicate water supplies, where one or another can carry the full load. Automobiles have multiple braking systems.

Autonomous functioning of coordinated items or systems: Human hearing and sight have duplicate components that function together to provide enhanced capability but each system can operate separately to provide substantial capacity. Headlights are a coordinated system.

Excess capacity: The human liver has such an excess of capacity that large portions can be removed without substantial effect on ordinary living. Similarly a woman's ovaries contain far more eggs than would be needed in a normal lifetime. Higher levels of strength in the frame of an automobile can be described as excess capacity.

High reliability systems: The human heart has evolved as a high reliability system with numerous internal checks and carefully supervised functioning. Similarly the electronic controls of cars are designed for very high reliability.

Automatic reprogramming: The human brain has some capability of "rewiring" itself in response to injury or illness. Smart systems that can "reprogram" might fall in this category.

Spare units: Some species reproduce at a high rate so that spare "units" are always available. Like spare tires they ensure that the species as a kind of overall system will keep functioning.

Functional redundancy: In addition to structural redundancy humans also have a certain amount of "functional" redundancy, where persons missing arms, sight or other organs develop "functional" substitutes to accomplish the same task. More importantly humans evolved the ability to make and use complex tools that add redundancy to their functional environment. Human factors engineering can identify functional substitutions.

5.5 EVALUATING THE RISK ENVIRONMENT[3]

When the space shuttle Challenger exploded during launch on January 28, 1986, a Presidential Commission [2] was established to:

1. Review the circumstances surrounding the accident and determine the probable cause or causes for the explosion.
2. Develop recommendations for corrective or other actions, based on the Commission's findings and determinations.

Richard Feynman, a Nobel Prize winning physicist from the California Institute of Technology, was appointed to the Commission. Dr. Feynman, near the end of his career and his life, took the appointment very seriously and devoted his entire time and energies to the investigation. One of his many contributions during the review was to determine the probability of failure of the space shuttle system.

During the investigation, Dr. Feynman asked three NASA engineers and their manager to assess the probability of failure P_f^s of a mission due to the failure of one component—the shuttle's main rocket engine. Secret ballots by the engineers indicated two estimates with $P_f = 1/200$ and one with $P_f = 1/300$. Under pressure, the program manager finally estimated the risk at $P_f = 1/100,000$. There was such a large difference between the manager's and the engineer's assessment that Dr. Feynman concluded that "NASA exaggerates the reliability of its products to the point of fantasy" [3].

The safety officer for the firing range at Kennedy Space Center, who had been under considerable pressure to remove the remotely controlled destruction charges from the shuttle, did not believe the reliability figures cited by NASA. He had collected data for all of the 2,900 previous launches using solid rocket boosters. Of this total, 121 had failed. This data provide a very crude estimate of $P_f = (121)/(2900) = 0.042 = 4.2\%$ or about one chance for failure in every 24 launches. The safety officer considered this high risk of failure to be an upper bound, because improvements made since the early launches had improved the reliability of the solid booster motors. Also, the pre-launch inspections on the shuttle were more thorough. His estimate of risk accounting for taking these improvements was $P_f^s = 1/100$. Dr. Feynman [3], after extensive interviews with engineers, reliability experts, managers from NASA and many subcontractors, concluded that the shuttle "flies in a relatively unsafe condition with a chance of failure of the order of 1%."

After the Challenger accident NASA began a series of comprehensive assessments of risk of flying the shuttle fleet in space. In 1995 the probability of a catastrophic failure was assessed at 1 in every 145 flights. After a series of safety related upgrades to the shuttle system, a new assessment in 1998 placed the probability at one in 245 flights. In October of 2002, NASA most recent assessment placed the probability at one in 265 flights or about 0.39%. How accurate are these assessments? The disintegration of the shuttle Columbia over Texas as it reentered space on February 1, 2003 was the second shuttle failure in 112 flights. This relatively high rate of failure $(2/112) = 0.01786 = 1.786\%$ raises questions about the validity of NASA's safety assessments [4, 5].

Clearly, risk assessment is a difficult task. Each component in a complete system must be evaluated to ascertain its failure rate. Then the components are placed in either a series or parallel arrangement, or some combination of the two, to model the system prior to determining its reliability. The system reliability may be lower or higher than the component reliability depending on the number and the arrangement of the components. Testing to determine component reliability is possible in some instances when the components are relatively inexpensive. However, establishing reliability estimates by testing requires a very large number of trials that often destroy the component. If you want to show a probability of failure less than $P_f = 1/1,000$,

---

[3] On June 1, 2014 the New York Times released a video titled Challenger, Columbia and the Nature of Calamity. We suggest you Goggle the title and watch the video to study the failure of a system and a more significant failure of management at NASA and at Morton Thiokol . Also read the Retro Report by Clyde Haberman.

you must test considerably more than 1,000 components to failure. Obviously, it would not be possible to test more than 1,000 booster rockets when the cost of a single test is several million dollars.

Often the probability of failure of a component must be estimated based on previous experience with similar applications. In some instances, it is possible to compute the probability of failure; however, these calculations require considerable knowledge of the spectrum of loading and the ability of the material to resist fracture. When analytical methods are inadequate and engineering judgment is required to assess the probability of failure, the estimate should be made by a senior engineer with considerable experience and expertise. Even then, the estimate should be pessimistic rather than optimistic. A frank and honest estimate of P_f, based on all of the data and knowledge available, is much better than unrealistic appraisals that give an unwarranted feeling of safety.

5.6 RECOGNIZING RISKY ENVIRONMENTS

When designing a product, there are usually risks involved in either the production of the product or in its use in the marketplace. What can you do to minimize the risk? In previous sections, the concept of safety factor, component and system reliability and evaluating the risk were briefly discussed. There is another component to minimizing risk — developing an ability to recognize the range of hazards involved. You must clearly recognize potential hazards before taking the necessary precautions in your designs to minimize the risks associated with them.

Human Factors

One of the most important and often forgotten areas of safety engineering is human factors. As an example, in 2005 an ATR 72 turboprop aircraft crashed near Palermo Italy killing 16 people. The plane ran out of fuel because the wrong fuel gauges had been installed. A manufacturer had created totally interchangeable fuel gauges for two different aircraft, the ATR 72 and the smaller ATR 42. In addition, the system had no redundancy; the low fuel indicator ran through the same gauge. No technical precautions were taken to make sure that a mechanic installed the correct gauge. It is critical that all engineers design products that can be safely used by the people who will use or install them. Some artificially defined "expert user" who never makes mistakes, reads all instructions and has the eye of an eagle and the touch of an artist does not exist.

Human factors engineering is a discipline in itself. In consumer products, reasonably foreseeable misuse of the product is a design criterion. In industrial products care must be taken to avoid confusion, mistakes and information overloads. A mass of warning labels and fine print added at the last minute often indicates poor design and inadequate system safety analysis.

5.7 A LISTING OF HAZARDS

Mowrer describes a complete list of hazards and provides an extended discussion of each in reference [6]. You are encouraged to read this chapter and to use the extensive checklists incorporated in his coverage. A brief excerpt from Mower's reference is provided in this section to assist you in recognizing the many hazards that should be considered when designing a product. This list of hazards includes:

- Dangerous chemicals and chemical reactions
- Exposure to electrical circuits
- Exposure to high forces or accelerations
- Explosives and explosive mixtures

- Fires and excessive temperature
- Pressure
- Mechanical hazards
- Radiation
- Noise

Dangerous Chemicals

Chemicals can be nasty and you must appreciate the extreme dangers of exposures to certain toxic substances. Have you read about the leak that developed in a storage tank in Bhopal, India, in 1984? The storage tank contained the toxic chemical methyl isocyanate used in the manufacture of pesticides. It was a significant leak with about 80,000 pounds of the chemical released to the environment. Three thousand nearby inhabitants were killed, 10,000 permanently disabled and another 100,000 injured. The cause of the failure was not in the design of the tank, but in the training of the individuals responsible for the plant maintenance and operation. Nevertheless, a catastrophic accident occurred, because several workers and managers in positions of responsibility did not adequately understand the dangers of this very toxic chemical.

Exposure to Voltage and Current

What about exposure to electrical circuits? Almost everyone has been shocked by the standard 120 volt, 60 cycle electrical power supplied by the local utility company. Why worry? You should worry because electrical shocks are **dangerous**. **Yes!** Even the 120-volt supply can cause major problems. Your body acts like a resistor and limits the current flowing from the voltage source through your hands, arms, legs, etc. The problem is that the resistance of your body is variable. It depends on the moisture on your hands, the type of soles on your shoes and even the moisture on the ground. If your hands are dry, your shoes have rubber soles and you are standing on a dry floor when you touch one wire of the circuit with only one hand, then you will probably not be harmed because you have a very high resistance path to ground. Consequently, the current flow through your body will be very small. However, if your hands are wet and you touch both of the wires (the white and the black) from the electrical supply, one with the left hand and the other with the right hand, you have placed 120 volts across your heart. You can be electrocuted with only 120 volts under these conditions. The moisture on your hands greatly reduces the effective resistance of the body.

Do not take chances with electricity. Insulate the operator from the circuits preferably with two independent layers of insulation. High voltages are even more serious than low voltages because the currents flowing through a human body increases dramatically. When the current flow through one's body increases to about 10 to 100 mA, serious consequences to the respiratory muscles result. Higher currents of 75-300 mA cause problems with a person's heart function.

The electrical current I flowing through the body can increase in two ways: first, by increasing the applied voltage V, and second, by decreasing the resistance R offered by the body. Ohm's law gives the relation among the voltage V, current I, and the resistance R as:

$$I = V/R \qquad\qquad\qquad (5.18)$$

Using adequate electrical insulation in the design of products using electrical power, markedly increases the resistance R and decreases the current I to negligible amounts.

Some people believe that low voltages (5 or 10 volts) are safe because they barely feel a tingle when touching a low voltage circuit. However, some low voltage circuits particularly on high performance processors carry substantial (100 A or more) currents. If you short a circuit with high current flow, an arc occurs which generates significant amounts of heat. Also, the flash of the arc can damage a person's eyes and the heat may produce serious burns. In your designs, insulate and/or shield low voltage circuits if they carry significant currents.

The Effects of High Forces and Accelerations

High forces and high accelerations (or decelerations) go hand in hand. Newton's second law requires the connection ($\mathbf{F} = m\mathbf{a}$). If the brakes are suddenly applied in an automobile, the car decelerates quickly and the passenger (without seat belt or air bags) is thrown into the windshield. You must always be concerned with acceleration or deceleration because of their effects on the human body. Military pilots are trained to withstand high accelerations (high Gs). A good pilot with a well-designed flight suit can pull 7 or perhaps 8 Gs before losing consciousness. However, a civilian will become irritated at less than 2 Gs. If you want to feel an acceleration thrill, go to an amusement park and ride the roller coaster. There is no need to incorporate high accelerations into the design of most new products. A reasonably hot sports car that accelerates to 60 MPH in six seconds requires an acceleration of only 0.46 Gs. You should be careful to keep both acceleration and decelerations low when you design products that move and accelerate.

Explosives and Explosive Mixtures

With the bombing of the U. S. S. Cole, the disaster at the Federal building in Okalahoma City and terrorists bombings at many locations, most people have become aware of the disastrous effects of large explosions. The gas pressures generated by blasts destroy substantial buildings, break glass over a wide region and kill or injure many people. The population is generally aware of the characteristics of common **detonating** explosives like dynamite and ANFO (ammonia nitrate and fuel oil). In most designs, you do not encounter a need to accommodate these traditional explosives. What you must recognize are other less apparent agents that act like explosives under special circumstances. Fuels such as natural gas, propane, butane, etc. can leak and combine with air to produce a **deflagrating** explosion when ignited. Boating accidents are common when gasoline leaks in an engine compartment. The resulting mixture of gasoline fumes and air can explode with a small spark, destroying the boat and killing or injuring the passengers. Still another unusual source of fuel for an explosion is dust. When handling large quantities of a combustible solid, dust (fine particles) is generated. If these particles are suspended in air, the resulting mixture will explode when ignited. A grain elevator exploded in Westwego, Louisiana, in 1977 killing 35 people when an explosive mixture of combustible particles from the grain and air was ignited.

Fires

Over a million fires occur in the U.S. every year. Some are vehicle fires (400,000) and others are structural fires (650,000). Many people die in the fires (4,700) and many more are injured (28,700). Clearly, fire is a serious problem. People are killed, injured or traumatized and property is lost (eight or nine billion dollars per year).

Engineers have to distinguish among the causes of fire **ignition**, which may be natural, negligent or even intentional and the cause of fire **disasters** which is often the result of the failure to take appropriate precautions in the design of products, structures and systems. One major problem that has been identified is the over reliance on inappropriate regulatory tests. Small scale **ignition** tests for example, are totally unsuited for evaluating the contribution of a material to an existing fire. Many serious tunnel fires have resulted from the improper classification of materials using such ignition tests.

Engineers have a responsibility to decrease the number of fires resulting from the products that they design. They need to understand the source of ignition. For example, did the fires start with an appliance or a motor that overheated for some reason? Determine the reason for the overheating and redesign the product so that excessive heat will not be generated. They also need to understand why and how a fire spreads. Some people believe that we are far too casual about fires. Carelessness in personal practices, poor design in products intended for the home and business and fraud (arson) to collect fire insurance reimbursement for lost property are routine.

Pressure and Pressure Vessels

Pressurized fluids are used for many good reasons, and in most cases, pressure vessels perform very well. It was not always the case. In Boston during the winter of 1919, a huge tank about 90 feet in diameter and 50 feet high fractured. It contained two million gallons of molasses that flooded the local area. Twelve people died and another 40 were injured in this accident.

It is relatively easy to design a pressure vessel today. In fact, the American Society for Mechanical Engineers (ASME) has developed a code that engineers follow in their design to produce pressure vessels that are certified as safe for service. However, on occasion, tanks fail in service. The difficulty is usually with the steel plates that are welded together to manufacture the tank. The steel employed in both the plates and the welds have to exhibit high fracture toughness at low temperatures, and the welds must be free of significant flaws. If you have the responsibility of designing a pressure vessel, follow the ASME code, make certain the welding procedures followed produce crack-free welds, and specify steel for the plate and welding rods that is sufficiently strong and fracture resistant. If you intend to become a mechanical engineer, you will have the opportunity to learn how to design pressure vessels, select suitable materials for their construction and specify welding procedures in courses presented later in the curriculum.

Mechanical Hazards

Mechanical hazards are features, which exist on products that may cause injury to someone. Consider as an example an electric fan used to cool a room. Is the fan blade adequately guarded, or is it possible for someone to stick his or her finger into the rotating blades? Suppose the cabinet you recently designed to hold a special tool has a sharp corner at hip level. Can someone walk into the cabinet and break skin or bruise a hip on that corner? You have designed a center post crane to lift material and move it over a 25-foot diameter area. Will the center post be stable under all conditions of loading or will it collapse? You have designed a new pizza machine that rolls the dough into sheets exactly 2 mm thick. Have you provided protection for the entrance to the rolls that prevent the operator from inserting his or her fingers into the rollers? You have designed a wonderful guard that prevents a person from exposing their hands and arms in operating a punch press. However, the guard is attached to the machine with two small screws. Have you used locknuts and/or safety wires to insure that the screws will not loosen during the operation of the press? Are the screws large enough to resist failure by shear? Is there any incentive to remove the guard?

There are many mechanical hazards encountered in designing equipment and products. Always examine each component, and look for sharp points, cutting edges, pinch points, rotating parts, etc. Do not count on peoples' good judgment. If it is possible to insert a finger into the machinery, even if it is a foolish act, you can be certain that sooner or later someone will do so.

Radiation Hazards

Radiation hazards are due to exposure to electromagnetic waves. The damage produced depends on many different factors such as:

- The strength of the source
- The degree to which the emitting radiation is focused
- The distance from the source
- The shielding between the source and the object being radiated
- The time of exposure to the radiation

The radiation spectrum is divided into several different regions—very short-wave length, visible light, infrared and microwave radiation. Short-wave length radiation is the most dangerous (x-rays, gamma rays, neutrons, etc.) with serious health risks (cancer) for overexposures. The hazards due to UV, visible light and IR are usually due to excessive exposure where serious burns to the skin occur. For very intense radiation, even short exposures are detrimental to the eyes. Microwave radiation is absorbed into the body and may result in the heating of one's internal organs. Because the means used by organs to dissipate this heat is not known nor is the effect of the localized increase in temperature known, it is prudent to avoid exposure to microwaves. In America, the Offices of Safety and Health Administration (OSHA) has issued a regulation limiting the power density of microwave energy to 10 mW/cm^2 for an exposure of six or more minutes. In designing products, where operators may be exposed to microwave radiation, shielding should be employed to reduce the power density well below the regulatory levels.

Noise

Noise levels that occur in the environment may be damaging to your hearing, may interfere with your work or play, and may degrade the quality of your life-style. Most noise is man-made, although occasionally a storm and Mother Nature provides the sounds of thunder and wind. If you listen occasionally to a band playing rock and roll music, there is a temporary shift in the threshold of audibility to a higher level of pressure. However, if you play in the rock and roll band almost every night for an extended period, the shift in the threshold of audibility becomes permanent and your hearing becomes impaired.

Noise is a pressure disturbance that propagates through some medium, usually air. The velocity of propagation through air is 344 m/s at room temperature. The pressure disturbance is oscillatory usually with many different frequencies. The frequency content of the pressure waves depends on the source of the sound. A note from a violin will have much higher frequencies that a note from a tuba. The intensity of the noise is measured with a sound level meter that consists of a microphone, amplifier and a display meter that provides a reading in decibels (dB).

As a general guideline, the threshold for audibility is less than 25 dB, before a person is considered handicapped. In addition, there is a threshold for feeling noise-generated pressure at about 120 dB and another threshold for pain between 135 and 140 dB. When designing a product, the noise level is a serious consideration. Clearly, the feeling and pain levels of noise intensity must be avoided, but what levels are satisfactory? The U.S. Environmental Protection Agency (EPA) has established standards, which provide guidance to engineers in the design of products. For example, the results presented in Fig. 5.6 show the relation among the sound pressure level, the communicating distance and the degree of speech intelligibility.

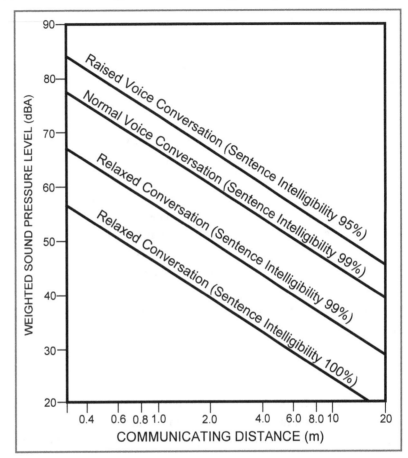

Fig. 5.6 Outdoor distances for intelligible conversation with a background of steady noise [8].

5.8 PRODUCT LIABILITY

Products liability is the legal doctrine that imposes responsibility on designers, manufacturers and sellers of products. The legal standard for product design in the USA is very high. The most typical formulation is provided by Section 402A of the Second Restatement of Torts:

(1) One who sells any product in a defective condition unreasonably dangerous to the user or consumer or to his property is subject to liability for physical harm thereby caused to the ultimate user or consumer, or to his property if:
   (a) the seller is engaged in the business of selling such a product, and
   (b) it is expected to and does reach the user or consumer without a substantial change in the condition in which it is sold.
(2) The rule stated in Subsection (1) applies although:
   (a) the seller has exercised all possible care in the preparation and sale of his product, and
   (b) the user or consumer had not bought the product from or entered into any contractual relation with the seller.

Suffice it to say that the failure to design for safety can be both financially and professionally disastrous.

5.9   <u>SUMMARY</u>

When a product is introduced in the marketplace, there is usually some risk involved to the workers fabricating the product, the customers and society-at-large. Hopefully, the risks will be small and acceptable to all concerned because of the significant benefits produced for the customers and/or the sociotechnical system. In determining if the risk is acceptable, the public needs an accurate and honest assessment of the consequences of a failure and the probability of the occurrence of a serious accident. It is the responsibility of the engineering profession to assist business administrators and governmental agencies in these assessments.

This chapter defined both risk and hazard and introduced the distinct problem of designing both intrinsic and extrinsic safety systems. It emphasized that ultimately society decides what risks will be acceptable. But while the public must accept risk if it is to have the benefit of certain products, engineers must make every attempt to minimize the probability of harm within socially acceptable constraints on cost and performance. There are many excellent engineering methods for ensuring safe design. The concepts of stress, strength, safety factor and margin of safety illustrate an approach to minimizing risk due to failure by fracture. Recommendations for a range of commonly accepted safety factors have been given. Issues to consider in selecting the safety factor to employ when sizing components have been described.

Sometimes it is understood that failures will occur in service. In these cases, it is necessary to determine the probability of the failure event. To illustrate methods for determining probability of failure, the concepts of failure rate and mean time between failures were introduced. It was shown that the failure rate could be determined by keeping records of the failures of components as a function of time after they were placed in service. These concepts are important, but they should not be confused with the probability of failure or survival.

Component reliability was introduced, and a method for computing reliability from the failure rate was shown and demonstrated. In developing this reliability equation, Eq. (5.12), the failure rate was assumed to be constant over the service life. It is important to observe that the probability of survival decreases as the service life is increased and the probability of failure increases with time in service. For reliable service for a long period of time, the failure rate must be extremely low.

System reliability is different than component reliability. A system is usually composed of many components. If the components are arranged in series, they must all function for the system to operate correctly. The reliability of a series connected system is determined from Eq. 5.13. It is evident from the data presented in Table 5.2 that the number of components arranged in a series markedly lowers the reliability of a system. Sometimes a system must function all of the time. If this is the case, a system failure cannot be allowed, because of its negative consequences. In these instances, engineers design with a number of redundant components arranged in a parallel system. Redundancy improves system reliability, as shown in Table 5.3, but at increased cost and the requirement for more power, weight and size.

The explosion of the space shuttle Challenger and more recently the disintegration of Columbia were described to illustrate the importance of evaluating the risk. It is often a very difficult problem and an analytical solution for the risk is often not possible. Nevertheless, it is a professional responsibility to prepare an intelligent, accurate, honest and frank estimate of the risk, and to insure that all of the principals involved are aware of the consequences of a failure.

Finally, a number of hazards that cause injury or death have been introduced. Unfortunately the listing is relatively long. It is important that you recognize the hazards and be vigilant in your designs to avoid them. Avoidance often does not require extensive calculation from elaborate formulae, but rather a detailed assessment of each component in the system and a good measure of common sense.

REFERENCES

1. Dally, J. W, P. Lall and J. Suhling, Mechanical Design of Electronic Systems, College House Enterprises, Knoxville, TN, 2007.
2. Lewis, R. S. Challenger: The Final Voyage, Columbia University Press, New York, NY, 1988.
3. Feynman, R. "Personal Observations on the Reliability of the Shuttle," Appendix to the Presidential Commission's Report, Ayer Co., Salem, 1986.
4. Mowrer, F. W., Introduction to Engineering Design: ENES 100, McGraw Hill, New York, NY, 1996, pp. 149-172.
5. Whoriskey, P., "Shuttle Failures Raise a Big Question," Washington Post, February 10, 2003, p. A-9.
6. Gugliotta, G and R. Weiss, "Dangers of Gauging Space Safety," The Washington Post, February 17, 2003, p. A-15.
7. Cunniff, P. F., Environmental Noise Pollution, Wiley, New York, NY, 1977, pp. 101-115.
8. "Information on Levels of Environmental Noise Requisite to Protect health and Public Welfare with an Adequate Margin of Safety," EPA Report, March 1974.
9. Powers D. G. and J. Proctor, Editors, Lockheed L-188 Electra, World Transport Press, April 1999.
10. Brannigan, G. P. Corbett, Brannigan's Building Construction for the Fire Service, 4th Edition, National Fire Protection Association, 2007.
11. Tuninter, "The ATR 72 Accident: Safety Recommendations," Addressed to European Aviation Safety Agency, EASA.

EXERCISES

5.1 Have you been involved in one or more automobile accidents since you began driving? Were you or anyone else injured or worse yet killed? Estimate the total mileage you have driven over this period and calculate your accident rate (number/mile). What were the reasons for the accident or accidents? Comment on your driving behavior and its influence on the accident rate.

5.2 Are the benefits of driving worth the risks of injury or death? Determine your probability of being killed in a fatal accident while driving this year. Hint: Statistics on fatal accidents are always listed in the World Almanac and on the web site for the NHTSA. Make any assumptions necessary for your analysis, but justify each of them.

5.3 If a component is designed with a safety factor of 3.75, what is its margin of safety?

5.4 NASA's space shuttle utilizes two booster rockets fueled with solid propellant to provide the thrust required for launch. Is this a redundant system? State your reason for this conclusion.

5.5 If the probability of failure of one of the solid propellant, booster rockets on the space shuttle is 1/1,000, determine the reliability of the solid booster rocket system.

5.6 You are designing a very large computer system to contain the database for a world-wide reservation system. It is estimated that at any instant 900 operators will be accessing the database, and another 2,100 operators will soon be ready with their requests for computer availability. The mainframe computers that you plan to employ each have a MTBF of 10,000 hours. Present a design of the computer system that will insure that all 900 operators have a 99.8 % probability of being served. In your design show all calculations and carefully list your assumptions. Justify the costs involved if each mainframe computer employed in the system is valued at $270,000.

5.7   Janet is an engineer in the transmission department of the Fink Motor Corp. Her job is to record data on the mileage prior to failure of the new lightweight transmission that has been placed in 114,000 of the new 2014 model of the Clunker III. She records the following data:

Mileage (1000 miles)	No. Failures, N_f
0 - 5	75
5 - 10	58
10 - 20	40
20 - 30	25
30 - 40	29
40 - 50	45
50 - 60	116
60 - 70	581
70 - 80	1,594
80 - 90	7,724
90 - 100	16,569

Determine the failure rate as a function of service life expressed in terms of mileage, and prepare a graph showing your results.

5.8   Using the data from the table in Exercise 5.7, determine the probability of survival and the probability of failure of the transmissions as a function of service life measured in terms of mileage. Prepare a graph of your results.

5.9   If the transmissions in the Clunker III were under warranty for 100,000 miles, what would be the consequences for the Fink Motor Corporation?

5.10   Prepare a listing showing the ratio of fatal events per flights for the airlines in the U. S. and Canada since 2000. The data needed to prepare this listing is at http://airsafe.com/airline.htm.

5.11   You are designing a cabinet-mounted, self-cleaning oven that requires very high power levels to heat its interior surfaces until they are free of all the splattered, burnt-on grease. As the lead engineer on this design team, what precautions should you take to insure that the oven would not be the source of a fire during its anticipated 15-year life?

5.12   There are many fatal accidents each year involving crashes with large trucks. The Federal Motor Carrier Safety Administration (FMCSA) is the government agency responsible for regulations involving trucks. Write a paper describing possible rules the agency could impose on the trucking industry to reduce the number of fatal truck crashes. Cite reasons for the new rules and predict the benefits that would result if they were implemented. Discuss the economic consequences for your suggestions.

CHAPTER 6

DEVELOPMENT TEAMS

6.1 INTRODUCTION

The development of quality, leading edge, high-performance products requires the coordinated efforts of many skillful individuals from different disciplines over an extended period of time. This group of individuals is organized within a company to act in an integrated manner to successfully complete the development process. The organizational structure employed varies from one corporation to another; it also depends on the size of the company engaged in the development. For relatively new firms in the entrepreneurial stage, team structure is usually not an issue. These firms are small with a single product and only ten to twenty individuals are involved in the development process. Everyone on the payroll is deeply committed to this product, and communication is usually accomplished around the lunch table. For very large corporations, where the number of employees may exceed 100,000, the company is often organized into divisions or operating groups. These large companies have many products or services that are offered by each division, and in some instances, one division may actually be a direct competitor of another division. For example, General Motors with its three automotive and two truck divisions produces many different models of each of its lines that compete for the same market segment.

In a large corporation, the total number of new products that are introduced each year may exceed 100. Communication is often very difficult because hundreds or even thousands of miles sometimes separate personnel involved in the development of a given product. Sometimes different divisions are in different countries, and business is conducted in different languages. Clearly, in a large firm, the organization of the product development team and the physical location of its team members become extremely important.

In the past twenty years, there have been many examples [1] demonstrating that a cross disciplinary team provides a very effective organizational structure for product development. The idea of forming a cross disciplinary team to design a new product appears simple until one examines the existing organizational structure of all but the smaller of the product oriented corporations. Larger corporations are usually organized along functional lines. The organization chart presented in Fig. 6.1 shows many of the functional departments that provide personnel needed to staff a typical product development team. Each department has a functional manager. Also, several departments are often grouped to form a division and each division has a director. Titles for the division leaders vary from firm to firm—but Director or Vice President are commonly employed.

The functional organization with its leadership structure and its ongoing operations presents a management problem when forming product development teams. As a team member, do you report to the team leader (product manager) or to the functional manager? When important decisions must be made, are they made by the product manager, by one or more of the functional managers or by the division directors? Clearly, the establishment of a product development team within a company that is organized along functional lines creates several operational problems for management. Functional organizations may also create difficulties for engineers attempting to serve several different managers. In the next section, these problems are addressed, and a successful team structure employed in many corporations that are organized along functional lines is described.

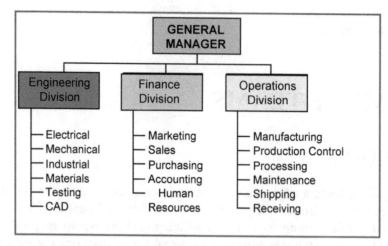

Fig. 6.1 Typical organization chart showing the disciplinary functions. This organization includes three divisions and three levels of management.

6.2   A BALANCED TEAM STRUCTURE IN A FUNCTIONAL ORGANIZATION

There are several organizational arrangements frequently used in forming teams within companies organized along functional lines. They include:

- Functional teaming
- Modified functional teaming
- Balanced teaming
- Independent team organization

Each of these techniques for forming product development teams in a functional organization has advantages and disadvantages. A complete discussion of these four different team structures is given in reference [2]. To introduce you to the concept of forming teams within a functional organization, the balanced team structure will be described.

The balanced team structure is represented in the organization chart presented in Fig. 6.2. The team is formed with members drawn from the functional departments; however, in most situations, the team members are located in close proximity in a separate room or a building wing reserved for the team. The team members are usually dedicated to a single project, and their responsibilities are coordinated and focused. The team members retain their reporting relationship to the functional manager, but the contact is less frequent than in functional teaming arrangements, and the relationship is less intense. Members of a balanced team often consider the product manager to be more important than their functional manager. The product manager is frequently a senior technical administrator with more experience, status and rank than the typical functional manager. The status of the product manager is dependent on the product development costs and the size of the development team. The product manager reports to the general manager, and usually is equal in status to the division directors in the organization. The senior program manager, with significant status and clout, leads the team and insures that resources will be available as required to meet schedules. The team members recognize the freedom from functional constraints that the balanced team structure implies. They assume ownership of the product specifications and commit fully to the success of the development activities. Communication is accomplished through the use of division and department coordinators and daily meetings. The team makes most of the routine decisions and only the higher-level decisions require the approval of the appropriate division directors.

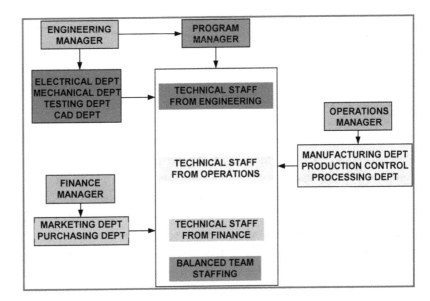

Fig. 6.2 Organization with a balanced team structure.

The primary disadvantage of the balanced team structure is the dedication of the team members to a single project. Many talented team members are needed by the functional managers to support design activities across several product lines. With some of the functional resources dedicated to a single product development, other products may not receive adequate attention. The balanced team structure enhances the development capability of a given product at the cost of reducing the technical capability that can be applied across the company's entire product line. A second disadvantage is that some team members from a given department or division may not have sufficient expertise to perform adequately. With loose and distant functional supervision and review, this fact may not be apparent. This problem often results in designs that may not be at the cutting edge of technology.

6.3   DEVELOPMENT PLANNING

As the development team becomes more independent and moves away from the functional departments, planning becomes more important. Team planning documents are often prepared by a small group of senior team members with talents in business, marketing, finance, engineering design and manufacturing. These planning documents incorporate detailed descriptions of the product, market analysis, business strategy for capturing market share, product performance specifications, development schedules and budgets, team staffing requirements, material selections, design strategies, technology prerequisites, manufacturing processes, tooling requirements, inspection procedures, test specifications, facility availability and distribution capabilities. An element of the plan covers techniques and personnel that are to be employed to insure quality and product reliability. Finally, the plan includes a section specifying the deliverables and other sections on methods to measure performance of the product and the productivity of the team.

   The development planning documents are extremely important, because they outline strategies for achieving the company's goals and objectives in developing a new product. Management may use some of these documents during periodic design reviews to monitor progress and to judge the team's performance. The plan is a well-defined contract between the team and the divisional managers involved in the product development. The team members are expected to sign-on; to commit to the level of effort required to meet the schedule, achieve the goals and to produce the deliverables. Management is expected to fund the program, to provide adequate staffing and to provide the necessary facilities, computers, software, tooling and equipment in a timely manner.

6.4 STAFFING

Cross-disciplinary teams are necessary in staffing a product development team regardless of the structure employed by the organization. Usually, at least five disciplines or functions should be represented on the team including: marketing, finance, design engineering, operations/manufacturing and quality control. The level of participation of each team member depends on the product being developed and the stage of the development. In some cases, team members may be assigned to two or more different projects. In other instances, several engineers from the same discipline are assigned to a single project to complete complex tasks necessary to meet the schedule for a single product.

Staffing on a product development team usually changes over the duration of the development period. Initially the team is small and staffed with senior personnel with demonstrated skills and talent. This small cadre may stay with the project for its duration. As the development proceeds, additional design engineers are required as technical strategies are converted into design concepts, then to design proposals and finally into detailed engineering drawings. Manufacturing engineers are needed in larger numbers when the early prototypes are being produced. After release of the design by engineering, the number of designers is reduced, and the emphasis is shifted to production where staffing from operations (production control, quality control, purchasing and plant maintenance) increases sharply. Clearly, the staffing of the development team is dynamic. Usually only a small fraction of the total team is committed to the project on a full time basis from its initiation to its completion. In some companies, individuals are committed to the development team for the duration of the project, but at varying levels of participation. This approach enhances ownership of the project by the individual team members.

6.5 TEAM LEADERS

The experience and leadership skills required of the product manager are strongly dependent on the team structure employed as well as the size and cost of the product development. Some products can be developed with a relatively small commitment of staff (say less than 10 members). In these circumstances, the team established is often structured along functional lines, and the product manager is often one of the functional managers. The functional managers are experts in their respective disciplines, skilled in managing personnel and knowledgeable (but not expert) in the other disciplines. They are also familiar with the company's entire product line.

Other product developments require a larger commitment of personnel (say a team numbering 10 to 50), and the balanced team structure, shown in Fig. 6.2, is probably the most suitable organization. The product manager is a seasoned staff member well-known for his or her expertise and respected for performing at a level that exceeds expectations. The appointment as a product manager (equal in rank and status to the functional department managers) often represents a promotion for the individual involved and a first experience in managing a technical group. The functional managers and the divisional directors provide close support through periodic reviews that monitor progress and reveal problems early in the development process.

As the size of development team becomes even larger (100 or more members), team leadership assumes major importance. The product manager in this situation is a senior member of the company's executive management with a well-established record of accomplishments. He or she is senior to the managers of the functional departments, and is at least equal in status to the division directors. The authority and responsibility given the product manager is significant, and his or her career depends strongly on the success of the product in the market and the productivity of the development team.

6.6   TEAM MEMBER RESPONSIBILITIES

Members of a product development team have two different sets of responsibilities—one to his or her disciplinary function and the other to the team. A team member, from say the electrical engineering department, provides the technical expertise in his or her specialty and ensures that the product is correctly designed with leading edge, state-of-the-art technology. The team member acts to bring all of the important functional issues that affect product performance to the attention of the team. The team member also represents the functional department to ensure that the product development team maintains the principles, goals and objectives of the functional discipline.

The most important team activity for the individual member is to share responsibility. Highly effective teams that routinely succeed have team members that hold themselves mutually accountable for the overall success of the product. Next, the team member must recognize and understand all of the product features, and fully participate in the techniques employed to meet the design objectives. Individual team members must assess team progress and participate in improving team performance. A team member must cooperate in establishing all of the reporting relationships required to maintain the communication among the team and functional departments. In many of the team structures, an individual may report to three different managers—the Division Director, the Department Manager and the Program Manager. The matrix organization structure (Fig. 6.2), inherent in forming development teams, creates complex reporting relations. Flexibility and cooperation on the part of the individual is required for the team structure to be effective.

Finally, the team members must be able to communicate in a clear concise manner in all three modalities: **oral, written, and visual (graphical).**

6.7   TEAM MEMBER TRAITS

When a group of individuals work together on a team to achieve a common goal, they can be extremely effective [3]. The team interaction promotes productivity for several different reasons. First, meeting together is synergistic in that one's ideas, when freely expressed, stimulate additional ideas by other team members. The net result is many more original ideas than would have been possible by the same group of individuals working independently. Another advantage is in the breadth of knowledge available in the team. The team is cross-disciplinary and the very wide range of skills necessary for the product development process is included within the team. Each member of the team is different with some combination of strengths and weaknesses. Acting together the team can build on the strengths of each member and compensate for any weaknesses. The grouping of individuals provides social benefits that are also very important to the individuals and to the corporation. There is a bonding that occurs over time and a support system develops that team members appreciate. The team develops a sense of autonomy valued by each of its members. The team develops solutions, implements them and then monitors the results with little or no direction from management. Empowered teams assume ownership of the project. They commit to the product development process and consistently perform beyond expectations.

While cross disciplinary teams have been employed in many different companies for at least three decades, little has been done by the educational system to develop team skills. On the contrary, both the secondary school and many college systems tend to encourage competitive attitudes and independence. Students compete for the "A" grades that might be given in a course and are discouraged from cooperating on assignments. This system prepares students to work independently, but team-building skills are not addressed.

The industrial workplace is much different. Team members compete, but not within the team. They compete with similar development teams from rival corporations. Team members bond, cooperate and consistently help each other by sharing assignments. Competition between team members is discouraged. Recognition and rewards go to the team as a whole much more often than to select individuals.

There is a set of characteristics that describe a good team member and another set that depicts an individual who can destroy the efforts of a team.

The characteristics of a good team member are:

1. Treat every team member with respect, trust their judgment and value their friendship.
2. Maintain an inquiring attitude free of predetermined bias so that team members will all participate and share knowledge and opinions in an open and creative manner. Listen carefully to the other team members.
3. Pose questions to those hesitant team members to encourage them to share their knowledge and experience more fully. Share your experiences and opinions in a casual, easy-going manner. Help other members to relax and enjoy the interactive process involved in team cooperation. Participate but do not dominate.
4. Observe the body language of the other team members, because it may indicate lack of interest, defensive-attitudes, hostility, etc. Act with the team leader to defuse hostility and to stimulate interest. Disagree if it is important, but with good reason and in good taste.
5. Emotional responses will occur when issues are elevated to a personal level. It is important to accept an emotional response even when you are opposed. It is also important to control your emotions and to think and speak objectively. You should be self confident in your discussions, but not dogmatic.
6. Allocate time for self-assessment so the team can determine if it is performing up to expectations. Make suggestions during these assessment periods to build team skills.
7. Be comfortable with your disciplinary skills, but continue to study to improve your capabilities. Communicate effectively by speaking clearly, writing clearly and concisely, and illustrating effectively with modern graphical tools.

The characteristics of a destructive team member are:

1. No member of the team should ever participate in a conversation that is derogatory about a person on the team or somewhere in the corporation. If you cannot make complimentary remarks about a person, keep the negative thoughts to yourself. Respect, trust and friendship are vital elements in the founding of a successful product development team. Derogatory remarks destroy the foundation necessary for respect, trust and friendship.
2. Arguments among team members should be avoided. You are encouraged to introduce a different opinion and participate in a discussion with a different viewpoint, but the discussion should never degenerate into a hostile argument. Every member on the team is responsible for quickly resolving arguments that arise among the team members.
3. All members of the team are responsible for on-time attendance at meetings. If a member is absent, the leader should know the reason beforehand and explain it to the team. You must be certain that all members appreciate and respect the reason for everyone's role on the team and the importance of every member attending every meeting.
4. Team progress is hindered when one member dominates the meeting. These people are often overly critical, intimidating and stimulate confrontations. If you have these characteristics, work hard to suppress them.
5. The team leader must be extremely careful about intervention. If the team is moving and working effectively, the leader should be quiet, because intervention in this instance is counter productive. When the team is having difficulty, intervention may be necessary depending on the problem. In the event a team member becomes hostile, intervention should

be quick, and the disagreement producing the hostility should be dealt with immediately. When a team is seeking consensus, the process may require time and the leader should wait five to ten minutes before intervening.

6. No member of the team can be opinionated including the leader. He or she is not a judge determining the correct solution. The leader seeks to facilitate to enable the team to reach a consensus on the most suitable solution. It is only when the team accepts the solution that implementation can begin.

Compare your traits with those listed above. You probably exhibit both good and bad characteristics. To be effective team members, you must work to enhance the favorable traits and to suppress those that are destructive to the efforts of the team.

6.8   <u>EVOLUTION OF A DEVELOPMENT TEAM</u>

If a number of workers or students are assigned to a new team, it is possible to observe several behavioral phases as the team bonds, melds and matures [4]. In the very beginning (Phase I), the team members usually exhibit both excitement and concern. They are excited about working on a high-profile project with a new group of talented people. It should be an opportunity to learn, make new friends, advance in position and/or stature and have fun in the process. At the same time, the team members exhibit signs of concern. They are worried about meeting and understanding other team members. The tasks assigned to the team are probably extremely vague at this stage of the development, and vagueness leads to uncertainty and anxiety. The skill levels and areas of expertise of fellow team members have yet to be determined. The personality of the team leader is unclear. There is worry over the new reporting relations that new team membership implies. In this orientation phase, team members are searching for their role in the team and evaluating their possibilities for success or failure. It is a tense time.

Dissatisfaction is the next phase in the evolution of the team (Phase II). You have met your fellow team members, and the news is not good. You recognize the differences in personalities and in work habits. A few of the team members do not understand how to tell time and never arrive on time when attending team meetings. Member schedules make it very difficult for the entire team to meet. Some team members lack social graces. Some members are inexperienced and unable to cope with their assignments. You have met the team leader and he or she is very difficult (stressed, demanding, exacting, impatient and abrupt). The schedule calls for the team to leap over mountains on a daily basis. The budget for the project is totally inadequate. Surely you are in a lose-lose situation. Your thoughts are dominated by schemes to transfer to a better team. During this phase, the progress of the team toward scheduled milestones is exceedingly slow.

With time progress is made, and the team members start learning how to work together. This period is the resolution phase (Phase III). Everyone agrees to attend meetings and arrive on time. Personality conflicts are smoothed over. Team members may never be true friends, but everyone manages to get along and show mutual respect. Team members have come to terms with the schedule, and have committed to the extra effort that it requires. The manager has mellowed, or team members have learned how to handle his or her impatient demands. Solutions for reducing development costs have been found. There appears to be a good possibility of meeting the product development plan. The team is beginning to perform well, as more members commit and contribute.

The next phase of the development team is the production phase (Phase IV). At this point, the team has melded into an efficient unit. Conflicts rarely arise. Lasting friendships begin to be formed. The strengths of each team member are fully utilized and the manager recognizes the talents of the team. Tasks are completed ahead of schedule and under budget. You now find yourself in a win-win situation.

Termination is the final phase in the evolution of the team (Phase V). The project has been completed and the ribbons have been cut. The team prepares to disband. In the termination phase, you reflect on your experiences, good and bad, over the duration of the project. You examine your individual performance and evaluate various factors that improved the effectiveness of the team. You meet with the manager for a performance evaluation. The team members share a sense of accomplishment in achieving the goals and objectives of the development.

6.9 A TEAM CONTRACT

Teams are effective because they meet together to focus their wide range of disciplinary skills and natural talents toward the solution of a set of well-formulated problems. The team meeting is the format for synergistic efforts of the team. Unfortunately, not all teams are effective because some members are so disruptive they destroy the cohesiveness of the team. Bonding of team members is vital if the team is to successfully solve the multitude of problems that arise in the product development process.

Teams **fail** because of three reasons. First, they deviate from the goals and objectives of the development, and cannot meet the milestones on the development schedule. Second, the team members become alienated and the bonding, trust and understanding critical to the success of the team never develops. Third, when things begin to go wrong and adversity occurs, **"finger pointing"** begins and an attempt to fix blame on someone replaces creative actions.

Clearly, many problems will arise to impair the progress of a development team. It is recommended that each team establish a set of simple rules it considers necessary to govern the team. Once established, each team member should sign and date this contract.

6.10 EFFECTIVE TEAM MEETINGS

Leadership is important to ensure that the team remains focused on the overall project goals/objectives and on the agenda at each meeting. Creating the correct team environment or atmosphere is essential so that the team members cooperate, share ideas and support each other to achieve solutions [5]. It is important for the team leader to assess the performance of the entire team at each meeting. When the team fails to make progress, it is critical for the team leader or possibly an individual member to make changes as necessary to enhance the team's effectiveness. Some typical difficulties encountered during team meetings are identified in the following paragraphs, along with suggested corrective actions team members can take to enhance the team's productivity.

One common problem impairing team productivity occurs when the goals and objectives of the meeting are not clear or they appear to be changing. While in many cases team meetings are planned well in advance and are supposed to follow an agenda, new topics are introduced and the meeting drifts from one item to another. This team behavior indicates that either some of the team members do not understand the goals, or they do not accept them. Instead, they are marching to the beat of their own drum and trying to take the team with them. Unless corrective actions are taken to focus the team on the original goals, this team meeting will fail and time and effort will be lost. A focused team stays on the agenda, and its team members accept the meeting objectives. Team discussions should generate a wide set of solutions to the problem being considered, and should continue until team members reach a consensus solution. A meeting is successful when an issue has been resolved and all of the team members accept the solution. This allows the team to move forward by organizing its talents to address the next problem.

Leadership is an essential element for team effectiveness. The most effective teams are usually democratic with a large amount of shared leadership. While there is a recognized leader, with a considerable degree of responsibility and authority, the team leader is supported by each team member. A team member with the appropriate expertise (relative to the problem being considered) will usually lead the discussion and effectively act as leader of the team during this period. However, in some instances, the team leader maintains strict control of the meeting and never shares the leadership role with any team member. Some team members resent this style of leadership and may not participate as fully as possible. The result is that complete utilization of the resources of the team never occurs.

As the meeting progresses, it is important to assess the discussions that are occurring. For a meeting to be successful, the discussion should include everyone on the team. A discussion should stay focused on an agenda item with few if any deviations to unrelated topics. The team members must listen carefully to one another and give all of the ideas presented a serious hearing. Teams tend to accomplish less when the discussion is dominated by only a few of the participants. These members tend to intimidate others in their efforts to control and dominate the team. The result is disastrous, because dominating team members essentially eliminate the contributions of other members.

Teams do not operate with total agreement on every issue. There must be accommodation for disagreements and for criticism. All of the ideas or suggestions made by every team member at any meeting will not be outstanding. Indeed, some of them may be ridiculous; criticism of these ideas must occur. However, the criticism should be frank and without hostility. Personal attacks must not be a part of the critique of an idea. If the criticism is to improve an idea or to eliminate a false concept, then it is of benefit to the progress of the team. The criticism should be phrased so that the team member advancing the flawed idea is not embarrassed. When disagreements occur, they should not be suppressed. Suppressed disagreements breed hostility and distrust. It is much better for the team to deal with disagreements when they arise. The root causes of a disagreement should be ascertained, and the team should take the actions necessary to resolve them. On some occasions voting is a mechanism used to resolve conflict. This procedure must be used with care particularly if the vote indicates the team is split almost equally in their opinion. A better practice is to discuss the issues involved and attempt to reach a consensus. Reaching a team consensus may require more time than a simple vote, but the results are worth the effort. Consensus implies that all members of the team are in general agreement and willing to accept the decision of the team. When the team votes, a simple majority is sufficient to resolve an issue. However, the minority members may become resentful if they are always on the short end of the vote. Over time, these members will stop accepting the outcome of the vote. These members will disengage from the project and will not commit to the actions necessary to implement the team's decision. The team meetings must be open. The agenda should be available in advance of the meetings and subject to change with added topics introduced as "new business". A sample agenda is shown in Table 6.1.

Team members should believe that they have the authority to bring new topics before the team. They should feel free to discuss their concerns associated with the team's operation. Secret meetings of an inside group should be carefully avoided. When the fact leaks that secret meetings are being held and that issues are prejudged by a select few, the effectiveness of the team is destroyed.

At least once during a meeting it is useful to evaluate the progress of the team. Is the team following the agenda? Is someone dominating the meeting? Is the discussion to the point and free of hostility? Is everyone properly prepared to address his or her agenda items? Are the team members attentive, or are they bored and indifferent? Is the team leader leading or is he or she pushing? If a problem in the operation of the team is identified during the pause for self-appraisal, it should be resolved immediately through open discussion.

Finally, the team must act on the issues that are resolved and the problems that are identified. To discuss an issue and to reach a consensus is part of the process, but not closure. Implementation is required for closure, and implementation requires action. When decisions are made, team members are assigned **action items**. These action items are tasks to be performed by the responsible individual; only when the tasks are completed is an issue considered closed. It is important that the action items are clearly defined, and the role of each team member in completing their respective tasks is understood. A realistic date should be set for the completion of each action item. The individual responsible should be clearly identified; he or she must accept the assignment without objection or qualification. A checking system must be employed to follow up on each action item to insure timely completion. If there is a delay, the schedule for the entire product development cycle may be at risk. It is important to deal with delays immediately and for the team to participate in the development of plans to eliminate the cause of the delay.

Table 6.1:
Weekly meeting of the Mighty Terrapin Team
Friday, February 2, 2018 8:00 AM

Weekly status report	Team Leader
Review of outstanding action items	
Action item #8	Member responsible
Action item #9	Member responsible
Action item #N	Member responsible
Report on progress	
Subsystem – Drive Train	Member responsible
Subsystem – Electrical	Member responsible
Subsystem – Structure	Member responsible
Subsystem – Control	Member responsible
Identify new problems	All member participate
Assignment of action items	Team Leader and volunteers
New business	All member participate
Summary	Team Leader
Adjourn at 9:00 AM or sooner	

6.11 PREPARING FOR MEETINGS

Effective team meetings do not just happen. It is necessary to prepare for the meeting and to execute post-meeting activities to insure success. The preparation usually involves selecting, arranging and equipping the meeting room, scheduling the meeting so that the necessary personnel are in attendance, preparing and distributing an agenda in advance of the meeting and conducting the meeting following rules established by the team. Equally important are actions taken by the team members following the meeting to implement the decisions that have been reached.

The meeting room and its equipment also affect the team's progress. The room should be sized to accommodate the team and any visitors that have been invited. Rooms that are too large permit the members to scatter and the distance between some members becomes too long for effective and easy communication. The seating arrangement should be around a table so that everyone can observe each other's face and body language. It is very difficult to engage anyone in a meaningful conversation if they have their back to you. Water, coffee, tea or soft drinks should be available if the meeting duration exceeds an hour. Smoking is strictly prohibited. The equipment that will be needed for presentations should be available, and its operation should be checked prior to the start of the meeting. A computer with projection capability is an essential feature in well-equipped meeting rooms. The availability of a computer during the meeting permits one to draw from a large database and to modify the presentation in real time. Additionally, it allows results of analyses or experiments to be clearly displayed in graphical form for the entire team to review.

Prior to the scheduled meeting, it is important to make careful preparations. Minutes from the previous meeting should be distributed with sufficient time for review before the next scheduled meeting. A detailed agenda, similar to the one shown in Table 6.1, is distributed to inform team members of the topics and/or problems that will be addressed in the next meeting. Individual team members responsible for specific agenda items are identified. Details of the meeting should be clear to all with assignments for individual members clearly defined. In some instances, information will be needed from corporate employees, suppliers or visitors that are not members of the team. In these cases, arrangements must be made to invite these people to the meeting so that they can provide the necessary expertise and respond to questions from all of the disciplines represented on the team.

In conducting the meeting, it is important to start on time. It is very annoying to the majority of the members to wait five or ten minutes for a straggler or two. The team leader must make certain that everyone involved understands that 8:00 AM means 8:00 AM and not 8:08 or 8:12 AM. The objectives of the meeting should be clearly understood and the time scheduled for each agenda item should be estimated. A team member should be assigned as the timekeeper, and another should act as a secretary to record notes and to prepare the minutes of the meeting. Team meetings are dynamic and these two roles often change over the duration of a project.

As the meeting draws to a conclusion, the team leader should take a few minutes to summarize the outcome of the discussions. This summary provides an opportunity to insure that assignments are understood, responsibility accepted and completion dates established. The meeting must be completed on time and everyone's schedule should be respected. If the meeting is not periodic (e.g. every Tuesday at 8:00 AM), then it is important that the time and place for the next meeting be scheduled.

After the meeting, it is productive for the team leader to make certain that the complete minutes have been prepared and distributed. If any member was not able to attend the meeting, the team leader should brief that person on the team's progress. Finally, the team leader should follow up on each assigned action item to ensure that progress is being made, and that new or unanticipated problems have not developed. It is clear that effective meetings do not happen by accident. Many members of the team work diligently before, during and after the meeting to make certain that the goals and objectives are clearly defined, that the issues and problems are thoroughly discussed, and that the solutions developed result in assignments that are executed with dispatch.

6.12   SUMMARY

Arguments for the importance of development teams in industry have been presented. Experience has shown that effective development teams are vital if a corporation is to develop highly successful products and introduce them in time to win a significant market share.

Functional organizations that exist within most corporations have been described. Reasons why a functional organizational structure often interferes with rapid low-cost product development have been given. The balanced team structure commonly employed in establishing development teams within large functionally organized corporations has been described.

The importance of locating team members to either enhance or inhibit communication has not been considered. However, you should be aware of the important role distance plays in effective and timely communication. The rule is simple. To enhance communication, minimize the distance between those needing to communicate with each other. The ideal situation is to place the entire development team in the same room and eliminate the walls and partitions.

Development planning and staffing of development teams has been covered. The success of the team depends strongly on generating a comprehensive plan. This plan is used initially to justify the development budget and its schedule. During the course of the development, the plan provides guidance for monitoring the progress and for judging both the team's performance and the product's attributes. Team staffing is cross-disciplinary with adequate representation from both the engineering and finance divisions. Engineers interact with the business personnel from marketing, sales and purchasing. Responsibility is shared between disciplines and functions.

The team's organizational structure and the team leader's management status have been discussed. This relationship is important because of executive authority that corresponds with a higher level of management. Higher-level managers carry more authority, many decisions can be made more rapidly, communication is more effective and team progress is often enhanced if the leader has status and clout associated with executive management.

Next, team member responsibilities and team member traits were discussed in considerable detail. Team members work in a matrix organization and often report to two or more managers. This arrangement is sometimes difficult particularly when you receive mixed signals from different managers. It is important to be flexible and remain cooperative with the program manager and functional managers. Team member traits are very important to your career. Evaluate your traits in an honest self-assessment. This is a critical first step in building team skills.

Most design teams evolve over the duration of a development project. Five phases usually experienced by the team in this evolution have been described. Not all of the phases provide pleasant experiences. Early in the evolution, a team encounters the dissatisfaction phase with very slow progress, low productivity and member gloom. To minimize the time in this phase, it is suggested that the team prepare and execute a contract that provides guidelines for member behavior.

Finally, guides for conducting successful team meetings have been provided. You will spend more time than you can imagine in team meetings. From the outset, learn the techniques for a successful meeting. Reasons for the failure of some meetings and the success of others have been presented. The role of the team leader and the behavior of the team members are described. The meeting room is more important than most engineers imagine. The progress of the project is dependent on the team's ability to control its meetings.

REFERENCES

1. Smith, P. G., D. G. Reinertsen, Developing Products in Half the Time, Van Nostrand Reinhold, New York, NY, 1992.
2. Schmidt, L. C. and Zhang, G., Dieter, G., Cunniff, P. F., and Herrmann, J. W., Product Engineering and Manufacturing, 2nd Edition, College House Enterprises, Knoxville, TN, 2002.
3. Barczak, G. and Wilemon, D., "Leadership Differences in New Product Development Teams," Journal of Product Innovation Management, Vol. 6, 1989, pp. 259-267.
4. Lacoursiere, R. B. The Life Cycle of Groups: Group Development State Theory, Human Service Press, New York, NY, 1980.
5. Barra, R. Putting Quality Circles to Work, McGraw Hill, New York, NY, 1983.
6. Clark, K. and Fujimoto, T., Product Development Performance, Harvard Business School Press, Boston, MA, 1992.
7. Wheelwright, S. C. and Clark, K. B., Revolutionizing Product Development, Free Press, New York, NY, 1992.

EXERCISES

6.1 Write an engineering brief describing why the team structure in a large corporation is so important in the product development process. Add a second paragraph indicating why team structure is much less important in a very small company.

6.2 Prepare an organization chart, like the one shown in Fig. 6.1, for the Clark School of Engineering. Describe the logic, as you see it, for this organizational structure.

6.3 List the advantages and disadvantages of the balanced team structure.

6.4 Outline a development plan that you would prepare if you were a senior team member representing an engineering function (discipline) on a newly formed development team. The product to be developed is a new communication device to be used (worn) by senior women living alone and requiring emergency assistance.

6.5 Write an engineering brief describing the staffing required for the team developing the new communication device described in Exercise 6.4. Give the reasons for changing the staff as the communication device evolves during the development process.

6.6 Describe the experience and status of the product manager for the balanced team structure.

6.7 Describe a matrix organization structure. Explain why this organization impacts an engineer assigned to a product development team with a balanced structure. If you were this engineer, how would you respond to inquires from your functional manager?

6.8 You (a male/female) are attending weekly team meetings and are always seated next to Patricia/Chip who is young and very attractive. Patricia/Chip is apparently interested in you because she/he frequently involves you in side conversations that are not related to the ongoing team discussions. Write a plan describing all of the actions you will take to handle this situation.

6.9 Oliver, the team leader, is a great guy who believes in leading his team in a very democratic manner. He encourages open discussion of the issue under consideration for a defined time period, usually five or ten minutes. At the end of the time period, he intervenes and calls for a team vote to resolve the issue. Write a critique of this style of leadership.

6.10 You, together with Ken and Joyce, are senior members on a development team with nine other members. The three of you are very knowledgeable and experienced. You get along famously, and have developed a habit of gathering together at a local pizza restaurant the evening before the scheduled team meeting. During the evening you discuss the issues on the agenda and arrive at some design decisions. Write a brief describing the consequences of continuing this behavior.

6.11 Prepare a list of ground rules that should be followed by all the members of a team in conducting an effective meeting.

6.12 Meet with your team and discuss a list of rules that will be used in guiding the team's conduct during the semester. Incorporate the team's list of rules in the form of a contract. As a final step, each member of the team is to sign the contract indicating his or her commitment to the rules of conduct.

CHAPTER 7

DESIGN BRIEFINGS

7.1   INTRODUCTION

You communicate by writing, speaking and with a number of different visual methods. All these modes of communication play a role as you attempt to convey your thoughts, ideas and concerns to others, and they all are important. Let's focus your attention on speaking in this chapter. You use speech almost continuously in your daily life. Why do you need to study about design briefings? There are several reasons. You usually speak with your friends and family in an informal style. You know them and feel comfortable talking with them. They know you. They are concerned with your well being, genuinely like you and are usually interested in what you have to say. A professional presentation is different. It is a formal event, which is usually scheduled well in advance. The audience may include a few friends, but usually it consists of peers (coworkers) and strangers. The time you have in which to convey your message is limited, and some in the audience may not be interested in you or your message. A person or two in the audience may be managers who control your advancement in the company. There are many reasons for tension headaches as you prepare for a professional presentation.

The design briefing is extremely important to both the product development process and to your career. Information must be effectively transmitted to your peers, management and external parties involved in the project. Clear messages accurately defining problems that the development team is effectively addressing are imperative. On the other hand, ambiguous messages are often misunderstood, hinder the definition of the problem and lead to delays in implementing timely solutions. Clearly, such messages should be avoided. The design briefing provides an opportunity to review the status of a specific product development. It also permits peers to share their ideas with you and affords management an open forum for assessing the quality of your work and the progress made by your development team. Because the professional presentation is critically important, you should become knowledgeable about some effective techniques for accurately delivering your messages to a mixed group of strangers, peers and acquaintances.

7.2   SPEECHES, PRESENTATIONS AND DISCUSSIONS

Let's distinguish between three formal methods of oral communication—speeches, presentations and group discussions.

Speeches

The speech is the most formal of the three modes of communication. In the year of a presidential or senatorial election, there are always many examples of speeches. Everyone is besieged with political addresses that clearly illustrate the key features of speeches. These speeches are given to large audiences in huge rooms, stadiums or coliseums. The setting is usually not appropriate for visual aids, although the speaker usually uses a teleprompter. The audience is diverse with a wide range of concerns. Those listening to the speech are of widely different ages with different interests and persuasions. The speech is carefully scripted with the speaker usually reading or closely following the script. Ad-libbing is carefully avoided. Communication is one way—from the speaker to the audience. Questions are usually not appropriate and generally not appreciated. Time is strictly enforced unless you are the President of the U. S. giving the State of the Union address. Fortunately, engineers are rarely called upon to make speeches; hence, this topic will not be developed further in this book.

Presentations

Professional presentations differ from speeches. Presentations are made to smaller audiences in smaller rooms that are usually equipped with electric power, light controls and projection devices. You may depend on visual aids, demonstrations, simulations and other props to help convey your message during a typical presentation. The audience is knowledgeable (about your topic) and usually has many common characteristics. The presentation is very carefully prepared because of its importance. It has order and structure, but it is not scripted. The flow of information is largely from the speaker to the audience; however, questions are permitted and often encouraged. The speaker is considered the expert, but discussion, comments and questions give the audience an opportunity to share their knowledge of the topic. Time is carefully controlled and is often insufficient from the speaker's viewpoint. The professional presentation is a vitally important mode of oral communication that you must quickly learn to master.

Group Discussions

Group discussions are also very important to the design engineer. The audience is smaller with much more in common. The group may all be members of a development team. The topic being discussed is narrowly focused. The speaker (central person) serves as a moderator, and he or she is an expert on the topic being considered. However, members of the discussion group (audience) may be as knowledgeable as the speaker (moderator). The moderator works in two modes. He or she may act as a presenter giving brief background information to frame an issue that prompts any member of the group to discuss the topic. The moderator may also direct questions to a group member known to be the most knowledgeable concerning the issue being addressed. The speaker (moderator) controls the flow of the information, but the flow is clearly multi-directional—from speaker to the group and from one group member to another. When a group (team) discusses a design issue, time is difficult to control and the content and range of coverage are strongly dependent on the skill of the moderator. Group discussion is very important in industry because the leader of a development team will often use this method of communication to identify problems and to initiate efforts directed toward their solution. Group discussion methods will not be described in this textbook; however, it is recommended that you watch **Washington Week in Review** on the PBS television network to gain some insights. While the participants are journalists, you can adopt many of moderator's clever techniques to guide engineering discussions.

Design Briefings

A design briefing is a type of professional presentation. It is the method you will use in speaking before the class and will be an educational objective emphasized in this course. Indeed, you probably will be required to make presentations describing your team's product development on at least two occasions—the preliminary and the final design reviews. In these two design reviews, you will employ visual aids to assist in delivering the information required to describe the status of your project with conciseness and clarity.

7.3   PREPARING FOR THE BRIEFING

A design briefing is too important to casually prepare. You should prepare very carefully to insure that you will accurately report the status of your team's development and to identify problems or unresolved issues. There are three aspects that you should consider in your preparations:

- Identify the audience
- Plan the organization of the presentation and your coverage of each topic
- Prepare interesting and attractive visual aids.

Identifying the Audience

In a typical college classroom, it is much too easy to identify the audience. You have your peers, the instructor and perhaps a visitor or two from industry. In an industrial setting, the audience will be more difficult to identify because it will be much more diverse. The size and diversity of the audience depends on the magnitude of the development project. A small briefing will have 10 to 20 people in attendance with most participants from within the company. A large briefing may include 50 to 100 representatives who are both internal and external to the company. The characteristics of the audience are important because you must adapt the content of the presentation to the audience. Classify your audience with regard to their status, interest and knowledge before you begin to plan the style and content for your presentation.

The status of the audience refers to their position in the various organizations they represent. Are they peers, managers or executives? If they are a mix, which is likely, who is the most important? If you are preparing a design briefing for high-level management, it must be concise, cogent and void of minor detail. Executives are busy, stressed and always short on time. High-level executives and managers are impatient and rarely interested in the technical details that engineers love to discuss in their presentations. Recognize these characteristics and adjust the content and the timing of your presentation accordingly. The executives are interested in costs, schedule, market factors, performance, risk or any other critical issue that will delay the development or increase projected costs of either the development or the product. Usually you will have only 5 to 10 minutes to convey this information at an executive briefing.

First and even second level managers have more time and are more interested in you and the designs that you are creating. If you prepare the presentation for this lower level of management, you can safely plan for more time (10 to 20 minutes). These managers are likely to be engineers and they share your knowledge of the subject. In fact, they probably will be more experienced, expert and knowledgeable than you. They will want to know about the schedule and costs because they share responsibilities with the higher-level management for meeting these goals. However, they are also interested in the important details, and you can engage them in a discussion of subsystem performance. The first and second level managers usually control the resources for the development team. If you need help, make sure that they receive the message in enough detail to provide the assistance required. Do not hide the problems that your team is encountering. Managers do not like surprises. If you have a problem, make sure they understand it and are able to participate in its solution. If you hide a problem from management and it causes a delay or escalation in costs, the manager will take the heat and you will be in a very cold and uncomfortable doghouse.

Peer Reviews

Design briefings for peers are often less tense and you have more time (20 to 30 minutes). You address schedule and cost issues because everyone needs to know if your team is meeting the milestones on the Gantt charts. However, the main thrust of your presentation pertains to design details. If you are working on a subsystem that interfaces with other subsystems, cover your subsystem in sufficient detail for all the team members to understand its geometry, interfaces, performance, etc. It is particularly important that the interface issue be fully addressed in the presentation. If four subsystems are to be integrated in a given product, many details about all four subsystems must be addressed to ensure that the integration goes smoothly. Suppose that your team is developing a power tool with a motor that draws a current of 10 amps, and some team member plans to employ a switch, rated at 6 amps, to turn the power on/off. The switch will probably function satisfactorily in the short-term tests with the prototype, but it will malfunction in the field with extended usage sometime after the product has been released to the market. Clearly, the two persons responsible for the interface between the motor and the switch did not communicate properly. The integration of the two subsystems failed.

The most important purpose of design briefings with peers is to make certain that all of the details have been addressed. You strive to integrate the subsystems without problems arising when the prototype is assembled and tested.

Peer reviews in the absence of management are very beneficial in the development process. The knowledge of most of the team members is about the same, although some members are more experienced than others. The topic is usually a detailed review of one subsystem or another. The audience is fresh and capable of providing a critical assessment of the technology employed. Your peers may check the accuracy of your analysis, comment on the choice of materials, discuss methods to use for manufacturing and assembly and relate their experiences with similar designs on previous products. Peer reviews afford the opportunity for the synergism that makes good products better. They also provide a forum for passing on important lessons from the more experienced designers to beginners in a friendly, stress-free environment.

The subject of the presentation is self-evident in a design review. You are developing a product, and the content of the design briefing will deal with one or more issues regarding this development. There is a choice of topics and considerable latitude regarding the content. Although advice has been given in previous paragraphs about matching content with the interests of your audience, you must understand the importance of knowing who will be in your audience. Executive reviews serve a different purpose than peer reviews, and the content and the time allotted for the presentation is adjusted accordingly.

For every type of review, there are two **absolute** rules that you must follow. First, **you must know the material cold**. It takes an audience about two seconds to understand that you are faking it. Also, some presenters fail to rehearse adequately, and the audience often misinterprets this lack of preparation with insufficient knowledge. In either event, when they realize that you are not the expert you allege to be, your presentation fails. You have **lost** the opportunity to effectively communicate your message. Second, **be enthusiastic**. It is all right to be calm and cool, but do not be dead on arrival. You must command the attention of your audience or they will quickly turn off. You control the attention of the audience only if you are enthusiastic and knowledgeable in presenting your material (story).

7.4   PRESENTATION STRUCTURE

There is a well-accepted structure for professional presentations. If you are familiar with PowerPoint®, a graphics presentation program (GPP) marketed by Microsoft, you are already aware of the templates included in this program for various types of presentations. An excellent structure for a design briefing is shown in the outline for a professional presentation presented below:

* Title
* Overview
* Status
* Introduction
* Technical Topics
  * First Topic
    * Background
    * Status
  * Second Topic
    * Background
    * Status
  * Final Topic
    * Background
    * Status
* Summary
* Action Items

Let's examine each of these topics individually and discuss the content that should be included on the visual aids that accompany your presentation. Recognize that the visual aids (electronic projection) control the flow and the information that you will present. For this reason, the topics listed above in the context of information included on the presentation slides will be addressed.

Title Slide

The title slide, illustrated in Fig. 7.1, obviously carries the title of the presentation. However, it also states your affiliation and the names of the team members that contributed to the work. Because you may be reporting on the work of the entire team, it is necessary for you to acknowledge their contributions on the title slide and in your opening remarks. A brief title—ten words or less—is a good rule to follow in drafting the title for your presentation. It is also a good idea to use descriptive words in the title. For example, you could use **PRELIMINARY DESIGN BRIEFING: VEHICLE DESIGN: DRIVE POWER SUBSYSTEM**. Eight words and you have informed the audience of the topic (a design briefing), the type of briefing (preliminary), and the product being developed (a drive power subsystem for your vehicle. A significant amount of information is successfully conveyed in eight words.

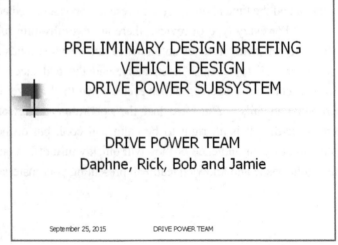

Fig. 7.1 Illustration of an informative and descriptive title slide.

Below the title, the team name is given together with the name of each team member involved in the development. It is essential that you share the ownership of the material covered in your presentation. Often a single member of the team covers the work of several people in a design briefing. It is not ethical to make the presentation without acknowledging the contributions of others. Students typically use first names in identifying team members, but in industry the complete names of the team members are used. If necessary, division and department affiliations are also stated.

It is also a good idea to date the title slide and to cite the occasion for the presentation. If you make many presentations, this information is useful when referring to your files, allowing you to easily reuse some slides after selective editing and/or updating.

Overview Slide

The second slide presents an overview of the design briefing. The overview for a presentation is like a table of contents for a report. You list the main topics that you intend to cover in the presentation. In other words, you tell the audience what you plan to tell them in the next 10 to 20 minutes. The number of topics should be limited. In a focused presentation, you can cover about a half dozen topics. If you press and try to cover ten or more topics, you will encounter difficulty maintaining the attention of the audience throughout the presentation. Constrain the tendency to tell everything—focus on the important issues. In a design review, the topics usually are organized to correspond with the subsystems involved in the product under development. For a component redesign, you concentrate on the issues guiding the component modification being considered. An example, of an overview slide for a preliminary design review of a drive power subsystem is illustrated in Fig. 7.2.

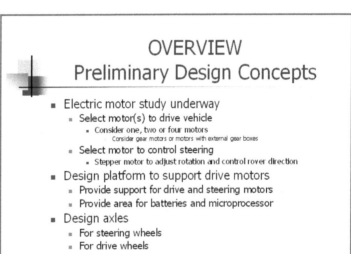

Fig. 7.2 The overview slide outlines the content in your presentation.

Status Slide

The third slide presents the product development status. You should report on the progress made by the team to date. It is a good idea to incorporate a Gantt chart showing the development schedule either on the status slide or a separate slide. The progress of the team on each task is defined on the Gantt chart and is immediately conveyed to the audience. If there are problems or uncertainties, this is the time to air the difficulties that the team is encountering. An example slide, shown in Fig. 7.3, illustrates uncertainties that the team has regarding

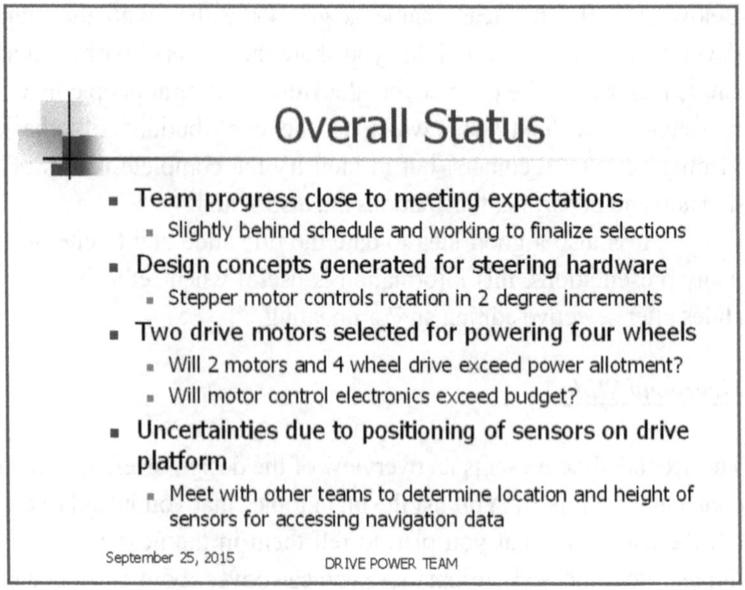

Overall Status

- Team progress close to meeting expectations
  - Slightly behind schedule and working to finalize selections
- Design concepts generated for steering hardware
  - Stepper motor controls rotation in 2 degree increments
- Two drive motors selected for powering four wheels
  - Will 2 motors and 4 wheel drive exceed power allotment?
  - Will motor control electronics exceed budget?
- Uncertainties due to positioning of sensors on drive platform
  - Meet with other teams to determine location and height of sensors for accessing navigation data

September 25, 2015   DRIVE POWER TEAM

Fig. 7.3 Status slide indicates overall progress and accomplishments while signaling potential problems.

Introductory Slide

The fourth slide covers the introductory material such as background, history or previous issues. It is important that the audience understand the product, the main objectives of the development, the role the team has in that development and the most pressing of the current issues. This introductory slide permits you to set the stage for the body of the presentation that follows. The audience responds better to your presentation when they know in advance the background and the topics that you will cover. They are better prepared mentally to receive your message.

Technical Topics

The body of the presentation usually deals with five or six technical topics. In most instances, the topics selected correspond to the more critical subsystems involved in designing the product. In the redesign of a component, the topics deal with the issues arising when considering new design concepts. Avoid trying to cover a large number of topics; it is better to report on a limited number of topics thoroughly than to rush through a dozen topics with incomplete coverage. It is suggested that you use two slides for each topic. The first slide describes the progress made by the team in generating design concepts and indicates the criteria employed in selecting the best concept. The second slide covers the status of developing the selected concept. The type of information reported on this slide includes accomplishments, outstanding issues, lessons learned, etc. An example of the status of the design of the drive system for a vehicle, which involves the use of two gear motors for speed and torque control, is presented in Fig. 7.4.

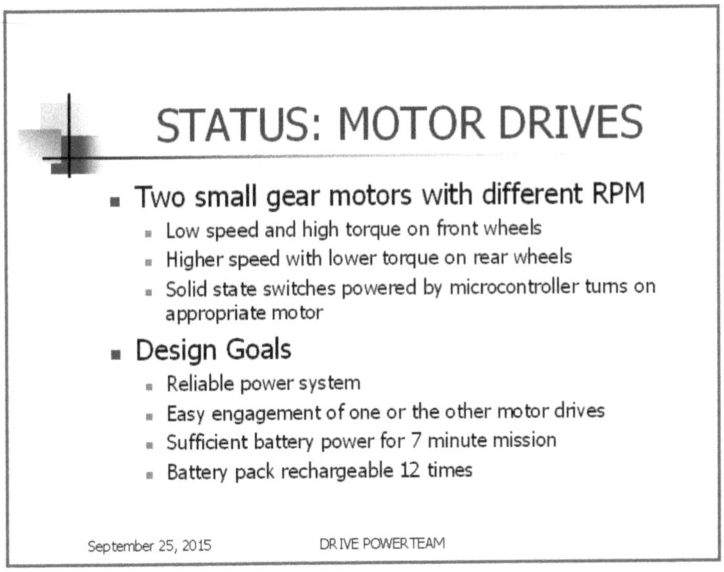

STATUS: MOTOR DRIVES

- Two small gear motors with different RPM
  - Low speed and high torque on front wheels
  - Higher speed with lower torque on rear wheels
  - Solid state switches powered by microcontroller turns on appropriate motor
- Design Goals
  - Reliable power system
  - Easy engagement of one or the other motor drives
  - Sufficient battery power for 7 minute mission
  - Battery pack rechargeable 12 times

September 25, 2015   DRIVE POWER TEAM

Fig. 7.4 Information presented in describing the early design of a drive subsystem for the vehicle.

Concluding Slides

After the technical topics have been covered, you conclude the presentation with a decisive pair of slides. Suggestions for two concluding slides are presented in Figs. 7.5 and 7.6. The next to last slide in the presentation, shown in Fig. 7.5, is titled "Key Issues." This is your opportunity to identify very important issues (questions) that your team has discovered. Do not hesitate. This is the time to introduce the uncertainties and to come forward and seek help. Design reviews are not competitive. In a design review, you seek help regardless of your status or role as a member of the audience. Managers will arrange support for your team if it is required. Peers will make suggestions and introduce fresh approaches to unresolved problems that the team may find useful. The design review is a formal process in a product development in which everyone participates. Take advantage of this review to identify uncertainties and to seek help whenever your team needs assistance.

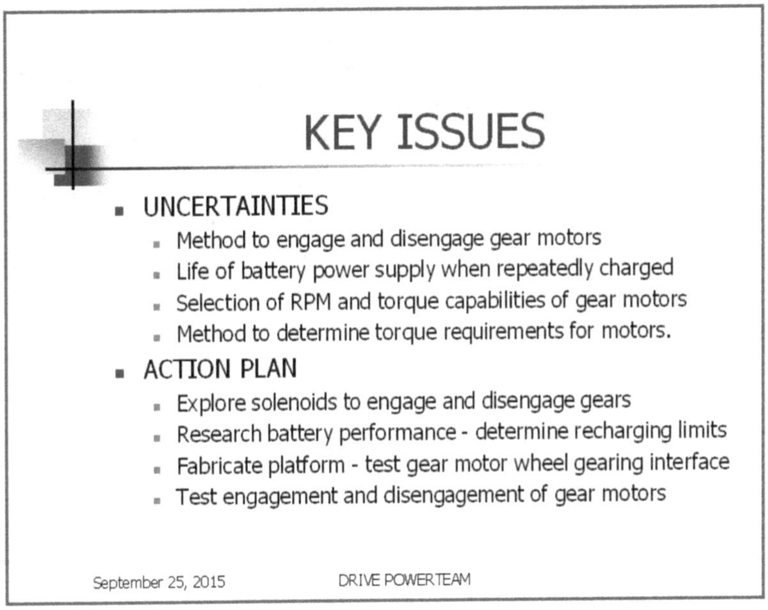

KEY ISSUES

- UNCERTAINTIES
  - Method to engage and disengage gear motors
  - Life of battery power supply when repeatedly charged
  - Selection of RPM and torque capabilities of gear motors
  - Method to determine torque requirements for motors.
- ACTION PLAN
  - Explore solenoids to engage and disengage gears
  - Research battery performance - determine recharging limits
  - Fabricate platform - test gear motor wheel gearing interface
  - Test engagement and disengagement of gear motors

September 25, 2015   DRIVE POWER TEAM

Fig. 7.5 The "key issues" focuses the team and management on your most significant problems.

The final slide presented in Fig. 7.6 is to present your plans for the future. Management is interested in how you intend to solve the issues that you have disclosed and the amount of time and money required for the solutions. Your peers will be interested in the technical approaches that you intend to follow. You should also define any requirement that you have for anyone in the audience. If you need help from your peers or more resources from the manager, be certain this assistance is clearly indicated on this final slide. You should distinctly identify those individuals who have agreed to provide the requested support. If action items are listed, clearly identify the responsible individual and estimate the time to completion. The planning incorporated in the final slide is very important. It is the prescription for your team's get-well program.

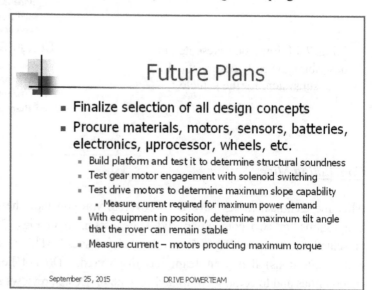

Fig. 7.6 The "future plans" slide gives you the opportunity to describe your approach for solving problems that the team is encountering.

7.5   TYPES OF VISUAL AIDS

Because visual aids control the content and the flow of the information conveyed to the audience, they are essential for a technical presentation. Slides provide visual information that reinforces verbal content. With visual aids you communicate using two of a person's five senses. Take time to carefully select the type of visual aid that will most effectively convey your message.

In selecting visual aids, you are usually limited. First, in your ability to produce the visual aids within the time and financial constrains imposed on the development team. Second with the availability of equipment and the room in which the presentation will be made. Is the room equipped with a computer-controlled projector? Another important consideration is the control of the light intensity in the room. Most rooms have light switches used to control the overhead lights. However, some rooms do not have blinds and bright sunshine will cause problems with the visibility of slides when they are projected.

Computer Projected Slides

Computer projected slides are an excellent choice if the room can be darkened. Computer projection of slides has several advantages:

- The presentation is easy to prepare using PowerPoint.
- The slides are in color and are more effective than black and white illustrations.

- Advanced visual effects (slide transition, slide building and video insertion) enhance the impact of the presentation.
- It is also easy to copy your presentation onto a CD or a memory stick.
- It is easy to transmit the presentation by attaching it to an email message.

A computer, usually connected to a digital projection system, stores your presentation file. You employ a mouse to click through your slide line by line. Finally, low-cost handouts with four or six slides per page can be prepared that aid the audience in following your presentation. The only disadvantage of computer-projected slides is that sometimes a relatively dark room is required for projection. With a darkened room, eye contact with the audience or even reading their body language is difficult. You lose visual contact with the audience that is so important in reading their response to your presentation.

Types of Visual Aids

There are several visual aids sometimes used in professional presentations. Video file are common; if you want to show motion or group dynamics, using video is clearly the best approach. Video cameras are readily available and after some practice you can become a reasonably good video producer.

Hardware, or other materials used in the product development, is sometimes passed around the audience during a presentation. The hands-on opportunity for the audience is a nice touch, but you pay a price with the loss of attention by some of the audience during the inspection period. It is recommended that you defer passing hardware to the audience during your presentation. Instead, after the presentation, invite the audience to inspect exhibits that are placed on a strategically located table. This approach maintains the attention of the audience throughout the presentation, while giving those interested in the hardware time for a much more thorough examination.

7.6 DELIVERY

Excellent presentations require well-prepared slides and a smooth, well-paced delivery. There are many aspects to delivering a presentation including: dress, body language, audience control, voice control and timing.

Attire

Let's start with dress. Broadly speaking, there are four levels of dress. The highest level, black tie for men and formal gowns for women, is a rare event and is never appropriate for design briefings. The next level, business attire for men and women, is sometimes appropriate for design briefings. A conservative Eastern company may have a dress code requiring business attire; whereas, a less formal Western company would encourage more casual attire. Casual attire should not be confused with sloppy attire. Casual attire is neat and tasteful, but without suits, ties and white shirts for men and suits for women. Although becoming more common in the workplace, sloppy attire (old jeans and a sweatshirt) is strictly taboo. You should select tasteful business or casual attire when you dress for the presentation. While it is recognized that most students these days prefer a relaxed, collegiate style of dress, you should resist the impulse to dress too casually.

Body Language

Posture is another important element in the presentation. In the words of former President Reagan, "stand tall." Your body language signals your attitude to the audience—you are the presenter and in control. Make sure you are calm, cool and collected. Nervous gestures with your hands, rocking on the balls of your feet, scratching any part of your body, pulling on your ear, etc. should be avoided.

Audience Control

Before you begin your presentation, take control of the audience. One approach is to pause a moment before projecting the title slide and immediately make eye contact with the entire audience. How do you look at everyone in the house? You scan the audience from left to right and then back looking slightly over their heads. Occasionally drop your eyes and make eye contact for a second or two with one individual and then another. Pause long enough for the audience to become silent (10 to 20 seconds). If someone is rude and continues to talk, walk toward them and politely ask for their attention. When you have everyone's attention, project the title slide and begin your delivery.

If you have rehearsed, you will not need notes. The slides carry enough information to trigger your memory. If they do not, you have not rehearsed long enough to remember the issues. Scripting the presentation is not recommended. People who script will eventually start to read their comments, which is a deadly practice. Rehearse until you are confident. Make notes to help you rehearse, but do not use them during the presentation.

Voice Control

When you begin speaking, insure that everyone can hear you. If you are not certain of how far your voice carries, ask those in the back of the room if they can hear. Use care with a microphone and amplifier, because the tendency is to speak too loudly. Try to control the loudness of your voice within the first minute of the presentation when you introduce the topic with the title slide. The audience will read that slide with anticipation, and they will bear with you as you adjust the volume of your voice.

Do not mumble. Speak slowly and clearly enunciate each word. It is better to present your material too slowly than too rapidly. If you can improve your enunciation, you will be able to increase the rate of your delivery without losing the audience.

Don't gasp for air while speaking. Learn to complete a sentence, pause for breath (without gasping) and then continue with the next sentence. There is nothing wrong with an occasional pause in the presentation, provided the pause is short.

Timing

If you forget some detail, do not worry. Skip it, and move to the next point that you are trying to make. If the detail is critical, count on one of your team members to raise the issue in the question and answer period. Most speakers forget some of the material that they intended to present and introduce some additional items on an ad-hoc basis. Usually, the audience will never be aware of your omissions or additions.

Studies have clearly shown that people can listen to conversation faster that the presenter can speak. The trouble with listening to someone who speaks very rapidly is not their rate of speech, but their enunciation. Many people who speak rapidly tend to slur their words, and the audience has trouble understanding poorly pronounced words. The trouble you have in listening to a person speaking too slowly is to stay with his or her message. While you are waiting for the next word, your mind goes off on a mental tangent. You anticipate that your mind will return to the speaker in time for the next word, but unfortunately your mind is often tardy. You miss keywords or even complete sentences before your attention returns to the speaker.

Listening is a skill that many people have not developed. As a speaker, it is your responsibility to keep the listener, even the poor ones, on the topic. You may use several techniques to maintain the attention of the audience. Employing the screen is probably the most effective tool for this purpose. As you project the slide, walk to the screen and point to the line on the slide that corresponds to the topic that you are addressing. If a few members in the audience are coming back from their mental tangents, you reset their attention on the current topic.

When referring to the image projected on the screen—and good speakers use this technique—stand to the left of the screen as shown in Fig. 7.7. Face the audience and maintain eye contact. When you must look at the slide, turn your head to the side and read over your shoulder. The 90° turn of your head and the glance at the slide should be quick because you want to continue to face the audience. **Under no circumstance should you stand with your back to the audience and begin reading from the slide as if it were a script.**

As you develop the discussion, point to key phrases to keep the audience focused on the topic. Engage both of their senses. Deliver your message through their eyes and ears simultaneously. When you point to a line on the slide, point to the left side of the line. People read from left to right—so start them from the initial point on the left side.

As you speak, modulate your tone and volume. Avoid both a monotone and a singsong delivery. A continuous tone of voice tends to put the audience asleep. A singsong delivery is annoying to many listeners. If you have a high-pitched voice, try to lower the pitch because a high-pitch tone is bothersome to many people

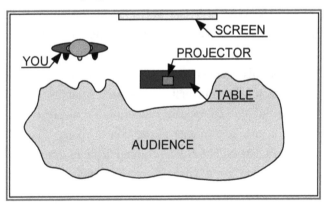

Fig. 7.7 Typical room arrangement for presentations. Position yourself to the left of the screen and face the audience.

If a question is asked, answer it promptly unless you intend to cover material later in the presentation that addresses this question. The answer should be brief, and you should try to avoid an extended dialog with some member of the audience. If a member of the audience is persistent with several related questions, simply indicate that you will speak with him or her off-line immediately after the presentation. The timing of your presentation can be destroyed by too many questions. Some questions are helpful because they permit audience participation, but an excessive number of questions cause the speaker to lose control of the topic, and the flow and timing of the presentation. Too many questions cause the format of the communication to change from a presentation to a group discussion. Group discussions have a purpose; however, they are different than a design briefing.

7.7   SUMMARY

The design briefing is extremely important to successful product development and your career. Information must be effectively transmitted to your peers, management and to external parties involved in the project. Clear messages, which accurately define the problems that the entire development team can effectively address, are imperative. The design briefing provides the opportunity to inform management and peers of your progress and problems.

There are three types of formal communication: speeches, professional presentations and group discussions. The characteristics of these three types of communication are described. For an engineer, the professional presentation is the most important and the formal speech is the least important. The main emphasis of this chapter is on the professional presentation with a focus on the design briefing.

Preparation for a presentation is essential. Accomplished speakers often spend the better part of a day preparing for a 20-minute briefing. There are three important elements in the preparation. First, know the characteristics of your audience, and insure that the content in the presentation corresponds to their interests. Second, organize the presentation in a manner acceptable to the audience. Make certain that the organization is efficient so that you use the allotted time effectively. Prepare high-quality visual aids (slides) that will help you convey your message and rehearse until you command the subject material.

A structure for the presentation is recommended that is commonly employed in design briefings. This outline includes four initial slides—title, overview, status and introduction. The body of the presentation is devoted to technical details using about eight to twelve slides. Two closure slides are employed for a summary and a listing of action items.

The type of visual aid selected is important because the visual transmission of information will markedly affect the outcome of the presentation. Computer projected slides have significant advantages.

The delivery makes or breaks a presentation. You have been provided with two pages of details about how to do this—and not that. However, the best way to learn to deliver a design briefing is by practicing your presentation. On your first two or three attempts, have a friend video the event. Then review your delivery style, technique and appearance during a trial presentation. You will identify many problems of which you were not aware. Another suggestion is to use one of your free electives to take a speech course. The experience gained in a typical communications course will not help much in crafting the content to include in a design briefing, but it will help you to develop very important delivery skills.

REFERENCES

1.   Wilder, L. Talk Your Way to Success, Simon and Schuster, New York, NY, 1986.
2.   Eisenberg, A., Effective Technical Communication, 2nd ed., McGraw Hill, New York, NY, 1992.
3.   Goldberg, D. E., Life Skills and Leadership for Engineers, McGraw Hill, New York, NY, 1995.
4.   Elbow, P. Writing with Power, 2nd Edition Oxford University Press, New York, NY, 1998.
5.   Struck, W. C. Osgood and R. Angell, Elements of Style, 4th Edition, Allyn & Bacon, Needham Heights, MA, 2000.
6.   Pickett, N. A. et al, Technical English: Writing, Reading and Speaking, Longman, Reading MA, 2000.

EXERCISES

7.1   Write a brief description of the characteristics describing your teammates.   Focus on those characteristics that will influence the language and content in your design briefings.

7.2   Prepare an outline for a preliminary design review. Include in the outline the titles of all of the slides that you intend to use.

7.3   Prepare the title slide for your preliminary design review.

7.4   Prepare an overview slide for your preliminary design review.

7.5   Prepare a status slide for your preliminary design review.

7.6   Prepare the key issues slide for your preliminary design review.

7.7   Beg or borrow a video camera. Record a practice presentation by a teammate. Critique his or her presentation with good taste. Then have a teammate record your presentation and listen carefully to their critique of your performance.

7.8   Practice the delivery of your presentation employing the screen of your laptop computer to display your slides.

7.9   Measure and record the time required to present your design briefing. As you rehearse, continue to measure the time. When the presentation becomes more polished, the time required for the presentation should decrease. Does it?

7.10   Prepare a group of slides for a design briefing that utilizes the slide transition and slide building features of PowerPoint.

CHAPTER 8

TECHNICAL REPORTS

8.1   INTRODUCTION

It is important that you enjoy writing, because engineers often have to prepare several hundred pages of reports, theoretical analyses, memos, technical briefs and letters during a typical year on the job. It is essential that you learn how to effectively communicate—by writing, speaking, listening and employing superb graphics. Communication, particularly good writing, is an extremely important skill. Advancement in your career will depend on your ability to write well. It is recognized that you will be taking several courses in the English Department and other departments in Social Sciences, Arts and the Humanities that will require many writing assignments. These courses should help you immensely with the structure of your composition and the development of good writing skills. Most of the assignments will be essays, term papers, or studies of selected works of literature. However, there are several differences between writing for an engineering company and writing to satisfy the requirements of courses such as English 101 or History 102.

In college, you write for a single reader—your instructor. He or she must read the paper to evaluate your work and to determine how well you are doing in class. In industry, many people within and outside the company, and with different backgrounds and experience, may read your report. In class, the teacher is the expert, but in industry the writer of the report is supposed to be the individual with the knowledge. Many of those working in industry do not want to read your report because they are busy: their phone is ringing continuously, meetings are scheduled back to back, and many important tasks must be completed before the end of the day. They read—if not skim—the report only because they must be aware of the information that it contains. They want to know the key issues, why these issues are important and who is going to take the actions necessary to resolve them. In writing reports in industry, you cut to the chase. There is no sense in writing a 200-word essay when the facts can be provided in a 40 to 50-word paragraph that is brief and cogent. An elegant writing style, so valued by instructors in the arts and humanities, is usually avoided in engineering documentation. In writing an engineering document, do not be subtle; instead be obvious, direct and factual.

In the following sections, some of the key elements of technical writing, including an overall approach, report organization, audience awareness and objective writing techniques will be described. Then a process for technical writing, which includes four phases: composing, revising, editing and proofreading is discussed.

8.2   APPROACH AND ORGANIZATION

The first step in writing a technical report is to be humble. Realize that only a very few of the many people who may receive a copy of your report will read it in its entirety. Busy managers and even your peers read selectively. To adapt to this attitude, organize your report into short, stand-alone sections that attract the selective reader. Three very important sections—the title page, summary and introduction are located at the front of the report so they are easy to find. At the front of your report, they attract attention and are more likely to be read. In college, you call the page summarizing your essay an abstract, but in industry it is usually called an executive summary. If it is prepared for an executive, perhaps a manager will consider it sufficiently important to take time to read it. Follow the executive summary with an introduction and then the body of the report. A common outline to follow in organizing your report is provided below:

- Title page
- Executive summary
- Table of contents, list of figures and tables
- Nomenclature—only if necessary
- Introduction
- Technical issue sections
- Conclusions
- Appendices

The title page, executive summary and introduction are the most widely read parts of your report. Allow more time polishing these three parts, because they offer the best opportunity to convey the most important results of your investigation.

Title Page

The title page provides a concise title of the report. Keep the title short—usually less than ten words; you will have an opportunity to provide more detail in the body of the report. The authors list their names, affiliations, addresses and often their telephone and fax numbers and e-mail addresses. The reader, who may be anywhere in the world, should be provided with a means to contact the authors to ask questions. For large firms or government agencies, a report number is formally assigned and is listed with the date of release of the report on the title page. The cover of a report describing a review of DOE's Nuclear Energy and Research and Development Program is shown in Fig. 8.1. This report was prepared by a committee of the National Research Council, which is the research branch of the National Academies. The committee members participating in the review and their affiliations are listed on a separate page early in the report.

Executive Summary

The executive summary should, with rare exceptions, be less than a page in length (about 200 words). As the name implies, the executive summary provides, in a concise and cogent style, all the information that the busy executive needs to know about the content in the report. What does the big boss want to know? Three paragraphs usually satisfy him or her. First, briefly describe the objective of the study and the problems or issues that it addresses. Do not include the history leading to these issues. Although there is always history, the introduction is a much more suitable section for historical development and other background information related to the problem. In the second paragraph, describe your solution and/or the resolution of the issues. Give a very brief statement of the approach that you employed, but reserve the details for the body of the report. If actions are to be taken by others to resolve the issues, list the actions, those responsible and the dates of implementation. The final paragraph indicates the importance of the study to the business. Cost savings, improvements in the quality of the product, gains in the market share, enhanced reliability, etc. can be cited. You want to convince the executive that your work was worth its cost, valuable to the corporation and that you and your team performed admirably.

Introduction

In the introduction, you restate the problem or issues. Although the problem was already defined in the executive summary, **redundancy is permitted in technical reports**. Recognize that you have many readers, so important information, such as a clear definition of the problem is placed in different sections of the report. In the introduction, the problem statement can be much more complete by expanding the amount of information describing the issues. A paragraph or two on the history of the problem is in order. Individuals reading the introduction are more interested than those reading only the executive summary; therefore, more detail is appropriate.

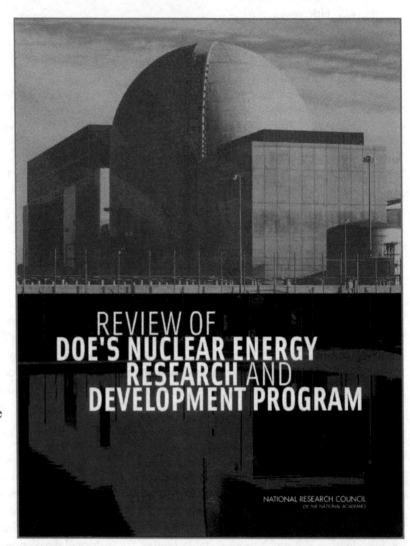

Fig. 8.1 Illustration of the cover page
of a professional report.

Following the problem statement, establish the importance of the problem to both the company and the industry. Again, this is a repeat of what was included in the executive summary, but your arguments are expanded. You can cite statistics, briefly describe the analyses that lead to the alleged cost savings, give a sketch of customer comments indicating improved quality, etc. In the executive summary, you simply stated why the study was important. In the introduction, arguments are presented to convince the reader that the investigation was important and worth the time, effort and cost.

In the next segment of the introduction, you briefly describe the approach followed in addressing the problem. You begin with a literature search, followed by interviews with select customers, a study of the products that failed in service, interviews with manufacturing engineers and the design of a modification to decrease the failure rate and improve the reliability of the product. In other words, tell the reader what actions were required to define and then solve the problem.

The final paragraph in the introduction outlines the remaining sections of the report. True, that information is already covered in the table of contents, but again, you repeat to accommodate the selective reader. Also, placing the report contents in the introduction gives you the opportunity to add a sentence or two describing the content of each section. Perhaps you can attract additional reader interest in one or more sections of the report and convince a few people to study the report more completely.

Organization

An organization for the report was suggested by the bullet list shown previously. However, you should recognize that an organization exists for every section and every paragraph in the report. In writing the opening paragraph of the section, you must convey the reason for including it in the report and the importance of its content. You certainly have a reason for including this section in the report, or you would not have wasted the time required to write it. By sharing this reason with your readers, you convince them that the section is important.

A paragraph is written to convey an idea, a thought or a concept. The first sentence in the paragraph must convey that idea, clearly and concisely. The second sentence should describe the significance or importance of the idea. The following sentences in the paragraph support the idea, expand its scope and give cogent arguments for its importance. State the idea in the first sentence and then add more substance in the remaining sentences. If the report is written properly, a speed-reader should be able to read the first two sentences of every paragraph and glean 80 to 90% of the important information conveyed in the report.

As you add detail to a section or a paragraph, develop a pattern of presentation that the reader will find easy to read and understand. First, tell the reader what he or she will be told, then provide the details of the story and finally summarize with the action to be taken to implement the solution. Do not try this approach in the theme that you have to write for English class, because the instructor will not appreciate your redundant style. Technical writing is not the same as theme writing. In technical writing, you will often reiterate facts to insure that the message is conveyed to the readers. Often one or more of the readers must take corrective actions for the benefit of the company and the industry.

You will frequently include analysis sections in engineering reports. In these sections, the problem is stated and then the solution is described. After the problem statement and the solution, you introduce the details. The reader is better prepared to follow the details (the difficult part of the analysis) after they know and appreciate the solution to the problem.

8.3   KNOW YOUR READERS AND THE REPORT'S OBJECTIVE

Another important aspect to technical writing is to know something about the readers of your report. The language that you use in writing the report depends on the knowledge of the audience. For example, I am writing this book for engineering students. I know your math and verbal skills from the SAT scores required for admission. I recognize that you have very good language skills and even better math skills. You are bright and articulate. I believe that the content in this book closely matches your current abilities. However, there are a few sections in the book that deal with topics usually found in the business school. Profits and costs are important in product development, and although they may not be as interesting as a computer programming or circuit analysis, they need to be clearly understood.

When you write a report for a class assignment, you understand that the instructor is the only reader with which you need to communicate. You also know that he or she is knowledgeable. In that sense, the writing assignments in a typical college class are not realistic. For example, suppose your manager asks you to write an assembly routing (a step by step set of instructions for the assembly of a certain product) for the production of your new widget in a typical factory in the U. S. What language would you use as you write this routing? If the plant is in southern Florida or the Southwest, Spanish may be the prevalent language. You should also be aware that a significant fraction of the factory workers in the U. S. are functional illiterates and many others detest reading. In this case, it might be a good approach to use fewer words and many cartoons, photographs and drawings to prepare an assembly routing document.

Know the audience **before** you begin to write, and adapt your language to their characteristics. Consider four different categories of readers:

- Specialists with language skills comparable to yours

- Technical readers with mixed disciplines
- Skilled readers, but not technically oriented
- Poorly prepared readers who may be functional illiterates

Category 1 is the audience that is the least difficult to address in your writings; the audience in the last category is the most difficult. Indeed, with poorly prepared readers, it is probably better to address them with visual presentations conveyed with television monitors and video files.

A final topic, to be considered in planning, is the classification of the technical document that you intend to write. What is the purpose of the task? Several classifications of technical writing are listed below:

- **Reports:** trip, progress, design, research, status, etc.
- **Instructions:** assembly manuals, training manuals and safety procedures.
- **Proposals:** new equipment, research funding, construction and development budget.
- **Documents:** engineering specifications, test procedures and laboratory results.

Each type of document has a different objective and requires a different writing style. If you are writing a proposal for development funding, you need persuasive arguments to justify the costs of the proposed program. On the other hand, if you are writing an instruction manual to assemble a product, arguments and reasons for funding are not an issue. Instead, you would prepare complete and simple descriptions with many illustrations to precisely explain how to accomplish a sequence of tasks required for an assembly.

8.4 THE TECHNICAL WRITING PROCESS

Whether you are writing a report as part of your engineering responsibilities in industry or as a student in college, you will face a deadline. In college, the deadlines are imposed well in advance and you have a reasonable amount of time to prepare your report. In industry, you will have less time, and an allowance must be made for the delay in obtaining approvals from management before you can release the report. In both situations, you have some limited period of time to write the report. The idea is to start as soon as possible. Waiting until the day before the deadline is a recipe for disaster.

Many professionals do not like to write; consequently, they suffer from writer's block. They sit at their keyboard hoping for ideas to occur. Clearly, you do not want to join this group, and there is no need for you to do so. A technique is described in this section that should help you avoid writer's block.

Define Your Task

Start your report by initially following the procedures described in the previous section. Understand the task at hand, and classify the type of technical document that you have to write. Define your audience; before starting to write, establish the level of detail and the language to employ. Prepare a skeletal outline of the report. Start with the outline presented previously and add the section headings for the body of the report. This outline is very brief, but it provides the structure for the report. This structure is important because it has divided the big task (writing the entire report) into several smaller tasks (writing one section at a time).

Gather Information

It is impossible to write a report without information. You must generate the information to be presented in the report. You can use a variety of sources: interviews with peers, instructors or other knowledgeable persons who are willing to help; complete literature searches at the library; read the most suitable references; conduct additional Internet searches for information. Take notes as you read, gleaning the information applicable for your report. Be careful not to plagiarize. While you can use material from published works, you cannot copy the exact wording; the statements must be in your own words. If the report has an analysis, go to work and prepare a statement of the problem, execute its solution and make notes about interpreting the solution.

Organize Data

When you have collected most of the information that is to be discussed in your report, organize your notes into different topics that correspond to the section headings. Then incorporate your topics in an initial outline that will grow from a fraction of a page to several pages as you continue to incorporate notes in an organized format.

Compose the Document

You are now ready to write. Writing is a tough task that requires a great deal of discipline and concentration. It is suggested that you schedule several blocks of time and reserve them exclusively for writing. The number of hours that should be scheduled will depend on the rate at which you compose, revise, edit and proofread. Some authors can compose about a page or so per hour, but most students need more time.

In scheduling a block of time for composing, you should be aware of your productive interval. Most writers take about a half an hour to come up to speed, and then they compose well for an hour or two before their attention and/or concentration begins to deteriorate. The quality of the composition begins to suffer at this time, and it is advisable to discontinue writing if you want to maintain the quality of your text. This fact alone should convince you to start your writing assignments well before the deadline date.

While you are writing, avoid distractions. Writing requires deep concentration—you must remember your message, the supporting arguments, the paragraph and sentence structure, grammar, vocabulary and spelling. Find a quiet, comfortable place and focus your entire concentration on the message and the manner in which you will present it in your report.

Naturally, some sections of a report are easier to write than others. The easiest are the appendices, because they carry factual details that are nearly effortless to report. The interior sections of the body of the report carry the technical details that are also easy to prepare, because describing detail is less concise and less cogent. Do not become careless on these sections because they are important; however, each sentence does not have to carry a knockout punch.

The most difficult sections are the introduction and the executive summary. It is advisable to write these two sections after the remainder of the first draft of the report is completed. Usually the introduction is written before the executive summary. Although the introduction contains much of the same information as the executive summary, it is more expansive. The introduction can be used as a guide in preparing the executive summary.

8.5   REVISING, EDITING AND PROOFREADING

Writing is a difficult assignment. Do not expect to be perfect in the beginning. Practice will help, but for most of us, it takes a very long time to improve our skills because writing is a complex task. Expect to prepare several drafts of a paper or report, before it is ready to be released. In industry, several drafts are essential because you often will seek peer reviews and manager reviews are mandatory. In college, you have fewer formal requirements for multiple drafts, but preparing several drafts is a good idea if you want to improve the report and your grade.

First and Subsequent Drafts

The first draft is focused on composition, and the second is devoted to revising the initial composition. Use the first draft for expressing your ideas in reasonable form and in the proper sections of the report. In the second draft, focus on revising the composition. Defer the editing process, and concentrate on the ideas and their organization. Make sure the message is in the report and that it is clear to all of the readers. Polish the message later in the process.

Several hours should elapse between the first and second drafts of a given section of the report. If you read a section over and over again, you soon lose your ability to judge its quality. You need a fresh, rested brain for a critical review. In preparing this textbook, the author composes on one day and revises on the next day. Revising is always scheduled for an early morning block of time. When you are rested, concentration is usually at its highest level.

Let's make a clear distinction between composing, revising, editing and proofreading. Composition is writing the first draft where you formulate your ideas and organize the report into sections, subsections and paragraphs. Unless you are a super talented writer, your first draft is far from perfect. The second draft is for revisions where you focus on improving the composition. The third draft is for editing where errors in grammar, spelling, style, and usage are corrected. The fourth draft is for proofreading where typographical errors are eliminated.

Revising

When you revise your initial composition, be concerned with the ideas and the organization of the report. Is the report organized so that the reader will quickly ascertain the important conclusions? Are the section headings descriptive? Sections and their headings are helpful to the reader because they help organize his or her thoughts. Additionally, they aid the writer in subdividing the task and keeping the subject of the section in focus. Are the sections the correct length? Sections that are too long tend to be ignored or the reader becomes tired and loses concentration before completely reading them.

Question the premise of every paragraph. Are the key ideas presented together with their importance before the details are included? Is enough detail given or have you included too many trivial items? Does the paragraph contain a single idea or have you tried to include two or even three ideas in the same paragraph? It is better to use a paragraph for every idea even if the paragraphs are short.

Have you added transition sentences or transitional phrases? The transition sentences, usually placed at the end of a paragraph, are designed to lead the reader from one idea to the next one. Transitional phrases, embedded in the paragraph, are to aid the reader place the supporting facts in proper perspective. You contrast one fact with another, using words like **however** and **although**. You indicate additional facts with words such as **also** and **moreover**.

Editing

When the ideas flow smoothly and you are convinced that the reader will follow your concepts and agree with your arguments, begin to edit. Run a spell checker, and eliminate most of the typographical errors and the misspelled words in your report. Search for additional misspelled words. Spell checkers do not detect the difference between certain words like **grate** or **great**, or say **like** and **lime** and between **from** and **form**.

Look for excessively long sentences. When sentences become 30 to 40 words in length, they begin to tax the reader. It is better to use shorter sentences where the subject and the verb are close together. Make sure that the sentences are actually sentences with a subject and a verb of the same tense. Have you used any comma splices (attaching two sentences together with a comma)? Examine each sentence and eliminate unnecessary words or phrases. Find the subject and the verb, and attempt to strengthen them. Look for redundant words in a sentence, and substitute different words with similar meaning to eliminate redundancy. Be certain to employ the grammar checking feature incorporated in your word processing program. It is not perfect, but it identifies many grammatical errors.

Proofreading

The final step is proofreading the paper to eliminate errors. Start by running the spell checker for the final time. Then print out a clean, hard copy to use for proofreading. Check all the numbers and equations in the text, tables and figures for accuracy. Then read the text for correctness. Most of us have trouble reading for accuracy because we read for content. We have been trained since first grade to read for content, but we rarely read for correctness.

To proofread, read each word separately. You are not trying to glean the idea from the sentence, so do not read the sentence as a whole. Instead, read the words as individual entities. If it is possible, arrange for some help from a friend—one person reading aloud to the other with both having a copy of the manuscript. The listener concentrates on each word and then checks the text against the spoken word to verify its spelling and grammatical correctness.

8.6   WORD PROCESSING SOFTWARE

Word processing software is a great tool to use in preparing your technical documents. The significant advantage of using this software is the ability to revise, edit and make the necessary changes without excessive retyping. You can mark, delete, cut, copy and move text. You can easily insert words, phrases, sentences and/or paragraphs.

While working with word processing software, it is difficult to revise, edit, and proofread on the screen because only a portion of the page is visible. Better results are obtained if you print a hard copy and view the entire page. With the entire page, you can see several paragraphs at once and can check that they are in the correct sequence. Use double spacing when printing the first few drafts to give adequate space between the lines for your modifications.

After you have completed the revisions on the hard copy, make the required modifications to the text using word processing software and save the results to a memory stick. It is recommended that you keep only the most recent version (draft) of the document. If you save several versions of the document, it is necessary to keep a logbook of the changes to each version. If the writing takes place over several weeks, it is easy to lose track of what changes you made and which version of the hard copy goes with which electronic copy. You will find it easier to keep one electronic file (on your memory stick) of the most recent draft with a hard (paper) copy to serve as a back up.

Word processing software has several features that are helpful in editing. The spellchecker finds most of your typographical errors and provides suggestions for correcting many misspelled words. The search or find command permits you to systematically examine the entire document so you may replace a specific word with a better substitute. A thesaurus is available to help you with word selection, but you must be careful when using it. Make sure you understand the meaning of the word that you select; do not try to impress the reader by using long or unusual words. Short words that are easily understood by the reader are preferred in technical writing.

Formatting the report is another significant advantage of word processing software. You can easily produce a document with a professional appearance. The formatting bar permits you to select the type of font, the point size (the height of the characters) and emphasis such as bold, underline or italic.

You can also format the page with four different types of line justification, which are commonly available. I am using word processing software to prepare this textbook and applying **justify** alignment for both the right and left margins. The word processing software automatically added the spaces required in each line and aligned both margins. In designing your page, use generous margins. One-inch margins all around are standard unless the document is to be bound; then the left margin is usually increased from 1.00 to 1.25 inch.

Tables and graphs can be inserted in the text. Take advantage of the ability to introduce clip art into the text or to transfer spreadsheets and drawings produced with other software programs into your report. Position the tables and figures as close as possible to the location in the text where they are introduced. Try to avoid splitting a table between two pages. If the table is longer than a page or two, place it in an appendix. Identify figures and photographs with suitable captions. The caption, placed directly beneath the figure, reinforces the message conveyed by the illustration.

8.7   <u>SUMMARY</u>

Writing is a difficult skill to master; most engineers experience significant problems when they prepare reports early in their careers. Unfortunately, the writing experiences in college do not correspond well with the writing requirements in industry. In college, you write for a knowledgeable instructor. In industry, you write for a wide range of people with different reading abilities. Moreover, the audience often varies from assignment to assignment. In both college and industry, you write to meet deadlines imposed by others. While writing may not be fun, there are many techniques that you can employ to make writing much easier and more enjoyable.

The first technique is organization; an outline for a typical report was suggested. Gather information for the report from a wide variety of sources and generate notes useful in refreshing your memory as you write. Sort the notes and transpose the information expanding the outline of your report.

When you organize the outline, but before you begin to write, determine as much as possible about your audience. They may be technically knowledgeable regarding the subject or they may be functionally illiterate. The language that you use in the report will depend on the reader's ability to understand. Also understand the objective of your document. Is it a report, an extended memo, a proposal or an instructional manual? Styles differ depending on the objective, and you must be prepared to change accordingly.

There is a process to facilitate the preparation of any document. It begins by starting early and working systematically to produce a very professional document. Divide the report into sections and write the least difficult sections (appendices and technical detail portions) first. Defer the more difficult sections, such as the executive summary and the introduction until the other sections have been completed.

A strategy for effective writing is to divide the effort into four different tasks—composing, revising, editing and proofreading. Keep these tasks separate:

- Compose before revising
- Revise before editing
- Edit before proofreading
- Proofread with great care.

Multiple drafts are necessary with this approach, but the results are worth the effort. Word processing software does not substitute for clear thinking, but it is extremely helpful in preparing professional documents. Word processing saves enormous amounts of time in a systematic editing process. It enables a mix of art, graphics and text neatly integrated into a single document. Word processing software has a thesaurus and word search features that are helpful in editing. It essentially turns a computer and printer into a print shop so that you have wide latitude in the style and appearance for your professional documents.

REFERENCES

1. Eisenberg, A., <u>Effective Technical Communication</u>, 2nd Edition, McGraw Hill, New York, NY, 1992.
2. Goldberg, D. E., <u>Life Skills and Leadership for Engineers</u>, McGraw Hill, New York, NY, 1995.
3. Elbow, P. <u>Writing with Power</u>, 2nd Edition Oxford University Press, New York, NY, 1998.
4. Alreb, G. T., C. T. Brusaw and W. E. Oliu, <u>The Handbook of Technical Writing</u>, 7th Edition, St. Martins Press, New York, NY, 2003
5. Struck, W. C. Osgood and R. Angell, <u>Elements of Style</u>, 4th Edition, Allyn & Bacon, Needham Heights, MA, 2000.
6. Pickett, N. A. et al, <u>Technical English: Writing, Reading and Speaking</u>, Longman, Reading MA, 2000.

EXERCISES

10.1 Prepare a brief outline of the organization for the final report, which describes the development of the autonomous vehicle that your team is designing.

10.2 Prepare an extended outline of the final report for the development of your team's autonomous vehicle.

10.3 Write a section describing one of the major subsystems for your team's autonomous vehicle.

10.4 Write a section covering the testing of a subsystem of the team's autonomous vehicle.

10.5 Write an Introduction for the final report on the team's autonomous vehicle.

10.6 Write the Executive Summary for the final report on the team's autonomous vehicle.

10.7 Revise the Introduction of the final report on the autonomous vehicle that was written by another team member.

10.8 Edit the Introduction after it has been composed and revised by other team members.

10.9 Proofread the Introduction after it has been composed, revised and edited by other team members.

10.10 Edit the final report after other team members have completed its composition.

10.11 Proofread the final report after other team members have completed editing it.

CHAPTER 9

ENGINEERING GRAPHICS

9.1 INTRODUCTION

Today engineers rarely use pencils to prepare formal engineering drawings. Communications among engineers working in design, manufacturing or operations is usually digital. Engineers communicate through the internet and the use of web tools.

Computers and several different software programs enable engineers to enhance their capabilities and significantly increase their productivity. Powerful computers, work stations and software programs are used to design, develop and manufacture most products. Software for engineering applications include: computer aided design (CAD), computer aided manufacturing (CAM) and computer aided engineering (CAE). To support your design efforts in this course, video tutorials for Creo Parametric, SolidWorks and Autodesk Inventor are available to assist you in becoming familiar with a CAD program of your choice.

9.2 COMPUTER AIDED DESIGN — CAD

Computer aided design involves the creation and preparation of engineering drawings representing the design of components, subsystems or complete assemblies. A computer with a CAD program allows designers to analyze their designs and test their ideas interactively in real time. We emphasize two equally important factors, the designer and the CAD program. The CAD program cannot replace the designer; however, these programs significantly improve an engineer's productivity. CAD programs are essential in implementing and tracking design changes that occur during the product development process. CAD programs and the designer are complementary. Most superior designs are produced with a balanced development team staffed with experienced personnel, who are proficient using advanced CAD systems.

An essential element in engineering design is to produce a clear definition of a component, subsystem or a complete assembly that will be manufactured. Producing engineering drawings of the component and/or the assembly defines the product. These engineering drawings establish the physical configuration of the component or assembly. The most significant advantage of a CAD design is the creation of solid models for all the components in an assembly. These models reside in a geometric database, and drawings of all the components and the assembly are derived from the solid models. Concurrently engineering analyses of the components or the system can be performed from information drawn from this database. Also, the database facilitates communications among the engineering departments.

Major characteristics of CAD systems include:

1.   A creative method of representing the design
2.   An effective method of communicating design information
3.   An approach to facilitate design modifications and implement redesign
4.   An effective method for executing engineering analyses
5.   A technique for integrating design with manufacturing and operations

9.3 THE DESIGN PROCESS

Design is a process for creating a product that **meets** a well-defined set of requirements. In general, a design has multiple solutions that are constrained by the available resources and the market. In essentially all cases, the final design of a product must be completed within a budget and on schedule. With advancements in communications, and increased global pressure to be the first to market, a concept known as concurrent engineering has been introduced. Concurrent engineering minimizes the time required for product development thereby enhancing profit potential. Concurrent engineering integrates manufacturing and product design processes reducing the time required to develop a new product. It also aids in balancing the conflicting requirements between the design and manufacturing. Concurrent engineering may be viewed as an integration of four stages of product development that include: conceptual design, preliminary design, detailed design and product prototyping, as shown in Fig. 9.1.

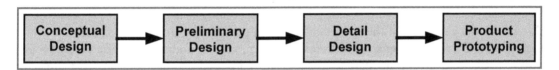

Figure 9.1 Essential stages in the development process

Conceptual design begins with the product specification, followed by generating design concepts and identifying critical product features. The process involves the definition of objectives, identification of alternative designs, and the evaluation of the merits of each design in terms of technical and financial feasibility. If a design does not meet the specifications, it is modified to generate another design that does meet specifications. The process involved in generating a conceptual design is illustrated in Fig. 9.2

Fig. 9.2 The conceptual design
process

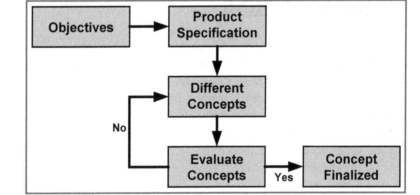

When the conceptual design is finalized a preliminary design is generated. Preliminary design involves visualization. Isometric drawings of the components and the assembly are constructed to inspect geometry and to facilitate the review of each component. Engineering drawings of components and assembly are drawn to ensure spatial compatibility. Material for each component is specified. Finally, analyses and tests are conducted to insure that the product will meet the product specifications. Component availability and cost are determined and an estimate is made for the total cost to produce the product.

During preliminary design, changes to the drawings are made without constraints. However, when the design is finalized a detailed design is generated. At this point in the design process, design changes follow a formal process that includes an engineering order, an engineering notice and an engineering change order.

Detail design involves specifying dimensional tolerances, an engineering analysis of the system and its components, confirmation of cost and availability of components and a critical review of assembly and component drawings. Graphics generated with CAD programs provide exceptional capabilities for engineering documentation. Upon completion of the detail design, a team meeting is convened to **release** the design. After a thorough design review by the entire team and management, the design is approved and released.

After the detailed design is released, a prototype representing the first article build is fabricated. It is necessary to build scaled or full-sized working models to verify the product performance. The first article build is also useful in demonstrating the company's ability to manufacture and assemble the product. Building a prototype aids in identifying design deficiencies not detected prior to the release of the design. Corrections to the design must be made and verified prior to initiating full-scale production. Minimizing the time required to fabricate these prototypes is extremely important in reducing the time and cost in developing a new product.

Rapid prototyping is a relatively new method for quickly fabricating prototypes of components. Rapid prototyping involves additive processes, whereas machining is a subtractive process with material removed to create the geometry of the component. Rapid prototyping involves the successive deposition of thin layers to build a three dimensional component. A number of additive processes are now available, which differ in the way layers are deposited to create parts and in the materials that are employed. Selective layer sintering and fused deposition melt or soften material to produce the layers. Stereo-lithography involves curing liquid polymers using different methods to solidify thin lines of plastic. The third method, called laminated object manufacturing, involves cutting thin layers of material to shape and joining these layers together to form a three dimensional object. Each method has advantages and disadvantage. The main considerations in choosing a machine are generally speed, cost of the equipment, cost and choice of materials and color capabilities. Additional descriptions of the 3D prototyping are presented in Section 9.16.

9.4 MANFACTURING

Process planning, which precedes manufacturing, includes choosing the type and sequence of each production operation, tooling requirements, production equipment and estimating the overall cost of manufacturing based on these selections. If the product is to be assembled, process planning includes establishing the number and sequence of steps involved in the assembly operation. Process planning often requires the design of an assembly line or a sequence of assembly cells. Production engineers, in charge of process planning, must understand production processes and the limitations of their equipment and the capabilities of the workforce. Components or subassemblies that cannot be produced in-house must be purchased from external suppliers. When the manufacturing process has been finalized, routing sheets are prepared that document the details of each step involved in manufacturing components and in assembling the product.

Computer-aided manufacturing (CAM) has grown in importance with improvements in digital equipment and software. Computer-aided process planning systems have been developed to automate process planning. For example, the digital files in a database contain the process plan for a particular component based on previous experiences. Any new process plan will be based on the existing process plan, saving time and reducing costs.

Significant progress has been made in developing the interface between the CAD and CAM systems. This interface enables digital data exchange between design and manufacturing departments. For example, the shared database permits numerical control (NC) codes for milling machines to be generated directly from the CAD files with additional information on tool selection and parameter settings, such as feed rate, depth of cut, etc.

When the planning process is complete, components are fabricated using one or more of several well-established manufacturing processes, as shown in Fig. 9.3. Components are collected and then assembled using an assembly line or a sequence of assembly cells. Quality control is achieved by inspecting components and the assembled products. Some of the components and finished products may be tested to insure performance.

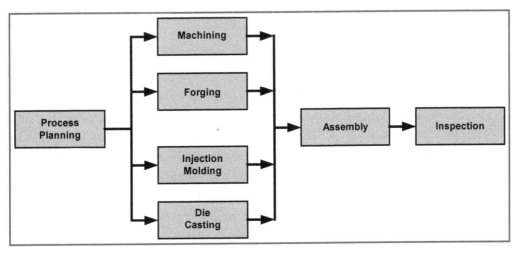

Fig. 9.3 Manufacturing processes

9.5 CAD SYSTEMS

During the past three decades, the development of both computer hardware and software has enhanced design tools and greatly facilitate communication. In the 1970s, CAD systems were developed using mainframe computers because of the major requirements for memory capacity and CPU speed. With advances in microelectronics, computing has shifted from mainframes, to minis, to work stations, to personal computers (PCs). Software companies have responded with new CAD programs having improved computing speeds and expanded memory capacity. Today, design engineers employ effective Windows-based CAD programs that operate on reasonably priced PCs.

The complexity involved in the development process for large complex products, such as an automobile, requires system integration, automation and optimization. These large complex products require massive memory capacity and high processor speed. To meet these requirements, the mainframe computer based CAD systems are still employed. On the other hand, PC-based CAD systems with more limited capabilities are employed by smaller businesses because of their lower costs. CAD programs, which are commercially available and operate effectively on PCs include AutoCAD, CATIA, I-DEAS, Creo Parametric, SolidWorks and UNIGRAPHICS.

9.6 ENGINEERING GRAPHICS

When communicating in writing, we follow strict grammatical rules for spelling, sentence construction and punctuation. When preparing engineering drawings, we also follow specific rules. For example, a three-dimensional view of a shaft is presented in Fig. 9.4. Its shape is cylindrical, which is 30 mm in diameter and 150 mm in length. Note that a symbol of ϕ is used on the drawing to indicate its cylindrical shape. The dimension given as M20-2.5 indicates it is a metric thread and with a diameter of 20 mm and a thread pitch of 2.5 mm. The centerline characterizes the symmetry of the cylinder.

Fig. 9.4 Dimensioning a cylindrical shaft.

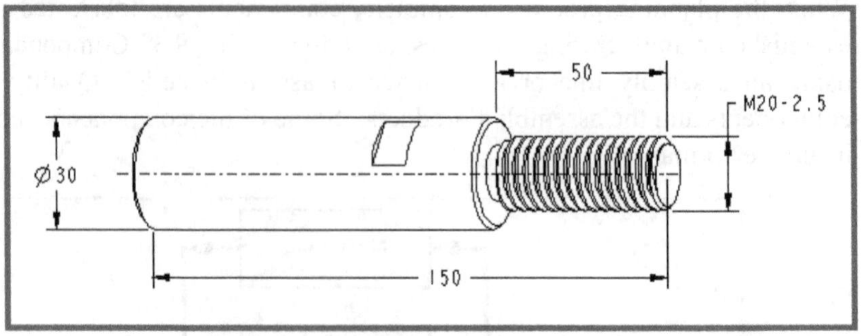

The American National Standards Institute (ANSI) establishes the standards for engineering documents in the U. S. The International Organization for Standardization (ISO) is responsible for international standards. The ISO consists of a network of national standards institutes from 154 countries. Engineering graphics is used to communicate by following the standards set by ANSI and/or ISO.

To facilitate communication in the engineering and manufacturing communities, engineering graphics employs three-dimensional spatial visualization with descriptive geometry and engineering drawings. Descriptive geometry is used to represent objects in the three-dimensions. Engineering drawings are blueprints required to execute the design and to guide operations in manufacturing. All the CAD software systems incorporate the engineering graphics rules adopted by ISO and ANSI. It is important for engineers to know and understand these rules.

9.7   VISUALIZATION AND DESIGN DETAIL

A three-dimensional view of the shaft is shown in Fig. 9.4. However, some information has not been provided, such as the chamfers at the two ends of the shaft and the location of the slot near the middle of the shaft. Two projection views of this shaft are presented in Fig. 9.5. Design engineers prefer the projection views over three-dimensional depictions. One of the major reasons for this preference is that features are represented in true lengths. For example, the dimension of φ30 and the dimension of 150 characterize the total volume of this component. The dimension of M20 indicates a standard metric thread with a 20 mm diameter. The dimensions of 26 and 10 characterize the size of the two flats, and the dimension of 75 gives the position of these two flats. It will become more evident why engineers prefer projection drawings as the additional content is described later. However, it should be recognized that projection views are more difficult to visualize. Fortunately CAD software enables one to combine project views and three-dimensional view(s) so that the communications between design engineers is easier and visualization of the geometry is facilitated.

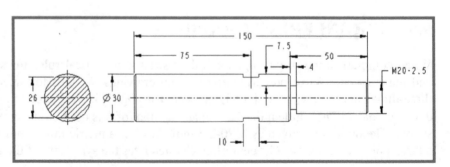

Fig. 9.5 Front and side projection views of a cylindrical shaft with detailed design information.

9.8   THREE-DIMENSIONAL SPATIAL VISUALIZATION

Because of the extensive usage of projection view drawings, the ability of an engineer to visualize an object in the three-dimensions is critical. Individuals acquire the visualization skills as they mature. Children learned topological spatial visualization by discerning the topological relationship of one object relative to others. For example, how close are the people relative to one another? Where is one person relative to others within a group? Later, in middle school children usually learned projective representation. They are able to recognize objects viewed from different perspectives. Finally, engineers learn to combine projective recognition with the concept of measurement.

An example to demonstrate different aspects of three-dimensional spatial ability is shown in Fig. 9.6. At the top of this illustration, the orientation of object AA is changed to a different orientation. On the lower portion, object BB is shown with one orientation. Five different orientations of object BB are shown at the bottom of the figure. Can you identify one of the five views that follows the same transform as the rotational sequence shown with AA.

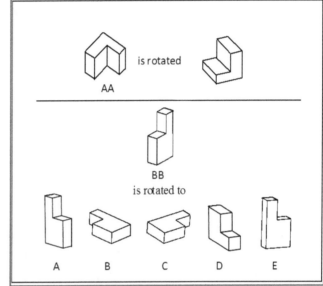

Fig. 9.6 An example of illustrating spatial visualization

To facilitate the visualization process, let's introduce a Cartesian coordinate system with its origin at a corner of the object BB, as shown in Fig. 9.7. By inspection, we observe that there are two rotations associated with the orientation of the object shown in AA. The first rotation of 90° is about the X-axis in the clockwise direction. The second rotation of 90 is about the Z-axis in the counterclockwise direction. Based on these observations, view D is the corresponding view in the lower portion of Fig. 9.6, because that is the orientation that results when object BB undergoes two similar rotations.

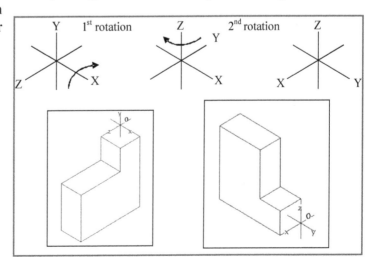

Fig. 9.7 Orientation of object BB after two rotations about the x and z axes.

9.9 PREPARING PROJECTION ENGINEERING DRAWINGS

Consider the three dimensional drawing of the object presented in Fig. 9.8. The three maximum dimensions of this object are (50 + 25 = 75) mm, 30 mm and 40 mm, respectively. Suppose you are to draw a projection view of this object.

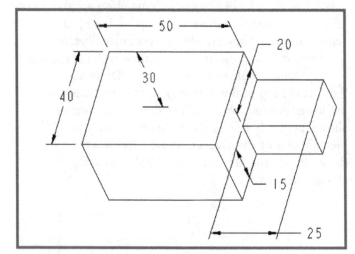

Fig. 9.8 An three dimensional drawing of an object

There are several steps usually employed in drawing a projection view.

Step 1: Layout

Layout the drawing by constructing four horizontal lines and four vertical lines, as shown in Fig. 9.9. Measure the distances between the lines to correspond with the three maximum dimensions (75, 40 and 30 mm) in the X, Y and Z directions, respectively.

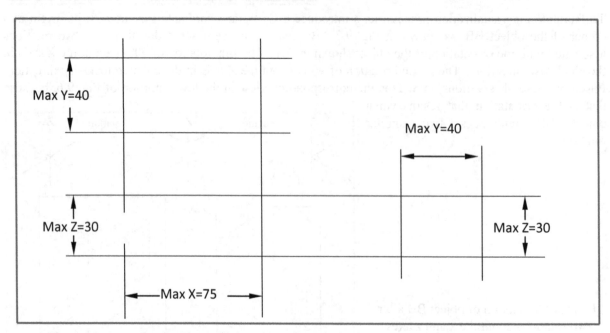

Fig. 9.9 Layout the drawing with boundary lines in both horizontal and vertical directions.

Step 2: Draw the front view

Sketch two rectangles, as shown in Fig. 9.10. The left side of the large rectangle and the right side of the small rectangle are aligned side by side; hence, the two rectangles share a common vertical line. Also note that the maximum X and Z dimensions control the size of the two rectangles and that the maximum Y dimension is not a factor in drawing the front view.

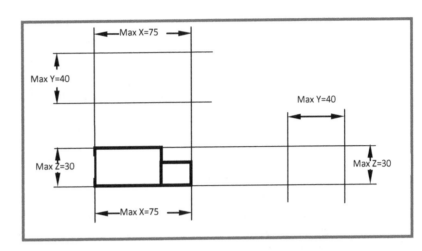

Fig. 9.10 Drawing the front view

Step 3: Draw the top view

Sketch the two rectangles, as shown in Fig. 9.11. The left side of the large rectangle and the right side of the small rectangle are aligned side by side; thus, the two rectangles share a common line. Also note that the maximum X and Y dimensions control the size of the two rectangles, and that the maximum Z dimension is not a factor in drawing the top view.

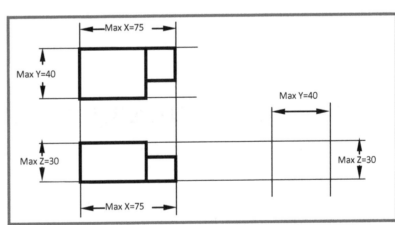

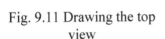
Fig. 9.11 Drawing the top view

Step 4: Draw the right-side view

Sketch two rectangles, as shown in Fig. 9.12. The large rectangle and the small rectangle are aligned side by side; hence the two rectangles share a common vertical line. Also note that the maximum Y and Z dimensions control the sizes of the two rectangles, and that the maximum X dimension is not a factor when preparing the right side view.

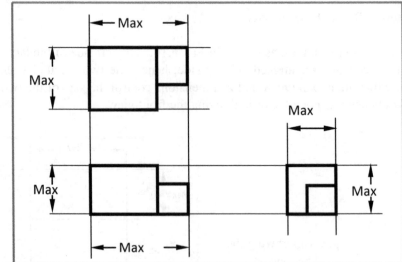

Fig. 9.12 Drawing the right side
view

The projection drawing in Fig. 9.12 shows the front, top and right side views. This layout is the standard adopted by ANSI for preparing an engineering drawing. The CAD software systems developed in the U. S. all follow this convention. Creating a projection view drawing with a CAD system is easy. However, it is important to recall that the front view is considered the parent view and the top and right sided views are considered child views. Because of this standard, the top view can only shift vertically, and the right side view can only shift horizontally when they are relocated, as illustrated in Fig. 9.13.

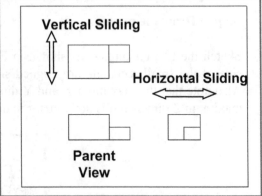

Fig. 9.13 Restrictions on shifting the child views created
with CAD systems.

9.10 **PRINCIPLES OF ORTHOGRAPHIC DRAWINGS**

Why are engineering drawings with projection views standard, when machines and structures are three-dimensional? The reason is because engineers communicate by using drawings displayed on a monitor or a sheet of paper both of which are two-dimensional. Preparing drawings for manufacturing is most effective when represented in two-dimensions to insure clarity and accuracy. For defining a three-dimensional component with a two-dimensional format (paper), orthographic projection is used.

Orthographic projection is based on the assumption that the component remains fixed as the observer moves to view it from different directions. Drawings prepared using the orthographic projection provide the shape and size of each component in an assembly. Dimensions and tolerances are added to these drawings to complete the definition of a component.

A method for preparing an orthographic projection is illustrated in Fig. 9.14, which shows the front view of a three-dimensional object. The front plane is drawn and parallel lines are projected forward to locate the points necessary to draw the front view of the object. It is evident that the front view does not completely specify the shape of the object, because the depths of the object's features are not specified. Additional projections are required to completely describe the stepped block.

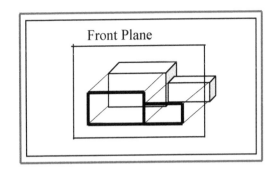

Fig. 9.14 Drawing a front view of a component with orthographic projection

The technique for drawing the top view is presented in Fig. 9.15. First, a horizontal plane is drawn to provide a surface upon which the top view can be displayed. The front and horizontal planes are perpendicular to each other. The top view shows the dimensions from the front to rear of the object.

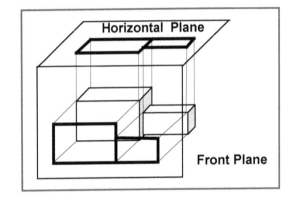

Fig. 9.15 Developing the top view with orthographic projection

In order to arrange these two views in two-dimensions, the horizontal plane is rotated through 90° to become co-planar with the frontal plane, as illustrated in Fig. 9.16.

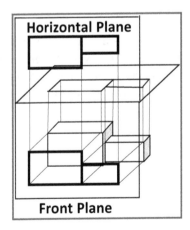

Fig. 9.16 Rotating the horizontal plane 90° to combine the front and top views on the front plane

A third plane is introduced to draw the right side view, as shown in Fig. 9.17. This plane is perpendicular to both the frontal and the horizontal planes. The right sided view is projected onto this plane. This view shows the shape of the object when viewed from the right side and provides the height and depth dimensions. To arrange the three views on the surface of a screen or paper, the right side plane is rotated 90° onto the frontal plane. The three views depict the three-dimensional shape of the object, as shown in Fig. 9.18.

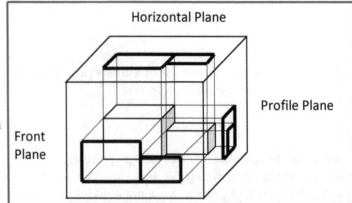

Fig. 9.17 Drawing the right side view with orthographic projection

In an orthographic projection, the three picture planes first introduced are called **planes of projection**, and the perpendiculars are called **projecting lines**. In examining these projections, consider the three planes as windows through which you observe the object.

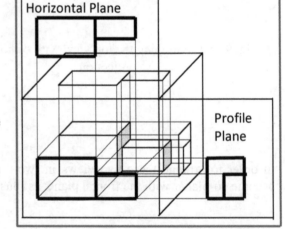

Fig. 9.18 Rotations to display the front, top and right side views on a single plane.

The projection layout, presented above, represents the convention widely adopted by design engineers in the U. S. However, design engineers in Europe and Asia employ a different layout and there are some differences. The alternative layout convention, illustrated in Fig. 9.19, shows four projection quadrants. Each quadrant is available for projecting a view of the object onto a picture plane.

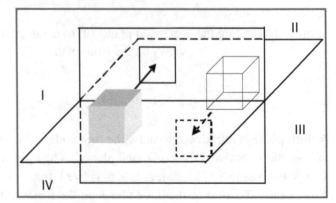

Fig. 9.19 Placement of cubes in Quadrants I and III illustrating a second reference system

If the method of orthographic projection is extended, the maximum number of principal views that can be drawn is six (top, bottom, left, right, front and back). As the observer changes position at 90^0 intervals, the object is completely boxed by a set of six projection planes. Four of the six principal projection planes are illustrated in Fig. 9.20. In each view, two of the three dimensions of height, width, and depth are evident.

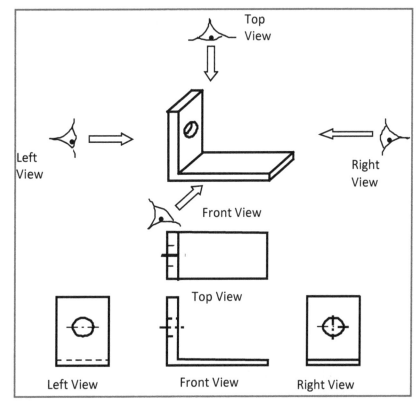

Fig. 9.20 Four of the six possible views of an angle bracket

The process of drawing six views is analogous to opening a six-sided box onto a flat surface. Because the front plane has been selected for display on this surface, the front view automatically appears. The other sides are hinged at the corners and rotated onto the flat surface. The projection onto the horizontal plane provides the top view. The projection onto the right profile plane gives the right side view. The projection onto the left profile plane is the left side view. By reversing the direction of observation, a bottom view is obtained instead of a top view, or a rear view instead of a front view. In some cases, a bottom view, rear view, or both may be required to show the detail of a complicated component.

In practice, three views usually are sufficient to represent the shape and dimensions of typical components. For objects with relatively simple geometric shapes, one or two views may be sufficient to completely define their shape and dimensions. A single view is sufficient to represent a simple cylindrical shaft, as shown in Fig. 9.21.

Fig. 9.21 A single projection view is adequate to define the dowel pin.

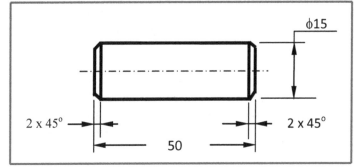

9.11   LINE VISIBILITY

To correctly representation many features in an orthographic projection, it is necessary to indicate whether a line is visible or hidden. Any view of an object is always bounded by visible lines; hence, only those lines that fall within the external outline are invisible, as shown in Figure 9.22.

Establishing line visibility is a requirement in preparing orthographic drawings representing many complex three-dimensional components. Fortunately, CAD systems correctly establish the visibility of the lines within an object's boundaries. With a CAD program, the command **show hidden** is usually available. Switching on the **show hidden** command results in displaying of all the hidden lines as dashed lines, and turning it off eliminates them.

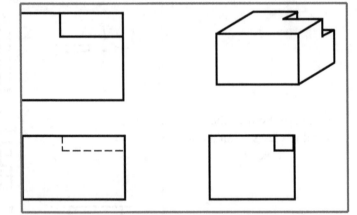

Fig. 9.22 Example of visible and hidden lines

9.12   TYPES OF LINES

Engineering drawings contain lines and symbols, with each type used for a specific purpose. The different types of lines and their purpose are shown in Fig. 9.23.

Visible or Object Line	——————	Dimension Line	⊢————⟶
Center Line	— · — · — · —	Phantom Line	— · · — · · —
Hidden Line	- - - - - - - - -	Cutting Plane Line	— · · — ·
Section Line	——————	Break Line	⟋⊢——⟋⊢

Fig. 9.23 Types of lines and their usage

A description of the various lines are listed below:

Visible lines or object lines: Solid, thick lines for outlines or visible edges.
Hidden lines: Dashed lines of medium weight for edges that are not visible in a view.
Centerlines: Thin lines composed of alternating long and short dashes used to represent the axes of symmetry.
Section lines: Solid, thin lines used for sectional views drawn at an angle to the edges.
Dimension lines: Same style as section lines, but used for dimensioning feature sizes.
Cutting plane lines: Solid, thick lines used to locate a section cut.
Phantom lines: Similar to centerlines except with double dots, used to represent adjacent parts.
Long break lines: Similar to centerlines except with break signs to indicate the end of a partially illustrated feature.

Different types of lines are used in preparing an engineering drawing to show the component's details. A CAD system may automatically select the proper line types for the drawing. However, it is important for you to understand the correct type of line when preparing a drawing.

9.13 DIMENSIONING

When preparing an engineering drawing, the first step is to create the projection view(s) and then dimensions are added to provide the sizes of the features. An example of the three dimensions used to specify the width, height and depth of a block is presented in Fig. 9.24.

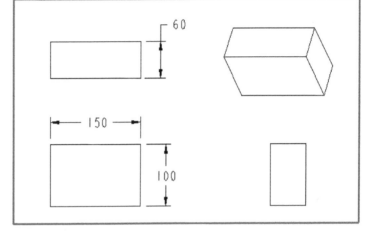

Fig. 9.24 Dimensions specifying the size of a rectangular block

Dimensioning includes specifying the size and position of an object's features, providing notes relevant to manufacturing, and giving information for purchasing, process planning, etc. With the addition of dimensions and notes, engineering drawings serve as a essential part of the documentation required for assembly, manufacturing, inspection, testing, etc. In fact, engineering drawings are often included as part of legal contracts. The American National Standards Institute (ANSI) has established standards titled: **Dimensioning and Tolerancing for Engineering Drawings**.

8.13.1 Size and Position Dimensions

There are two types of dimensions. The **size dimension** as the name implies is associated with the size of a feature, as illustrated in Fig. 9.24. The **position dimension** locates a feature with respect to another feature(s) or to a reference(s) line. To illustrate the differences, let's examine the Fig. 9.25. The component, fabricated from sheet metal, has dimensions of 150 and 100 specifying its width and depth. The thickness of the plate, 6 mm, is specified with a note. The dimensions of $\phi20$ and $\phi25$ define the sizes of the two holes. These dimensions all specify size.

The dimensions of 24 and 35 are position dimensions, because they define the location of the center of the hole with a 25 mm diameter. The dimensions of 20 and 30 are also position dimensions, because they define the location of the center of the hole with 20 mm diameter. Position dimensions are as important as the size dimensions, because without dimensions defining the locations of the holes, the part cannot be fabricated.

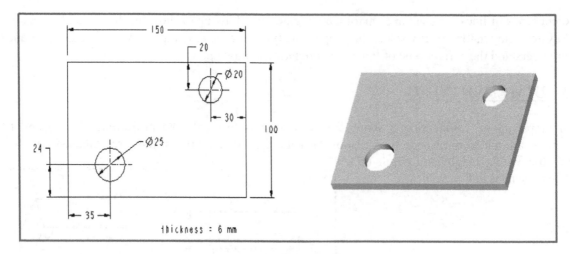

Fig. 9.25 Example drawing showing size and position dimensions

9.13.2 Using Reference Lines to Specify Dimensions

It is common practice to use reference lines when specifying dimensions. To understand the use of references, let's examine the dimensions shown in Fig. 9.25. Consider the two position dimensions used to define the center location of the 20 mm diameter hole. It is evident that the lines defining the right and top edges are the two references to specify the 30 and 20 mm dimensions, respectively. In the same way, the line defining the left and bottom edges of the plate serve as the two references for specifying the 35 and 24 mm dimensions, respectively. It is necessary to select reference lines, when specifying position dimensions.

The importance of selecting a reference before specifying a position dimension is illustrated in Fig. 9.26, which shows an assembly of three components. The sheet metal part (depicted in Fig. 9.25) is assembled with two other components. One (blue) is located at the upper right corner, and the other (green) at the lower left corner, as shown in Fig. 9.26. To ensure the alignment of the two components with the sheet metal part requires that the position dimensions for the holes on the two mating components have the same references, as shown in Fig. 9.27.

Fig. 9.26   An exploded view of components to be assembled with the sheet metal part.

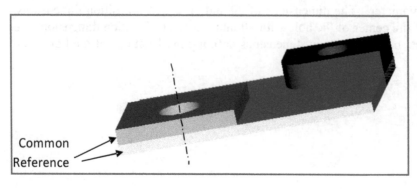

Fig. 9.27 Common reference lines among the components ensures smooth assembly.

To facilitate the selection of reference lines, most of the CAD systems automatically select a minimum set of references, without notifying the designer. On the CAD display, the reference information is usually indicated by dashed-lines, but these references are not always suitable. When the references are not adequate, it is the responsibility of designer to redefine them.

A drawing of a sheet metal part, presented in Fig. 9.28, illustrates the use of size and position dimensions. The thickness SD3 of the sheet metal is given in a note. The size dimensions are given as SD1, SD2, SD3, SD4, SD5, SD6 and SD7. The position dimensions are given as PD1, PD2, PD3, PD4 and PD5. All of these dimensions are required to completely define this component.

There are always different ways to dimension a component. The criterion for selecting the most appropriate way is to make all of the dimensions clear so that manufacturing, assembly or purchasing can be performed without confusion. Another important aspect in dimensioning is to give the dimensions to aid the machinist. For example, in dimensioning a hole, specify its diameter, not its radius. Drill bits used to cut the hole are specified by their diameter.

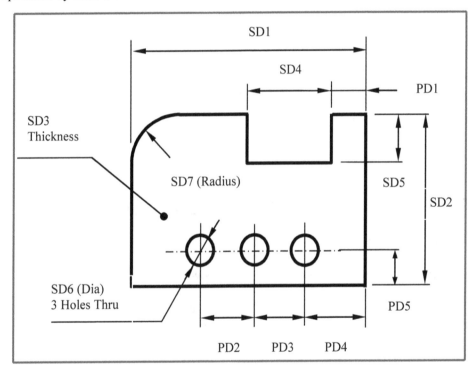

Fig. 9.28 Size and position dimensions.

For components that can be described with a single feature size dimensions are required, but not position dimensions. For example, Fig. 9.29 shows a wedge, which is an object that can be depicted with a single feature. Hence, three size dimensions are required, but no position dimensions are necessary.

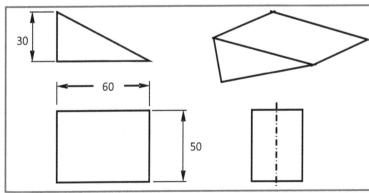

Fig. 9.29 A single feature object, where size dimensions are required but not position dimensions.

9.14 TWO SYSTEMS OF UNITS

There are two units of measurement employed when dimensioning. In the this country, both the U. S. Customary System (English) and the SI system (metric) are employed. The initials SI refer to the International System of Units. Numerical values shown in engineering drawings are assumed to be in millimeters in the SI system or inches in the U. S. Customary system. The conversion factor between inch (in.) and millimeter (mm) is 25.4 mm = 1 inch. A comparison of units used to dimension the front view of a block is presented in Fig. 9.30. Units of mm or inch are not included on the dimension lines. The length unit employed on the drawing is identified in the drawing block. All CAD systems have a unit conversion function, permitting automatic conversion between the two systems.

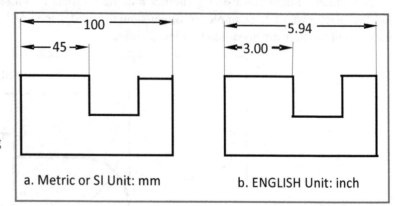

Fig. 9.30 Examples of dimensioning in SI and U. S. Customary units.

9.15 GUIDELINES FOR DIMENSIONING

Twelve guidelines for specifying dimensions on engineering drawings are listed below:

Guideline 1: Specify all dimensions in one view if possible.

This practice minimizes the number of views needed to represent an object. The example presented in Fig. 9.31 illustrates the use of a single view to completely describe a pin with rotation symmetry.

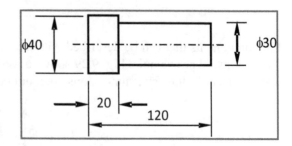

Fig. 9.31 All dimensions in shown in a single view

Guideline 2: Specify the three maximum dimensions.

This practice is important for production control to ensure availability of the stock required for fabrication. When the three maximum dimensions are specified it is easy to verify the availability of stock that is the correct size.

Guideline 3: Consider manufacturing processes when dimensioning.

Suppose an end mill with a diameter of 20 mm is used to machine a keyway, as shown in Fig. 9.32. The end mill is inserted into the block 30 mm from its left side and it traverses 40 mm in cutting the keyway. Hence, the left side line is used as the reference for the position dimension.

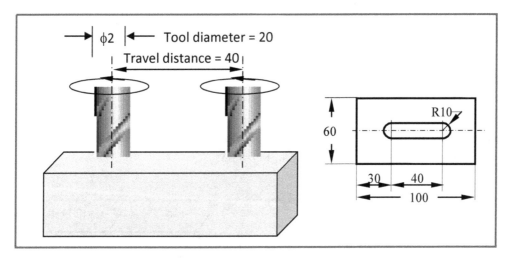

Fig. 9.32 Dimensioning a keyway to facilitate the manufacturing process

Guideline 4: Use section views to show hidden features and their dimensions.

An example of a section view, with dimensions, is presented in Fig. 9.33. All of the dimensions describing the size of the countersunk hole are included in a single view. Without using a section view, dashed lines must be used to size this hole. Best drawing practices reduce the number of dashed lines to represent hidden features; thus, enhancing the clarity of engineering drawings.

Fig. 9.33 Use a section view to show and dimension hidden features.

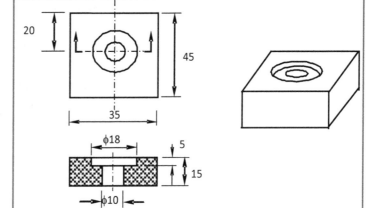

Guideline 5: Use centerlines for symmetrical parts.

Two different ways to dimension an identical part are shown in Fig. 9.34. The drawing displayed on the left has a centerline, and the one the right does not. It is evident a drawing with a centerline reduces the number of dimensions required. Also the presence of a centerline emphasizes that the right and the left sides be identical in size and shape.

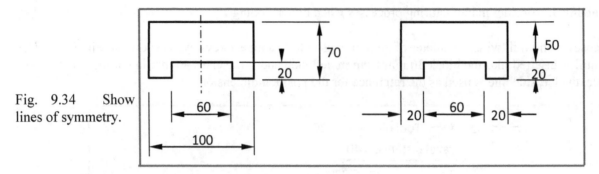

Fig. 9.34 Show lines of symmetry.

Guideline 6: Do not close a dimension loop.

In practice, redundant and unnecessary dimensions occur when all the individual dimensions are specified together with the overall dimension, as illustrated in Fig. 9.35.

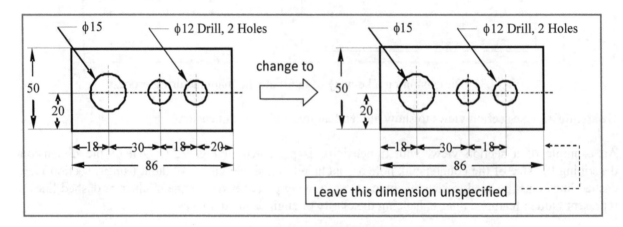

Fig. 9.35 Remove redundant dimensions to avoid forming a closed loop

Guideline 7: Consider reference(s) when specifying position dimensions.

Selecting dimensions for position usually requires more consideration than selecting dimensions of size, because there are often several options. Usually a datum reference should be determined before positional dimensions are specified. Datum references include finished surfaces, centerlines or some combination of them, as illustrated in Fig. 9.36.

Fig. 9.36 Use references lines to specify size and position dimensions.

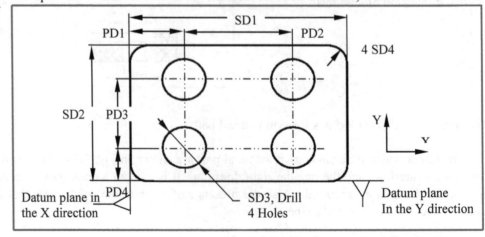

Guideline 8: Avoid intersections between dimension lines.

Two cases demonstrating the intersection of dimension lines are shown in Fig. 9.37. To avoid intersecting dimension lines place dimension lines for the smaller dimensions adjacent to the view.

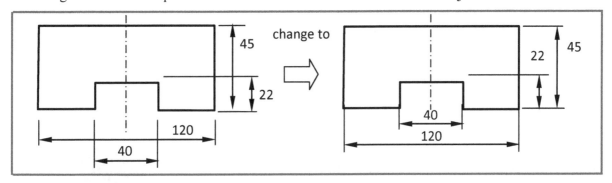

Fig. 9.37 Avoid intersections of dimension lines.

Guideline 9: Use national and international standards when specifying dimensions.

The method for showing threads on a cylinder is presented in Fig. 9.38. Note that M indicates a metric thread, which followed by the thread's major diameter and its pitch in millimeters. Engineers do not draw the shape of threads. Dashed lines are used to indicate the length of the threads. Most CAD systems follow the national and international standards, when dimensions are added to engineering drawings.

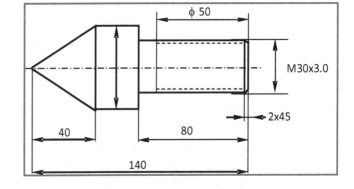

Fig. 9.38 Use dashed lines to represent the length of the threads.

Guideline 10: Add a chamfer to eliminate a sharp edge or a fillet to mitigate a reentrant corner.

The large chamfer (4 mm by 45°) on the end of a shaft, illustrated in Fig. 9.39, is to facilitate assembly. The fillet at the reentrant corner (R = 5 mm) is to reduce the stress concentration. Two small chamfers (2 mm by 45°) eliminate the sharp corners on the larger round. These chamfers eliminate sharp edges that abrade or cut the hands of assembly personnel.

Fig. 9.39 An example of chamfers and a fillet.

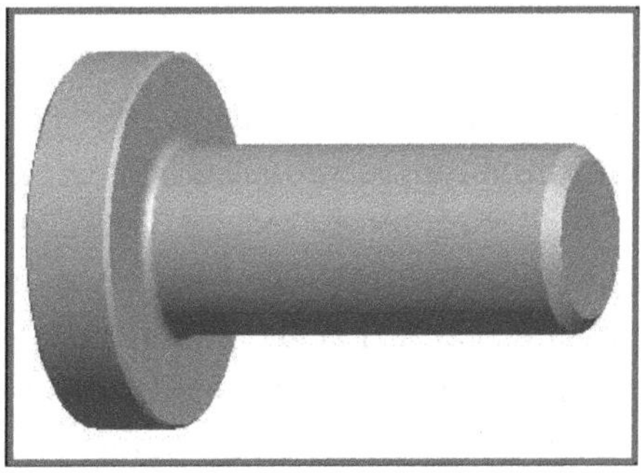

Guideline 11: Add tapers to critical surfaces when designing a mold.

When designing a mold or die for casting or injection molding, parting surfaces are required. As illustrated in the Fig. 9.40, the die is made from a cap (yellow) and a base (red). They have a parting surface that is geometrically identical on both portions. To strip a component (blue) from the mold, tapered surfaces, which are perpendicular to the parting plane must be specified.

Fig. 9.40 Tapered surfaces of the mold are perpendicular to the parting plane.

Guideline 12: Use an appropriate scale for engineering drawings.

The scale of an engineering drawing should be selected so that the projection views are clearly depicted, and dimensions, tolerances and notes are easily readable. A component may be drawn to any convenient scale, which is specified in the title block. The dimensions shown correspond to the actual size of the component not the size of the view on the drawing. The scale is a constant regardless of the value of the dimensions.

8.16 TOLERANCES

The American Society of Mechanical Engineers (ASME) defines tolerance as the total amount by which a specific dimension is permitted to vary. The numerical value of a tolerance is the difference between the maximum and minimum limits. Tight tolerances tend to increase the cost of production, while components with large tolerances usually cost less to fabricate. However, in some cases components with large tolerances may pose difficulties in assembly. A balance between precision and cost must be reached during the design phase.

Products used on a daily basis are designed to operate at different performance levels. Take as an example two very different products, each with a wheel that rotates on an axle. Suppose one wheel and axle assembly is installed on a child's wagon and another on a Boeing 787. In both cases the purpose of the wheel and axle assembly is to allow the vehicle to roll. Obviously, the performance levels are quite different for these two applications, and the degree of precision needed for each assembly differs greatly. The need for precision and tight tolerances is much greater for Boeing's 787 than for the child's wagon.

Businesses require that parts be specified with increasingly exact dimensions. Many parts, fabricate by different companies at widely separated locations, must be interchangeable. This requirement demands precise size specifications and achieving exacting tolerances in production. Dimensioning components, within a range of variation sufficiently narrow to ensure interchangeability, requires the designers to be knowledgeable about tolerances and assembly practices.

9.16.1 Tolerancing Components

Tolerances are the allowable deviations from a specified dimension. For example, the tolerance of a component's length with a dimension specified as 120 mm, can be defined as 120 ± 1 mm. This means that a component is acceptable if this dimension varies from 119 mm to 121 mm. Manufacturing costs usually increase as tolerances become tighter; hence, it is prudent to specify tolerances as large as possible to minimize costs without compromising the function of the component.

Tolerances allow the designer to control the maximum and minimum sizes of features and their positions relatively to each other. The component shown in Fig. 9.41 has a hole with a diameter that may vary from 50.50 mm to 50.00 mm. The range of allowed variation, 0.50 mm, is the numerical value of this tolerance.

Tolerance = Maximum − Minimum Dimension.

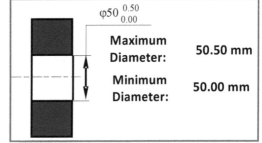

Fig. 9.41 Example of tolerance a hole.

The tolerance of a dimension often affects the tooling used to produce the component. Manufacturing components with large tolerances can be accomplished by using less precise tooling than those with tighter tolerances. The hole shown in Fig. 9.41 can be made with a sharp drill, but if the tolerance is reduced to 0.05 mm, a more expensive processes using milling and reaming are required.

9.16.2 Fitting Mating Parts

Clearance Fit

The contact between two mating components during an assembly is called a fit. A typical example is the fit between a hole and a shaft, shown in Fig. 9.42. The fit of a shaft in a hole is very common in a wide range of products. Wheels rolling on shafts and pistons sliding in cylinders are two examples. The performance and life of these systems dependents on the fit of the mating components. There are two common classifications used to specify different types of fit —clearance and interference fits.

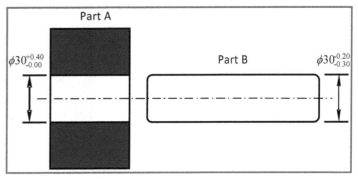

Fig. 9.42 Fit of a shaft in a hole with tolerances for clearance.

Let's assume that we are dealing with a mass production process, where a shaft is taken from a batch of shafts and a part containing a hole from a second batch of components. With a clearance fit, the diameter of the shaft is always smaller than the diameter of the hole. Hence, the shaft **clears** the hole, leaving an annular gap. The amount of clearance varies and depends on the tolerance placed on the hole and shaft.

The example presented in Fig. 9.42 indicates that the actual dimension of a shaft may vary from 29.70 mm to 29.80 mm, and the diameter of the hole varies from 30.00 mm to 30.40 mm. The tolerances are 0.10 mm on the shaft and 0.40 mm in on the hole, respectively. The maximum clearance is determined by subtracting the smallest shaft diameter from the largest diameter of the hole. The minimum clearance is determined by subtracting the largest shaft diameter from the smallest diameter of the hole:

$$\text{Maximum Clearance} = A_{max} - B_{min} = 30.40 - 29.70 = 0.70 \text{ mm}$$
$$\text{Minimum Clearance} = A_{min} - B_{max} = 30.00 - 29.80 = 0.20 \text{ mm}$$

Tolerance of the assembly for the clearance fit = 0.70 − 0.20 = 0.50 mm

Fig. 9.43 Clearance between mating parts A and B after assembly.

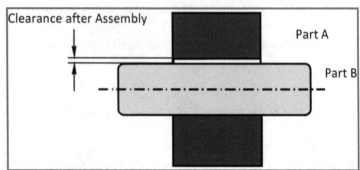

Interference Fit

Interference fit occurs when the shaft diameter is larger than the diameter of the hole. Interference fits are also known as force fits, because the shaft is forced into the hole during assembly. A light hammer blow of the end of the shaft may be sufficient to assemble a light force fit, but a heavy force fit often requires the shaft to be cooled and/or the hole to be heated to facilitate assembly. After assembly, an interference fit provides a permanent mating of the two components. Disassembly requires the shaft to be forced out of the hole, which is difficult. In severe cases, the housing containing the hole may have to be cut apart to remove the shaft. Interference fits are recommended when assembling mating parts, which are to remain fixed and not move relative to each other. A bearing fitted onto an axle is an example. The drawing of a shaft and a disk, presented in Fig. 9.44, illustrates an interference fit. The shaft diameter is larger than the diameter of the hole by 0.05 to 0.35 mm depending on the tolerances.

Fig. 9.44 Example of an interference fit.

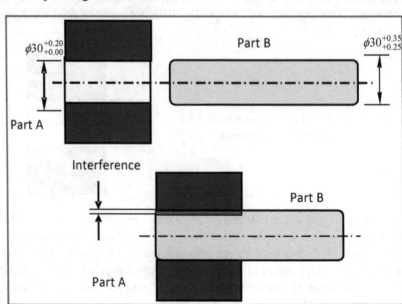

8.17 <u>RAPID PROTOTYPING</u>

In manufacturing and the machining operation material is removed from a block to fabricate a component to the size and shape specified on an engineering drawing. Machining is classified as a subtractive process because material is removed as a part is fabricated. During the past two decades significant progress has been made in developing additive manufacturing techniques. Manufacturing has included techniques that are additive for hundreds of years. These techniques included fastening together components by riveting, screwing, forge welding, or electric arc welding. However, forming parts by adding thin layers or strings of material is relatively new. This new technique is termed rapid prototyping or rapid manufacturing. Rapid prototyping depends on CAD files where the three dimensional geometry of a component is available. The digital data in these files provide the input for the rapid prototyping equipment.

Rapid prototyping software takes digital files representing the blueprints from CAD programs and slices them into digital cross-sections of the component. Each slice is input for the rapid prototyping machine that successively prints the slice. Depending on the rapid prototyping equipment, layer after layer of material is deposited on the machines base until the layering is complete. At this stage the final three-dimensional configuration has been printed. Results show that the digital model and the physical model are essentially the same.

Surface smoothness depends on the printer resolution, which includes layer thickness and x-y resolution in dots per in. or mm. Typical layer thickness is about 0.10 mm, although some machines can print layers as thin as 0.016 mm. Resolution in x and y is comparable to that of laser printers. The three-dimensional dots are about 0.05 to 0.1 mm in diameter. Fabrication of a prototype with traditional techniques may require times ranging from several hours to several days, depending on the size and complexity of the prototype. Rapid prototyping systems can usually reduce this time to a few hours, although it varies widely depending on the type of machine used and the complexity of the prototype.

Conventional manufacturing methods such as injection molding of plastic components are less expensive for manufacturing in high quantities, but additive manufacturing can be faster, more flexible and less expensive when producing relatively small quantities. Rapid prototyping machines provide designers and development teams the ability to produce concept models using a desktop size printer, as shown in Fig. 9.45.

Fig. 9.45 3D printers used by students
in an introductory design class.

Three different of additive processes are currently employed to develop prototypes. They differ in the manner layers are deposited and in the materials used. With selective laser sintering and fused deposition modeling the materials are melted to produce the layers. With sterolithography liquid polymers are cured using different techniques. With laminated object manufacturing, thin layers are of paper, plastics or metals are cut to shape and fused together. Each method has its advantages and disadvantage. The primary considerations in selecting a machine are speed, cost of the 3D printer, cost of the printed prototype, and cost and choice of materials and color capabilities.

9.17.1 Extrusion Deposition

The fused deposition process uses a plastic filament or metal wire that is fed from a reel to supply material to an extrusion nozzle. The nozzle serves to melt the material and to turn the flow on and off. Stepper or servo motors control the x, y position of the nozzle. The horizontal and vertical directions are controlled by a CAM program. The prototype part is produced by extruding small beads of thermoplastic to form layers. The layers fuse together and the thermoplastic hardens immediately after extrusion from the nozzle. The fused deposition process is illustrated with the cartoon presented in Fig. 9.46.

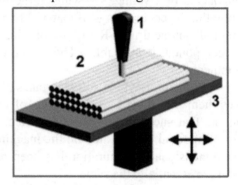

Fig. 9.46 The fused deposition process:
1— nozzle ejecting molten plastic
2— deposited material
3—controlled movable platform

Various plastics can be melted, including acrylonitrile butadiene styrene, polycarbonate, and polyphenylsulfone. There are restrictions on the shapes that can be fabricated, because stalactite-like structures cannot be supported as the polymer solidifies.

9.17.2 Granular Particle Binding

Another method for forming a three-dimensional part by adding material involves selective fusing of materials in a granular bed. After fusing the selected lines or areas, the base is indexed downward and another thin layer of granules is added. The sequence is repeated until the component is complete. A laser is usually employed to sinter the granules producing solid plastic parts. Some metal powders can be sintered with the laser, but their structure is porous. For higher strength prototypes fabricated from metals an electron beam is used to melt the powders. A high vacuum is required for electron beam melting, but unlike metal sintering methods that do not require melting temperatures to be achieved, e-beam formed parts are dense, without significant porosity and strong.

A similar approach involves an ink jet from a printer. The ink jet is modified to spray dots of liquid binder. The prototype is created one layer at a time by spreading a layer of powdered resin and then printing dots of binder to form a cross-section of the part. This sequence is repeated until all of the layers have been formed. Ink jet printing is compatible with colored resins and with elastomers. The strength of the prototypes can be increased by impregnating the structure with wax or thermo setting polymer such as epoxy.

9.17.3 Selective Liquid Curing

With selective liquid curing, a vat of liquid polymer is exposed to a narrow beam of light from a digital light projector, while the surrounding area is illuminated with light that will not affect the polymer. The liquid polymer exposed to the beam of light hardens. The base platform of the machine then moves down in small increments and the liquid polymer is again exposed to light. The expose, cure and move sequence is repeated until the prototype shape has been formed. The liquid polymer is drained, leaving the prototype.

A similar approach using an ink jet printer squirts very thin 16 to 30 μm layers of a light sensitive polymer forming a layer of the cross section. An ultraviolet light mounted adjacent to the ink jet cures the photopolymer as soon as it is deposited. The process is repeated layer after layer until the prototype is completed. The prototype is completely cured and may be handled immediately, without post-curing.

8.18 SUMMARY

Computers and many software programs enable engineers to enhance their capabilities and significantly increase their productivity. Major characteristics of CAD systems are:

1. A creative method of representing the design
2. An effective method of communicating design information
3. An approach to facilitate design modifications and implement redesign
4. An effective method for executing engineering analyses
5. A technique for integrating design with manufacturing and operations

Design is a process for creating a product that **meets** a well-defined set of requirements. In general, a design has multiple solutions that are constrained by the available resources. In essentially all cases, the final design of a product must be completed within a budget and on schedule. Conceptual design begins with the product specification, followed by generating design concepts and identifying critical product features. If a design does not meet the specifications, it is modified to generate another design that meets the specifications. When the conceptual design is finalized a preliminary design is generated. During preliminary design, changes to the drawings are made without constraints. However, when the design is finalized a detailed design is generated. At this point in the design process, design changes must follow a formal process that includes an engineering order, an engineering notice and an engineering change order.

Process planning, which precedes manufacturing, includes choosing the type and sequence of each production operation, tooling requirements, equipment and estimating the overall cost of production based on these selections. If the product is assembled, process planning includes establishing the number and sequence of steps involved in the assembly operation. When the planning process is complete, components are fabricated using one or more of several well-established manufacturing processes.

When communicating in writing, we follow strict grammatical rules for spelling, sentence construction and punctuation. When preparing engineering drawings, we also follow specific rules. The American National Standards Institute (ANSI) establishes the standards for engineering documents in the U. S. The International Organization for Standardization (ISO) is responsible for international standards. All the CAD software systems incorporate the engineering graphics rules adopted by ISO and ANSI. It is very important for CAD users to understand these rules.

Orthographic projection views represent the standard means for engineers to communicate design details. Drawings transmitted to manufacturing are most effective when represented in two-dimensions to insure clarity and accuracy. Orthographic projection drawings provide the shape and size of each component in an assembly. Dimensions and tolerances are added to these drawings to complete the definition of a component. Dimensioning includes specifying the size and position of an object's features, providing notes relevant to manufacturing, and giving information for purchasing, process planning, etc.

Twelve guidelines for specifying dimensions on engineering drawings were described:

1. Specify all dimensions in one view if possible
2. Specify the three maximum dimensions
3. Consider manufacturing processes when dimensioning
4. Use section views to show hidden features and their dimensions
5. Use centerlines for symmetrical parts

6. Do not close a dimension loop
7. Consider reference(s) when specifying position dimensions
8. Avoid intersections between dimension lines
9. Use national and international standards when specifying dimensions
10. Add a chamfer to a sharp edge or a fillet in a reentrant corner
11. Add taper to critical surfaces when designing a mold
12. Use an appropriate scale for engineering drawings.

EXERCISES

8.1 Use your personal experience to describe the essential steps involved in engineering design. Make sure that the following information is presented.

- What was the product you designed?
- Was the design process a team effort or your own effort?
- What was the design objective(s)?

8.2 Based on your personal experience or observation, describe the importance of documentation in engineering design. Why do computers and information technology play an important role in today's engineering design community?

8.3 Have you made any engineering drawings? If yes, how did you prepare your engineering drawings? Did you prepare them using a computer and a software system, or using a pen or pencil by hand?

8.4 What are the three engineering projections that are most commonly used? In what situations do you need more than three projections? In what situations do you only need two projections or one projection?

8.5 Prepare an engineering drawing of an object where one projection is needed.

8.6 Prepare an engineering drawing of an object where two projections are needed.

8.7 Prepare an engineering drawing of an object where three projections are needed.

8.8 Prepare an engineering drawing of an object where more than three projections are needed.

8.9 Tolerances can be specified in three formats depending on their nature. They are unilateral format, bilateral format, and limit format. For each of the tolerances listed below, write the other two forms.

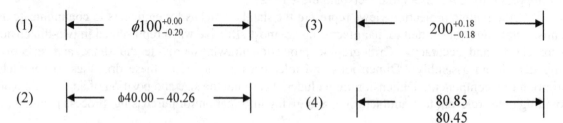

(1) $\phi 100^{+0.00}_{-0.20}$ (3) $200^{+0.18}_{-0.18}$

(2) $\phi 40.00 - 40.26$ (4) $\begin{array}{c} 80.85 \\ 80.45 \end{array}$

8.10 As illustrated in the following figure, the dimensions and tolerances for parts A and B are $50^{+0.12}_{-0.06}$ mm and $75^{+0.20}_{-0.04}$ mm, respectively. Parts A and B will be assembled into part C, which is a container. The maximum and minimum clearances between the assembly and the container are 0.50 and 0.02 mm, respectively. Determine the maximum and minimum dimensions of the container's width.

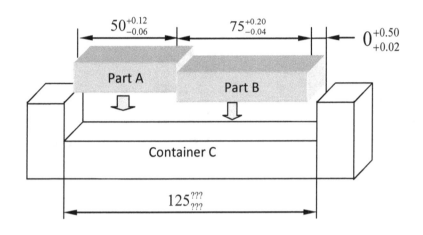

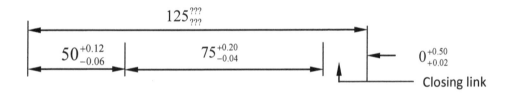

8.11 Thousands of ball bearings have been purchased. Determine the maximum and minimum diameters of shafts, which will be assembled with the bearings shown below. The nominal dimension of the inner diameter of the bearing is 50 mm, and the nominal dimension of the outer diameter is 85 mm.

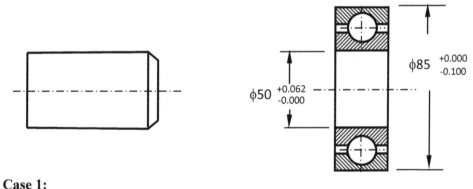

Case 1:
For clearance fit, the maximum clearance is 0.302 mm and the minimum clearance is 0.008 mm.
The maximum diameter is _____ mm and the minimum diameter is _____ mm.
Case 2:
For transition fit, the maximum interference is 0.091 mm and the maximum clearance is 0.006 mm.
The maximum diameter is _____ mm and the minimum diameter is _____ mm.

Case 3:

For interference fit, the maximum interference is 0.105 mm and the minimum interference is 0.014 mm. The maximum diameter is _____ mm and the minimum diameter is _____ mm.

8.12   An assembly consists of five components. They are Container, Part1, Part2, Part3, Insert1, and Insert2. The assembly process belongs to "clearance fit." The range of the clearance in the horizontal direction after the assembly is between 0.03 and 0.30 mm. The range of the clearance in the vertical direction after the assembly is between 0.02 and 0.22 mm.

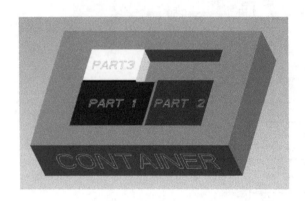 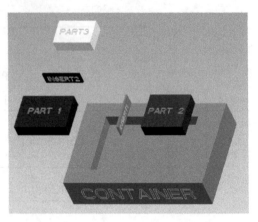

The dimensions and tolerances of Container, Part1, Part2 and Part 3 are given. They are shown below.

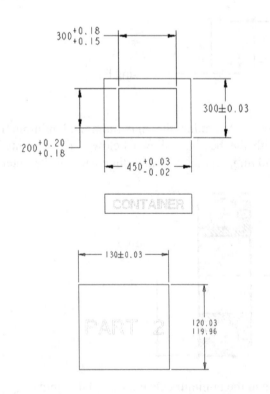

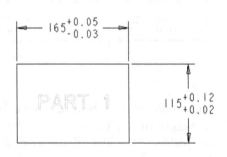

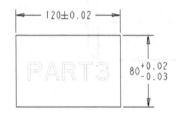

What are the maximum and minimum dimensions of Insert 1, which is used to control the clearance in the horizontal direction during this assembly process? What are the maximum and minimum dimensions of Insert 2, which is used to control the clearance in the vertical direction during this assembly process?

CHAPTER 10

MICROSOFT EXCEL

10.1 INTRODUCTION

Microsoft® Excel, a spreadsheet program, is an extremely powerful tool, because it is useful in many different applications. In addition to providing an introduction for using the Excel spreadsheet, techniques for three important applications will be described in detail in this chapter. These applications include:

- Preparing tables and graphs
- Making calculations
- Conducting design trade-off studies

While several different spreadsheet programs are commercially available, Excel® has been selected because it is the most popular. Many universities have adopted it and you will be able to access it in campus computer labs during your tenure in college. The content in this chapter is intended to both develop your entry-level skills in Excel and to encourage you to begin thinking like a practicing design engineer. Hopefully, you will find the spreadsheet tool important enough to develop a much higher skill level through independent study.

10.2 EXCEL BASICS

The version of Excel described in this chapter is included in Microsoft Office 2010 suite of programs. However, much of the content in this chapter is applicable to earlier versions of Excel, which you may still be using. Microsoft Office 2007 and later has been given a major facelift as compared to prior versions of Office. The primary change is that the menu-based software has been replaced with an all-purpose "ribbon" along the top of the screen. The ribbon provides a tabbed browsing environment that contains all of the items found under the familiar Microsoft menu headers (i.e., File, Edit, View, Insert, etc.). The use of these tabs minimizes the user's time in searching drop down menus. We suggest that you take the time to explore each of the tabs along the top ribbon and familiarize yourself with the new configuration for all the Microsoft Office programs. The Excel ribbon is presented in Fig. 10.1.

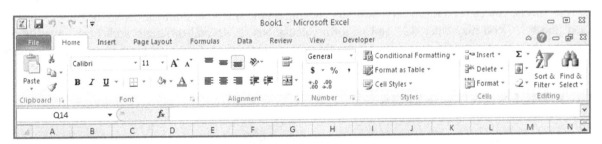

Fig. 10.1 The "ribbon" menu of Microsoft Office Excel

Most of the screen below the ribbon is occupied by the spreadsheet, which is simply a large table with many columns and rows as shown in Fig. 10.2. The columns are labeled across the top of the table with letters A, B, C,... Q and beyond. The rows are labeled down its left side with numbers 1, 2, 3,... 16 and beyond. Each small block, called a cell, is identified with its coordinates (e.g. in Fig. 10.2, the cell A1 is outlined with a border indicating it is active). The letter (column location) is placed before the number (row location). The cells are simply locations in a very large table (spreadsheet) where you can enter numbers, text or formulas.

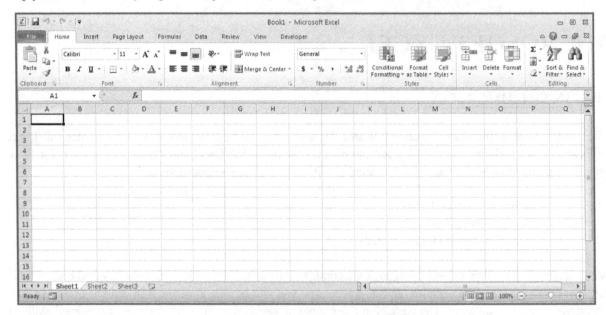

Fig. 10.2 The Excel spreadsheet showing cell A1 as the active cell

You may move about on the spreadsheet using either the mouse or the keyboard. When the mouse pointer is on the spreadsheet, its location is indicated with a large plus sign. When the large plus sign is pointed at the appropriate cell—click; the cell becomes active and is ready to receive the data you enter. You may also move the active cell by using the arrow keys, the tab key, the shift-tab keys, the enter key, the shift-enter key and the control-arrow keys. Try them and note how the active cell moves about the spreadsheet.

The spreadsheet is much larger than the table shown on the screen. In Fig. 10.2, only columns A through Q and rows 1 through 16 are shown. Explore the number of rows on the spreadsheet by holding down the control and the arrow-down key. Note that the number of the last row is 1,048,576. There are a total of 16,384 columns. You have a total of $(1,048,576)(16,384) = 17,179,869,184$ cells on the spreadsheet. Your spreadsheet is huge—let's hope you never have to use all of the available cells!

At the bottom of the spreadsheet, in Fig. 10.2, you will find three tabs identifying sheet numbers 1, 2 and 3. When you open Excel and create a new file, you immediately begin a workbook for a project. The program automatically establishes three sheets (pages) in this workbook. If your project requires more than three worksheets, you can add more very easily by clicking on the small tab immediately to the right of the tab labeled Sheet 3. You may enter data and perform calculations on several different worksheets in a file by clicking on the tabs to move from one sheet to another. Now that you are familiar with the layout of Microsoft Excel, let's explore how to use Excel to create tables, graphs, and to perform calculations.

10.3 TABLES AND GRAPHS

You will often collect large amounts of numerical data that must be presented at a design briefing or in an engineering report. At other times, you may have several different results from a mathematical relationship that must be expressed in a meaningful format. You have two choices in presenting large quantities of numerical data—show the numbers in tabular form or on a graph. Both methods of presentation, the table and the graph, have advantages and disadvantages. The method that you select depends on your audience, the message, and the purpose of the presentation.

In this section, the technique for arranging data in a table using a spreadsheet is described. Data entry in tabular format into a spreadsheet is also necessary when preparing graphs. Three common methods for representing data in the form of graphs will be described, including:

1. Pie charts
2. Bar charts
3. X-Y graphs

Each type of graph is used for a different purpose and selection of the correct type of presentation is essential in communicating effectively. For instance, the pie chart is used to show distributions, whereas the bar chart is used to compare one set of data with another. X-Y graphs, the most frequently used graphic in engineering, shows the variation of a dependent variable Y with an independent variable X. Regardless of the type of table or graphic required, Excel provides a technique for creating high-quality, professional graphics.

Tables

Very early in the design of your vehicle, your team will be faced with the requirement of making a special table called a parts list or bill of materials. The parts list provides an estimate of the overall weight of your vehicle, which is needed in the preparation of your design report. A complete parts list should include every component required to construct your team's vehicle. For each item listed, the parts list should include the quantity of each component required, the vendor, the part number, a weight estimate and its cost. An example of an incomplete parts list created in Excel is presented in Table 10.1.

Table 10.1
Sample of an incomplete vehicle parts list

	A	B	C	D	E	F	G
	C9 Excel 1						
1	**Item Description**	**Vendor**	**Quantity**	**Unit Cost ($)**	**Total Cost ($)**	**Weight (N)**	
2	Arduino Uno	Univ. of Maryland	1	$ 42.50	$ 42.50	0.31	
3	3 Speed Axle Gear Box	Hobby Engineering	1	$ 7.99	$ 7.99	0.28	
4	Gear Motor	Spark Fun Part #	1	$ 24.95	$ 24.95	0.20	
5	Stepper Motor	Spark Fun Part #	1	$ 6.95	$ 6.95	0.18	
6	Stepper Motor Drive	Spark Fun Part #	1	$ 14.95	$ 14.95	0.22	
7	Proximity Sensor	Spark Fun Part #	1	$ 13.95	$ 13.95	0.11	
8	Off Road Wheels (pair)	Spark Fun Part #	2	$ 14.95	$ 29.90	0.95	
9	Shaft (Stainless 1/4 D by 12 L)	Spark Fun Part #	1	$ 3.59	$ 3.59	0.10	
10							
11	Total				$ 144.78	2.35	
12							
13	Sample formulas used to creat the table						
14	Cell E2: = C2*D2						
15	Cell E11: = SUM(E2:E9)						
16							

A number of operations are required to construct Table 10.1. First, each cell must be formatted. For instance, you may have noticed that each cell along the top row of the table (row 1) is shown in bold font type. Cells can be made bold by highlighting the desired cells and clicking the standard Microsoft icon for Bold font found under the Home tab on the main ribbon. Under this tab, the font style, size, alignment, formatting, color, etc. may be changed for each cell and each character within the cell. Under the home tab, there is subsection called Alignment. The top right icon in this subsection is called "Wrap Text". This icon can be used to allow the descriptions written in a cell to wrap and create multiple rows within the cell instead of being cut off at the start of the adjacent cell text. Another important set of commands can be found under the Home ribbon tab in the Number subsection. These icons are used to set the number of decimal places shown in the highlighted cells and/or the style of number to be used.

Because the costs were given in U. S. dollars ($) at the top of the parts list, each of the numbers in columns D and E were formatted to display two decimal places, which gives a precision of $\pm$ one cent. Similarly, column F, showing weight is specified with two decimal places because the scale used to weigh these components was accurate to within $\pm$ 1 g. It makes no sense to list a weight to a higher level of certainty than it can be measured than it does to provide a cost to a higher precision than $\pm$ one cent. You should not list numbers in a table with a higher level of precision (number of decimal places) than your certainty in reporting.

There are a number of other formatting operations with which you must become comfortable to make professional looking tables. These include adding borders and adjusting the widths of each column/row (which is as simple as left clicking and dragging the line separating the column/row labels). You are advised to spend time early in your academic career familiarizing yourself with the commands required to make highly professional looking tables and graphs.

Another set of operations is required to create the parts list shown in Table 10.1. The use of mathematical formulae simplifies the creation of this table. For instance, because the quantity is provided in column C and the unit cost in column D, the total cost (column E) for a specified part is found by multiplying the numbers in columns C and D together. Excel has a built in mathematical library to make these types of repeated calculations efficiently. From row 8 of the parts list, the cost for a pair of off road wheels is $14.95 and the quantity required is 2 pair. Hence, the total cost is $14.95 \times 2 = \$29.90$. In Excel, this calculation is made very simply by typing "=C8*D8" into cell E8. The "=" sign indicates to Excel that a formula will follow and the "*" sign indicates the multiplication of the numbers in these two cells. The basic algebraic operations use the following symbols: + to add, − to subtract, * to multiply, and / to divide. For a much more complete set of formulae and functions available in Excel, click on an empty cell and type = and note the function box shown in Fig. 10.3. Note the functions listed and then explore More Functions.

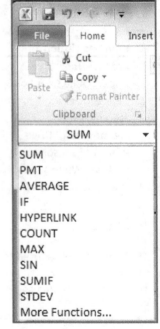

Fig. 10.3 The function box opens by placing an equal sign in any cell.

To enter equations more efficiently, you can select a cell to be used in a formula by simply clicking on it with the mouse instead of typing its location in a cell. Another very useful feature of Excel is that after the first equation has been typed into a cell, it can be used to automatically compute any number of similar calculations. For instance, let's automate the computation of the Total Cost column (column E). To accomplish this, copy the formula in cell E2 by pressing CTRL-C and then highlight the cells in column E from rows 2 through 9 where you are to perform a similar calculation. Next, paste this formula into each of these cells by typing CTRL-V. Excel automatically performs all of these calculations using an indexing system with the cells E# =C#*D#, where the # sign indicates the numbers of the row you have highlighted. To check the accuracy of these calculations, double click on any of the boxes in column E to show the formula hidden within that cell.

Excel has hundreds of different mathematical functions that can be used in equations, ranging from trigonometric functions like "=sin(Cell)" to statistical functions like "=average(Cell Range)". In creating Table 10.1, one additional computational technique was used in Excel. Instead of adding up all of the cells in the column showing the total cost ("=E2+E3+…+E9), the summation tool was used. To follow this procedure, type "=" select the sum function, as shown in Fig. 10.3, and highlight the column of cells to be summed. The summation will be performed automatically when you hit the enter key. If you double click on the cell showing the total cost, you will observe "=sum(E2:E9)" as the mathematical operation. Here, "E2:E9" provides the Cell Range over which the mathematical operation was performed. Similarly, by copying the formula for the total cost, and pasting it into cell F11, you can quickly compute the total weight of the entire rover. If you double-click on this cell, it will show "=sum(F2:F9)". Excel indexes to the F column from the E column when the copy/paste function is used across columns in a similar way that it indexed row numbers when calculating the Total Cost as the product of the Unit Cost and the Quantity.

Control of Indexing Column Letters and Row Numbers

You have control over the way in which Excel indexes when the copy/paste commands are employed. For instance, an entire column of numbers (let's say measurements in feet) can be converted into inches by multiplying each cell in feet by a conversion factor (12 in / 1 ft). To accomplish this conversion, the factor 12 can be placed in a cell, say B2. When calling this number, you must define it as B2 in the first equation written. The "$" before B locks the number 12 in column B and the "$" before 2 locks the number 12 in row 2. When mathematical operations are indexed across columns and rows, the quantity in cell B2 will always be used for the calculations if the $ sign precedes the column and row indicators.

Excel is an extremely powerful computational tool that can be used for much more than making simple tables like the parts list shown in Table 10.1. With practice, you will find that the time spent becoming proficient in Excel will be of great benefit to you in many of your engineering classes.

The Pie Chart

Excel is an extremely useful tool for quickly creating high quality graphics. The first graphic that we will create is a pie chart. A pie chart is usually employed to show distributions in data. To apply this to the your project, the data in a complete parts list has been rearranged to create a table showing the cost to construct each of the vehicles major subsystems as shown in Table 10.2.

Table 10.2
Vehicle subsystem cost breakdown among four subsystems

	A	B
	Subsystem	Cost ($)
1		
2	Wheels and Chassis	33.85
3	Motors and Gearing	41.65
4	Electronics & Batteries	21.45
5	Sensors and Controls	52.62
6		
7	Total	149.57

Using the data listed in Table 10.2, we will use Excel to create a pie chart illustrating the distribution of costs in amount and in percentage among the sub-systems. We will clearly identify the subsystem requiring the largest portion of the $149.57 budget. To begin, click on the Insert tab located on the main ribbon and observe the chart options, illustrated in Fig. 10.4. The options available include Column, Line, Pie, Bar, Area, Scatter and Other. Let's create pie chart by clicking on the Pie icon and from the drop down box select the 3-D pie graph to represent this data.

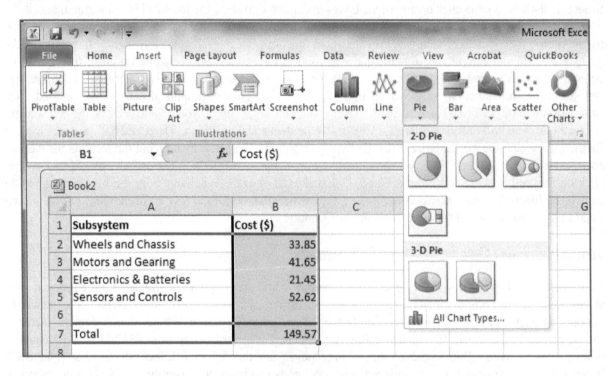

Fig. 10.4 Screen observed under the insert tab with the pie chart selected.

A screen view of this procedure to create a 3-dimensional pie chart is presented in Fig. 10.4. Mark the column of the four subsystems (A2:A5) and the column of the four costs (B2:B5) and click on the icon of the expanded pie chart. From the chart layout box select the version that provides legends to the right and a title at the top. The pie chart shown in Fig. 10.5 appears, which shows the percent of costs for each subsystem. We clicked on the title to activate it and added "Percentage of Costs for the Subsystem". There are several different options for displaying data in a pie graph that are evident when you click on the Chart Tools/Design Tab and locate the Chart Layout area. Make your selection.

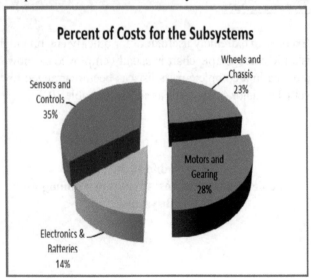

Fig. 10.5 Three-dimensional pie chart showing percentage costs for the vehicle.

The final information that should be added to every graphic you create includes the title of the graph, axis labels, legend labels, etc. To make these additions, switch to the "Layout" tab under the Chart Tools tab. The Labels subsection allows you to add the final information to the graph. Even for the simple pie chart constructed, it is necessary to add a title.

The procedure described above for inserting a pie chart can be used to insert many different types of charts and graphs. However, this chapter will limit the discussion to pie charts, bar charts, and X-Y graphs, but you should explore the other types of charts that can be created using Excel.

Bar Charts

Bar charts can also be used to show distributions, but they are better suited for illustrating comparisons. As an example of a bar chart, consider the tabulation of enrollment in the Mechanical Engineering and the Electrical and Computer Engineering Departments in the A. J. Clark School, University of Maryland at College Park over the period from 2008 to 2013, which is presented in Fig. 10.5. At a glance, you can note that the number of engineering student enrolled in both departments has in this six year period. On a second glance, the reader can observe that the growth in Mechanical Engineering was more rapid than in Electrical and Computer Engineering. The increase in Mechanical Engineering over this period was 68.6% compared to a growth of 35.0% in Electrical and Computer Engineering. Significant amounts of important information can be presented in bar charts, as illustrated in as Fig. 10.6.

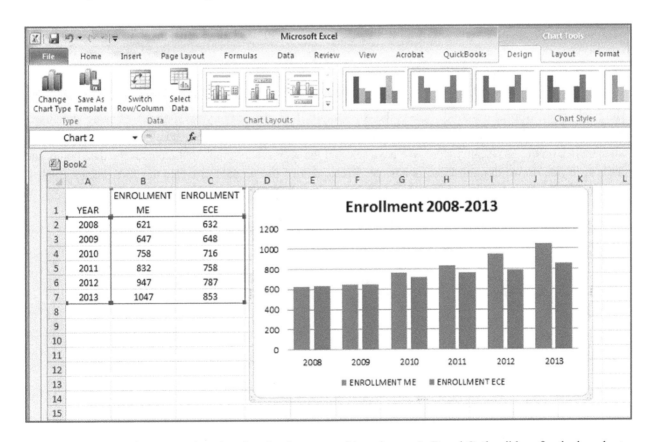

Fig. 10.6 A composite screen shot showing the data entered in columns A, B and C, the ribbon for the bar chart and finally the bar graph.

Examination of the ribbon for the bar chart with the Design tab active shows Chart Styles, Chart Layout and Select Data. The selection of the Chart Style and the Chart Layout are self-evident. However, Select Data is usually necessary in order to properly orient the presentation and to correctly label the axis. Clicking on the Data Source icon brings onto the screen a dialog box for arranging the data onto the two available axes. Inspection of Fig. 10.7 shows that Enrollment ME and Enrollment ECE are the legend entries. The data in columns B and C is used to establish the height of the bars in the Y axis. We clicked in the edit button for the horizontal axis label and marked the sequence of numbers from 2008 to 2013. The year of enrollment appeared along the X axis. Finally we clicked on the title box and added Enrollment 2008-2013 to complete the bar chart. When you begin to understand the many options available in Excel for creating charts, you will be able to create professional graphs and charts quickly.

Fig. 10.7 Selecting the data for placement on the bar chart.

The X-Y Scatter Graph

The final graphic format described here is the X-Y graph. It is inserted in your spreadsheet using techniques that are similar to those used to create pie and bar charts. First, create a table of data. Next, click on the insert tab and click on the X-Y scatter icon. We have selected the X-Y scatter graph that includes the data points and the best fit line drawn through these points. A screen shot of the process to create the X-Y scatter graph is presented in Fig. 10.8.

	A	B	C
1	Year	4 Year Graduation Rate	5 Year Graduation Rate
2	2005	42.7	57.9
3	2006	48.6	62.7
4	2007	51.9	64.4
5	2008	53.8	67.3
6	2009	59.6	

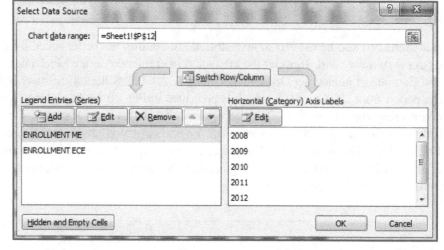

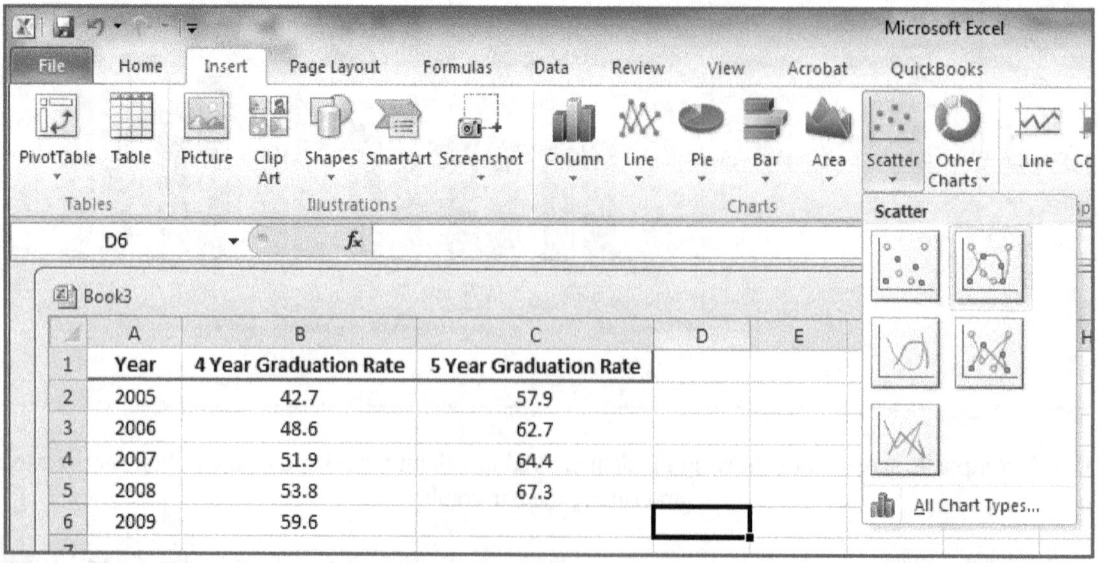

Fig. 10.8 Preparing an X-Y scatter graph showing graduation rates for the period from 2005 to 2009.

After selecting the X-Y scatter graph that includes the data points and the best fit line drawn through these points the Chart Tools screen appears. We select the Chart Type with the Blue and Red data points and lines. We open the Chart Layout icon and select Layout #9. Finally, we label the X and Y axes and titles the graph. The legend was added automatically by the selection of Layout #9. The result of these selections is presented in Fig. 10.9.

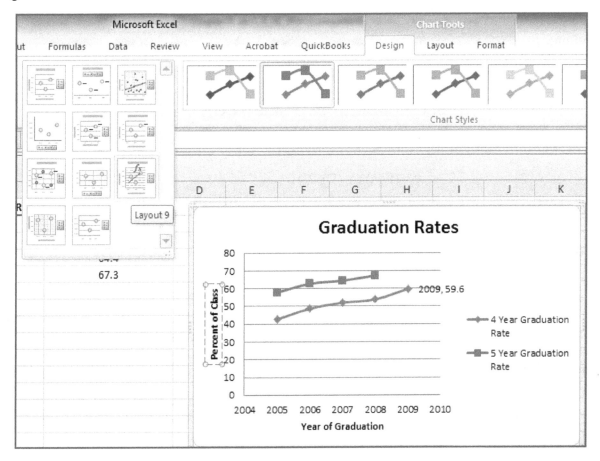

Fig. 10.9 A composite screen shot showing the selection of the chart style, chart layout and the graph showing graduation rate over the period from 2005 to 2009.

Before moving to the next topic, please note the improvement in the graduating rate. This is actual data for the A. J. Clark School of Engineering at the University of Maryland. The improvement is due to the Keystone Program initiated in 2005. The Keystone Program places the instructors with demonstrated teaching skills in the ENES courses that the students take in their first two years. The Keystone Program has created a culture for learning that includes excellent instruction, hands-on experiences and quality computer facilities and laboratories. The author is privileged to be able to contribute to this program by working with the Keystone faculty members to create textbooks for three introductory engineering courses.

10.4 EXCEL AS A DESIGN TOOL

You will soon realize that design is a highly iterative process. This fact is better understood when you consider the number of times you will be required to compute the vehicles weight, the force necessary to achieve a velocity and distance in a specified time and the torque needed to achieve the target values. Let's consider two examples to illustrate the use of a spreadsheet in interactive design.

EXAMPLE 10.1

Determine the net propulsion force required to achieve a velocity of 1.0 ft/s in 10 s and the distance travelled during this time period.

Solution:

Recall Newton's second law and write:

$$F = m\,a = m\,dv/dt \qquad \text{or} \qquad m\,dv = F\,dt \qquad\qquad (a)$$

Integrating the second of Eq. (a) yields:

$$v = (F/m)\,t = (Fg/W)\,t \qquad\qquad (b)$$

where $g = 32.17$ ft/s2 is the gravitational constant and W is the vehicle's weight in lbs.

Integrating Eq. (b) yields:

$$s = v\,t = (Fg/W)\,t2 \qquad\qquad (c)$$

Let's use Excel and explore solution space to acquire a sense of the magnitude of the net force required to achieve the specified velocity of 1.0 ft/s. We will make an assumption about the weight of the vehicle as 2.2 lb. but enter this value into the spreadsheet in such a way that it can be changes with a single cell entry. Our first trial for the solution is presented in the spreadsheet shown in Fig. 10.10.

	A	B	C	D	E	F
3	F = ma = m dv/dt					
4	m dv = F dt					
5	v = (F/m) t = (Fg/W) t					
6	s = v t = (Fg/W)t^2					
7						
8	gravitational constant g	Weight W				
9	32.17	2.2	Net force F (lb)	Fg/W	Velocity v (ft/s)	Distance s (ft)
10			0.1	1.46	7.31	37
11	Time t (Sec)		0.2	2.92	14.62	73
12	5		0.3	4.39	21.93	110
13			0.4	5.85	29.25	146
14			0.5	7.31	36.56	183
15			0.6	8.77	43.87	219
16			0.7	10.24	51.18	256
17			0.8	11.70	58.49	292
18			0.9	13.16	65.80	329
19			1	14.62	73.11	366

Fig. 10.10 Initial entries in the spreadsheet with the weight W = 2.2 lb. and the time t = 5 s.

A quick inspection of the results shows that the range of the net force from 0.1 to 1.0 lb. was much too large. Even with a net force of 0.1 lb., the vehicle achieved a velocity of 7.31 ft/s after only 5 s. Let's change the range for the net force to 0.01 to 0.1 lb., the time to 10 s and observe the results in Fig. 10.11.

8	gravitational constant g	Weight W				
9	32.17	2.2	Net force F (lb)	Fg/W	Velocity v (ft/s)	Distance s (ft)
10			0.01	0.15	1.46	15
11	Time t (Sec)		0.02	0.29	2.92	29
12	10		0.03	0.44	4.39	44
13			0.04	0.58	5.85	58
14			0.05	0.73	7.31	73
15			0.06	0.88	8.77	88
16			0.07	1.02	10.24	102
17			0.08	1.17	11.70	117
18			0.09	1.32	13.16	132
19			0.1	1.46	14.62	146

Fig. 10.11 Second net force entries with the weight W = 2.2 lb. and the time t = 10 s.

Again we note that the velocity is higher than the target value of 1.0 ft/s after 10 seconds with the applied net forces listed in Column C. We perform a final iteration and consider the net force ranging from 0.001 to 0.01 lb. The results, presented in Fig. 10.12, show that a net force of 0.007 lb. applied for 10 s results in a velocity of 1.02 ft/s and a distance travelled by the vehicle of 10.2 ft. The entries in cells A:12 and B:9 are locked in place using the $ before both the column letter and the row number. Hence you can change the weight or the time and observe the results for the velocity and the distance travelled.

8	gravitational constant g	Weight W				
9	32.17	2.2	Net force F (lb)	Fg/W	Velocity v (ft/s)	Distance s (ft)
10			0.001	0.015	0.15	1.5
11	Time t (Sec)		0.002	0.029	0.29	2.9
12	10		0.003	0.044	0.44	4.4
13			0.004	0.058	0.58	5.8
14			0.005	0.073	0.73	7.3
15			0.006	0.088	0.88	8.8
16			0.007	0.102	1.02	10.2
17			0.008	0.117	1.17	11.7
18			0.009	0.132	1.32	13.2
19			0.010	0.146	1.46	14.6

Fig. 10.12 Third set of net force entries with W = 2.2 lb. and t = 10 s yields a valid solution space.

We have solved the example problem, but at this stage we have not interpreted the solution. First we are dealing with very small forces unless you seek high velocities. A net force of 0.01 lb. produces a velocity 1.46 ft/s and travel distance of 14.6 ft. Considering the size of the track area is only about 10 ft by 10 ft, the vehicle will travel from one side to the opposite side in less than 10 s. Is this sufficient time for the steering mechanism on the vehicle to adjust its direction and to control its path? In the missions assigned in this class, you will find that control is more important than speed. Moreover control takes time and speed reduces the time available for adjustments to power and steering. Think about driving down a two-lane, winding road with curves and hills at 100 MPH and you have an analogy.

Finally in this example, we have used the term net force rather than force. There are two forces acting on the vehicle, namely the propulsion force that you control by the design of the drive motors and gearing, and the frictional forces that oppose motion. You have little control over the frictional forces as they depend on the state of the dirt that the vehicle will encounter. In loose sand the frictional force could become larger than the propulsion force and the vehicle will stall. On a stretch of hard surface the friction force will be relatively small and the vehicle will accelerate and run the risk of losing control. These facts imply that you will need to provide the vehicle with a means to quickly adjust the propulsion force to overcome the frictional force but not by a large amount.

EXAMPLE 10.2

You are to explore solution space to determine the torque T applied to the axle of a wheel with a diameter of D in order to generate a single wheel propulsion force of 0.25 N. Assume that the wheel does not slip when the torque is applied.

Solution:

Construct a free body diagram representing the wheel as shown in Fig 10.13.

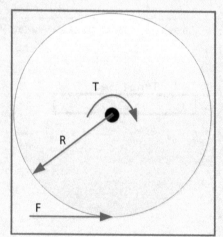

Fig. 10.13 Free body diagram of the wheel showing the application of the torque T at the wheel's center, the radius R of the wheel and the propulsion force F acting on the wheel.

If we write $\sum MO = 0$ and sum the moments about the center point on the wheel, it is evident that:

$$T = F \times R \qquad\qquad (a)$$

Let's use Excel to explore solution space. First open a new book in Excel and type FORCE as the heading in Column A. Below this heading list forces from 0.05 to 0.50 N in increments of O.05 N. Next create Columns

---

B though G with the heading of T (N-mm) to record the output. Check the spreadsheet in Fig 10.14 to note the procedure.

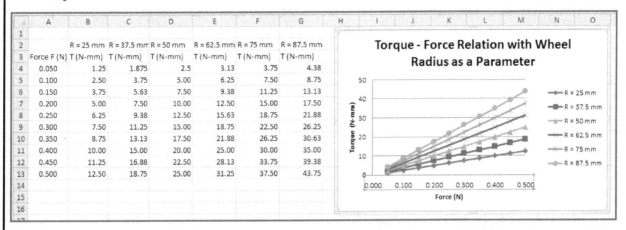

Fig. 10.14 Spreadsheet input data and resulting graph that shows the torque required to generate a propulsion force for different wheel radii. of

On Row 2 we have made entries for the wheel radius that varies from 25 mm to 87.5 mm in increments of 12.5 mm. The results for the required torque T were achieved by multiplying the radius by the force for each of Columns B to G. The graph was constructed by following the procedure already established for X-Y Scatter Charts. As expected the torque required increases linearly with the wheel radius.

For loose sand a large wheel radius is important to lift the chassis high enough to avoid contact with the sand. Using large diameter wheels will necessitate high torque motors, but this should not pose a problem because many gearhead motors are commercially available that are capable of delivering the required torque.

10.5 SUMMARY

Microsoft Excel has been introduced and procedures for creating tables and graphs have been described. Tables represent an excellent method for reporting numerical data. They are well suited for the presentation of precise numerical data, such as the data required in a parts list.

Many different types of graphs are used to aid in visualizing data. Graphs are much less precise than tables, but they show trends, comparisons and distributions much more effectively. The pie chart is used to indicate the distribution of some quantity. The procedure for creating a pie chart using Excel was described in detail. The bar chart, used to compare the magnitudes of two or more quantities, was introduced. The bar chart was demonstrated and the methods used in its construction were discussed.

Trends are best-illustrated using X-Y graphs. X-Y graphs effectively show the trend of one variable with respect to another. The procedure for constructing an X-Y Scatter Graph was described.

Finally, the use of Excel in the design process has been introduced. Two examples were presented to demonstrate the use of Excel when calculating parameters that affect the selection of components for the vehicle. In design of a new system there are many unknowns and valuable insight is gained by exploring solution space. The exploration process is enhanced with Excel where parameters can be changed and the boundaries of solution space modified with relative ease. The Excel spreadsheet greatly aids the iterative design process by providing a means for conducting automated design trade-off studies with numerical and graphical results that are displayed immediately.

EXERCISES

9.1   Using Microsoft Excel, create a parts list for the vehicle your team is developing. Remember to include the manufacturer, model number, description, cost, and weight for each component on your team's vehicle on the parts list.

9.2   Using Microsoft Excel, prepare a table similar to the one presented as Table 9.2 showing the breakdown of weight and cost between the main sub-systems of your vehicle design.

9.3   Use the data from Exercise 10.2 to construct a pie chart showing the distribution of your team's cost budget between each of the main sub-systems.

9.4   Use the data from Exercise 10.2 to construct a pie chart showing the distribution of your vehicle weight between each of the main sub-systems.

9.5   Use the data from Exercise 10.2 to construct a bar chart showing the distribution of your team's cost budget between each of the main sub-systems.

9.6   Use the data from Exercise 10.2 to construct a bar chart showing the distribution of your vehicle weight between each of the main sub-systems.

9.7   Using Excel, create a table to calibrate an ultrasonic proximity sensor. The data should include distance from an object as well as the associated Arduino analog input reading.

9.8   Prepare an X-Y scatter graph showing the data found in Exercise 10.7. What trend can be observed in the graph of the data?

9.9   Fit an appropriate trendline to the X-Y scatter graph created in Exercise 10.7. Include the R^2 value on the graph. Was the trendline you established a good fit of the data acquired?

9.10   Create a sizing calculator for your team's power sub-system that computes the required battery capacity based on the current flow required for each electrical component on your vehicle.

CHAPTER 11

BEGINNING MATLAB® for ENGINEERS

11.1 INTRODUCTION

MATLAB® is a high-level language and programming environment for technical computing. With MATLAB, you can perform basic numerical calculations interactively, visualize results, and write computer programs all within one environment. This versatility and ease of use has made MATLAB a standard tool for engineering computation, both in the university and in industry. MATLAB has many features in common with other computing environments such as IDL®, MathCAD®, and Mathematica®. Because of widespread adoption of MATLAB both by industry and academia and the availability of a large number of powerful toolboxes that extend its basic capabilities, we will describe key concepts for MATLAB in this chapter.

We will show you some of the basic capabilities of MATLAB and help you become comfortable in getting started using MATLAB to solve engineering problems. MATLAB has many advanced capabilities that are not covered in this chapter. Once you become familiar with the key concepts in MATLAB, you can easily build upon your expertise as you encounter more complex problems that MATLAB can solve. References are given at the end of the chapter to help you explore more advanced applications of MATLAB.

11.2 BASIC CONCEPTS IN MATLAB

11.2.1 Basic Data Structure

The basic variable structure in MATLAB is a matrix. In fact, the name MATLAB comes from "matrix laboratory". MATLAB was originally designed to provide an easy-to-use interface to matrix analysis and linear algebra software libraries. Since linear algebra is a fundamental area of mathematics that underlies many engineering fields, the linear algebra capabilities and notation of MATLAB have made it a natural choice of software environment for engineering. MATLAB is used by engineers to process signals, model system behavior, and solve other mathematical problems that involve systems of equations.

When starting MATLAB, you will encounter a prompt in the command window:

```
»
```

You type commands after this prompt followed by the Enter key.

A row vector can be created as follows:

```
» x = [ 4 8 3.2]

x =

    4.0000    8.0000    3.2000
```

By default, MATLAB prints out the results of the assignment. If a semicolon is included at the end of the line, the command is still executed, but the printout of the result will be suppressed.

A column vector can be created by

```
» x = [ 4; 8; 3.2]

x =

    4.0000
    8.0000
    3.2000
```

The semicolons denote the end of a line. A column vector can also be created by defining a row vector and then transposing it:

```
» x = [ 4 8 3.2]'

x =

    4.0000
    8.0000
    3.2000
```

A 3-by-4 matrix can be entered as:

```
» y = [1 2 3; 4 5 6; 7 8 9; 10 11 12]

y =

     1     2     3
     4     5     6
     7     8     9
    10    11    12
```

Complex values can be entered as follows:

```
» f = 4 + 3.5i

f =

    4.0000 + 3.5000i
```

The functions i and j are predefined to be the unit imaginary number. These functions can be used interchangeably. However, if an assignment is made to a variable i or j, the functions with the corresponding names will no longer return their pre-assigned values.

11.2.2 Creating Simple Matrices

Colon notation provides a convenient way of generating incremental sequences:

```
» m = [1:8]

m =

     1     2     3     4     5     6     7     8
```

In general, m = [start:increment:finish] generates a row vector beginning with start, increasing by increment, and ending with finish.

```
» m = [0:0.1:0.5]

m =

     0    0.1000    0.2000    0.3000    0.4000    0.5000
```

We can refer to a range of values in a matrix using colon notation:

```
» y(2:4,1:2)

ans =

     4     5
     7     8
    10    11
```

This prints all values in rows 2 to 4 and columns 1 to 2. If we want to refer to the **entire** range of one of the arguments, we can simply use a colon without a range:

```
» y(2:4,:)

ans =

     4     5     6
     7     8     9
    10    11    12
```

A matrix of zeros can be created with

```
» z = zeros(3,5)

z =

     0     0     0     0     0
     0     0     0     0     0
     0     0     0     0     0
```

Likewise, a matrix of ones can be created with

```
» q = ones(2,6)

q =

    1    1    1    1    1    1
    1    1    1    1    1    1
```

11.2.3 Strings

MATLAB can manipulate strings as well as numerical values. For example, the assignment

```
» fn = 'music.wav'

fn =

music.wav
```

assigns the string "music.wav" to the variable fn. This variable can then be used to refer to that string. String values are always enclosed in single quotes. To read in an audio (WAV) file called "music.wav", use either

```
» z = wavread(fn);
```

or

```
» z = wavread('music.wav');
```

11.2.4 Viewing Variables

Variable values can be viewed in multiple ways. If we want to examine the value of location (3,2) in the matrix y, we simply type

```
» y(3,2)

ans =

     8
```

Notice that the result is stored temporarily in a variable called ans. However, you must omit the semicolon at the end of the line; otherwise, the result will not be displayed.

The default format in which variable values are displayed can be controlled with the format command.

```
» format long
» w = 1/300

w =

   3.333333333333333e+002
```

This displays the value in scientific notation with 15 digits rather than the default 5 digits. The extra line-feeds can be suppressed as follows:

```
» format compact
» w = 1000/3
w =
    3.333333333333333e+002
```

Type "help format" for further options. More sophisticated formatting can be accomplished with the fprintf command. (Type "help fprintf" at the MATLAB prompt for details.)

11.2.5 Interactive Calculations

MATLAB can be used as a calculator. Suppose we want to find the length of a ladder leaning against a wall, where the bottom is 1.5 m from the wall and the top is 3.5 m up the wall. In MATLAB:

```
» z = sqrt(1.5^2 + 3.5^2)

z =

    3.8079
```

If we want to find the height against the wall of a 3 m ladder leaning at a 25-degree angle, we calculate:

```
» 3*sin(2*pi*25/360)

ans =

    1.2679
```

Note that the quantity pi is a predefined function in MATLAB. We can assign calculated values to variables in MATLAB and then use them in further calculations.

11.2.6 File Input and Output

Data can be loaded and saved in various formats. The save command can save one or many MATLAB variables in a single file. For example, the command:

```
» save stuff q w z
```

saves the variables q, w, and z in a file called stuff.mat.

The load command reads files containing numerical data previously saved by a save command in MATLAB. If we clear the MATLAB workspace:

```
» clear
```

all of the variables previously created in MATLAB will be erased. Then we load the stuff.mat file in with the command:

```
» load stuff
```

to input all the previously saved variables. To see the current workspace, use a `whos` command:

```
» whos
  Name        Size            Bytes  Class

  q           2x6                96  double array
  w           1x1                 8  double array
  z           1x1                 8  double array

Grand total is 14 elements using 112 bytes
```

The `load` command can also read text files containing tables of numbers separated by blanks. In this case, you must specify the entire filename, including the extension. For example, if `stuff.txt` contains a table, you load it with

```
» load stuff.txt
```

The result is that MATLAB creates an array variable called `stuff` (the filename without the extension) that contains the data in the table.

11.3 GRAPHICS

MATLAB contains a wide variety of options for displaying and annotating numerical data. We cover some of the more common options here.

11.3.1 Plot

The plot command plots the values in vector y versus the values in x. To plot the sine function shown in Fig. 1:

```
» x = [0:0.05:2];
» y = sin(2*pi*x);
» plot(x,y)
```

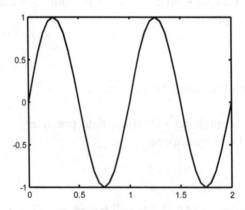

Fig. 1 Graph of a sinusoidal function.

Multiple curves can be displayed on the same graph using different line styles, and axes can be labeled as shown in Fig. 2:

```
» z = 0.5*sign(y);
» plot(x,y,':',x,z,'-')
» xlabel('time (s)')
» ylabel('voltage')
» legend('sine','square')
```

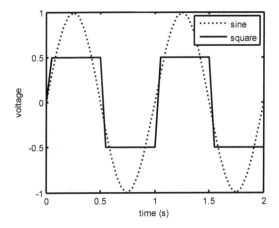

Fig. 2 Sine and sine squared functions are graphed and labeled.

Note that the arguments ' : ' and ' – ' control the line style for the two graphs. Many more options are available for plot. See the MATLAB **help for plot** to obtain more details.

11.3.2 Bar

Bar graphs, as illustrated in Fig. 3, are useful for comparing discrete scenarios. They can be created similarly to plots in MATLAB:

```
» x = [1:4];
» y = [ 3 3 3.5 4; 3.5 3 4.5 5]';
» bar(x,y)
» xlabel('Machine (before/after modification)')
» ylabel('Speed')
```

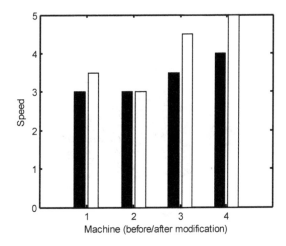

Fig. 3 Bar graph showing speed for various machines before and after modifications.

11.3.3 Stem

Stem (or lollipop) plots are often used to plot sequences, which occur when time signals are sampled at regular intervals. Consider sampling a sine wave having a frequency of 100 Hz (cycles per second) with a sampling period of T = 1/8000 seconds to obtain the graph shown in Fig. 4:

```
» f0 = 100;
» T = 1/8000;
» n = [0:50];
» x = sin(2*pi*f0*n*T);
» stem(x)
```

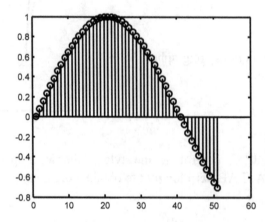

Fig. 4 A stem (or lollipop) plots used to represent sequences that occur when time signals are sampled at regular intervals.

11.3.4 Mesh

The meshgrid function allows us to define a set of grid points in two dimensions for evaluating and plotting functions of two variables as illustrated in Fig. 5. The first output variable (xx in this example) is a 2-D array that varies only in the x direction. The rows of xx are copies of the first input argument. Likewise, the second output variable is a 2-D array that varies only in the y direction, with columns being copies of the second input argument.

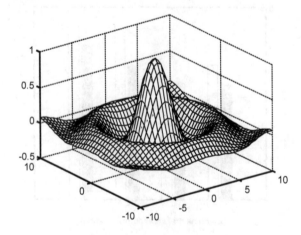

Fig. 5 A three-dimensional graph prepared using the meshgrid function.

The entries for plotting a function of two variables are presented below:

```
» [xx,yy] = meshgrid([-10:0.5:10],[-10:0.5:10]);
» r = sqrt(xx.^2 + yy.^2) + 1e-6;
» f = sin(r)./r;
» mesh(xx,yy,f)
```

The .^ and ./ operations are explained more fully in Sec. 4.

11.3.5 Printing Graphics

Figures in MATLAB can be printed in either of two ways. The first is to select File->Print on the figure menu. This brings up a dialog box that allows you to select specific options for printing. The second is to use the `print` command from the MATLAB prompt. This allows you to print figures from within a MATLAB script and provides greater flexibility than the Print menu. In either case, it is possible to print either to a printer or to a file. Printing to a file allows you to generate a figure that can be incorporated later into a document. You can choose, for example, to print to a TIFF image or to an EPS figure. One can also select Edit->Copy Figure and then paste the figure into a word processor or presentation software.

11.4 LINEAR ALGEBRA

Solving a system of equations is an important part of many engineering disciplines. A system of linear equations can be expressed using matrices and vectors, which can then be solved easily in MATLAB. Consider a plastic part consisting of three substances. We need to determine the volume of each substance needed to make the part. Substances X, Y, and Z are mixed together with a volume of x, y, and z cm^3. The density of the substances is:

Substance	X	Y	Z
Density (g/cm^3)	0.8	1.2	1.3

The overall weight (density times volume) of the part must be 7.15 g. The total volume of the part must be 6 cm^3. Finally, the weight of Substance X within the part must be 0.8 g. This implies the following set of equations:

$$0.8x + 1.2y + 1.3z = 7.15$$
$$x + y + z = 6$$
$$0.8x = 0.8$$

These equations can be written in matrix-vector form as:

$$\begin{bmatrix} 0.8 & 1.2 & 1.3 \\ 1.0 & 1.0 & 1.0 \\ 0.8 & 0.0 & 0.0 \end{bmatrix} \begin{bmatrix} x \\ y \\ z \end{bmatrix} = \begin{bmatrix} 7.15 \\ 6.00 \\ 0.80 \end{bmatrix}$$

In MATLAB, you enter the known matrix and vector quantities as:

```
» A = [0.8 1.2 1.3; 1 1 1; 0.8 0 0];
» b = [7.15 6 0.8]';
```

The solution to this system of equations is given by $x = A^{-1}b$. This can be done in MATLAB by computing the inverse of A directly:

```
» Ainv = inv(A)

Ainv =

         0          0     1.2500
  -10.0000    13.0000    -6.2500
   10.0000   -12.0000     5.0000
```

Then:

```
» x = Ainv*b

x =

     1.0000
     1.5000
     3.5000
```

However, the preferred way to solve system of equations in Matlab is to use the backslash operator:

```
» x = A\b

x =

     1.0000
     1.5000
     3.5000
```

This is preferred because it is much less sensitive to round off errors in the computations used to solve the equation. The vector x in MATLAB contains the solved values for the quantities x, y, and z. Suppose now that we have multiple b vectors for which we want a solution for the same A matrix. Let

```
» B = [7.15 6 0.8; 6.19 5.7 1.6]';
```

This defines two columns for B, the first the same as before and the second equal to [6.19 5.7 1.6]'. Because these two systems share a common A matrix, we can solve the systems for two different solutions simultaneously:

```
» X = A\B

X =

     1.0000     2.0000
     1.5000     2.2000
     3.5000     1.5000
```

The solution to the first system is found in the first column (same as before). The solution using the second b vector is now found in the second column.

MATLAB interprets "*" as a matrix multiply whenever the quantities are matrices or vectors (as in Ainv*b above). An error will result if the column dimension of the left matrix does not match the row dimension of the right matrix. In some cases, it is desirable to multiply two matrices together point-by-point instead of as a matrix multiply. This can be done using the point-wise multiply operator .*:

```
» x = [ 5 2 1];
» r = rand(1,3)

r =

    0.4860    0.8913    0.7621

» x.*r

ans =

    2.4299    1.7826    0.7621
```

The same principle can be used for point-wise division (./) as well as several other operators such as the power function (.^).

11.5 SCRIPTS AND FUNCTIONS

MATLAB allows command sequences to be recorded in a text file as an alternative to typing them in directly. These files can then be run as though the commands were entered at the command line. These command files fall into two basic categories--scripts and functions. These files are created with a .m extension. Because of this, MATLAB script and function files are called m-files.

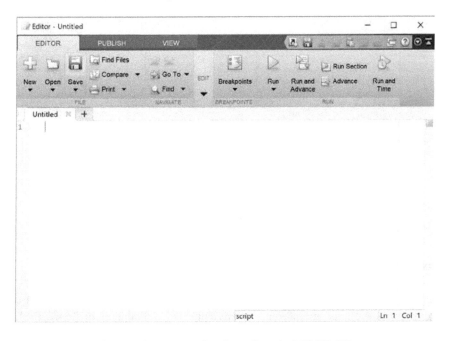

Fig. 6 The screen for the editor in MATLAB.

11.5.1 Scripts

MATLAB commands can be entered into a text file exactly as they would normally be typed at the command line. To run the sequence of commands contained in the script, simply type the base name of the file as a command at the prompt. For example, you can write a script to create a simple sequence and plot the result:

```
» edit
```

This command brings up the MATLAB Editor/Debugger as shown in Fig. 6. You enter the following commands into the editor:

```
n = [-5:5];
x = abs(n);
plot(n,x)
```

We then save this to file plotdemo.m in the current directory. At the MATLAB prompt, we type:

```
» plotdemo
```

this generates the graph shown in Fig. 7.

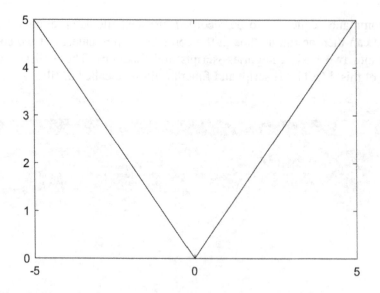

Fig. 7 A graph of the absolute value of n as n varies from −5 to +5.

11.5.2 Functions

New functions can be created in MATLAB with m-files. This means that MATLAB can operate as a programming language as well as an interactive environment. The capabilities of MATLAB can be extended simply by writing and adding m-files to the existing collection. MATLAB provides an extensive set of optional m-file packages called **toolboxes** in a number of different categories, such as signal processing, controls, data acquisition, and symbolic math.

A MATLAB m-file function has a specific form. Consider the m-file `sinewav.m`:

```
function [x,t] = sinewav(f0,per,dur)
%SINEWAV   Sine wave generator
%   [X,T] = SINEWAV(F0,PER,DUR) generates a discrete %   sine wave with
continuous frequency F0 Hz,
%   sampling period PER seconds, and duration DUR
%   seconds
%
%   The sample values are returned in X and the
%   sample times in T.

% this calculates the time sample locations
t = [0:per:dur];
% this computes the signal itself
x = sin(2*pi*f0*t);
```

This m-file illustrates several aspects of MATLAB functions. First, the file begins with the keyword "function". This keyword instructs MATLAB that the file is more than a script. It has an input argument (or arguments) and an output argument (or arguments). Second, the name of the function matches the base portion of the filename. Third, input arguments are passed to the function by separating the arguments with commas and enclosing the argument list in parentheses. Output arguments are separated by commas and enclosed in brackets if there is more than one output argument. Otherwise, the brackets can be omitted. Finally, the function contains comments. MATLAB ignores everything on a line after a "%" appears. Comment lines that appear immediately after the function line are treated as help information. If you create this function and then type "help sinewav", it will print the following statement to the command window:

```
SINEWAV   Sine wave generator
   [X,T] = SINEWAV(F0,PER,DUR) generates a discrete
   sine wave with continuous frequency F0 Hz,
   sampling period PER seconds, and duration DUR
   seconds
```

The function would be called from the command line as:

```
» [zsin,zt] = sinewav(500,44100,10);
```

This would create a sine wave of 500 Hz sampled at 44.1 kHz and lasting for 10 seconds. The samples are stored in `zsin` and the time locations in `zt`. Note that the input and output arguments do not have to have the same names as the variables used inside the function. The function actually makes copies of the input variables for internal use and does not modify the input arguments even if they have the same name as the variables used inside the function. (An exception to this is if a variable is declared to be `global`.)

The special capabilities of MATLAB allow programs (functions) to be written quite compactly compared to other high-level languages. Consider, for instance, the following C program, which also generates a sine wave.

```
#include <stdio.h>
```

```
#include <math.h>

main( argc ,argv)
int argc;
char *argv[];
{
    int i,n;
    double *sv, f0;

    n=5000;
    f0=100;

    sv = (char *) calloc(n, double);
    for (i = 0; i < 50000; i++) {
        sv(i) = sin(2*3.1415927*f0*I/44100);
    }
}
```

As you can see, the C program requires many more lines and is also much more difficult to understand (or debug!). Furthermore, this program does not prompt for input arguments or write to an output file. Adding these capabilities would make the program even longer.

11.6 ENGINEERING TOOLS

MATLAB is capable of solving a large range of engineering problems. In this section, we discuss the capabilities of MATLAB in two areas and summarize other applications.

11.6.1 Differential Equations

Differential equations provide an important mathematical description of many physical systems of interest in engineering. Liquid and airflow, mechanical motion, chemical reactions, and electrical current can all be described by differential equations. Differential equations are simply equations whose terms are **derivatives**. A derivative is a function that describes the rate of change of another function. A simple example of a derivative is a plot of vehicle acceleration as a function of time. Acceleration is the rate of change, or derivative (first derivative with respect to time) of the velocity. Velocity, in turn, is the rate of change, or derivative, of position.

The solution of a differential equation is a function whose derivatives fit the specified differential equation, subject to certain initial and boundary conditions. The analytical solution of a differential equation ordinarily requires a great deal of mathematical knowledge. Most engineering curricula require a course in differential equations, which usually follows several calculus courses. Fortunately, MATLAB can solve a wide variety of differential equation types without requiring a great deal of background on the part of the user. Furthermore, MATLAB can easily give numerical solutions to differential equations that cannot be solved analytically.

Consider the following example. Suppose we press the gas pedal in a car to the floor at a constant rate, taking 3 seconds to get the pedal to the floor. Then we keep the pedal floored for 3 seconds and finally let off for 3 seconds. The acceleration function will look something like the graph presented in Fig. 8.

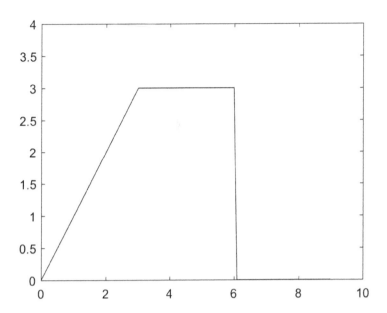

Fig. 8 Acceleration of a car as a function of time.

Knowing how the car accelerates, we wish to find the velocity as a function of time as well as the distance traveled as a function of time. To do this, we must write our own m-file function – called an **odefile** in this context – that describes the relationship among the various functions. (The "ode" stands for **ordinary differential equation**. This term contrasts not with **extraordinary** differential equations but with **partial** differential equations. Partial differential equations describe rates of change of functions of more than one variable.) The input of the odefile function must consist of two arguments – the time t at which the relationship is to be evaluated and a vector y whose elements are the (unknown) values of the functions we wish to find. The output must be a vector dydt the same size as y whose elements are the rates of change (the derivatives) of the corresponding elements of y. For this problem, we call the odefile **myode**, which we construct as follows:

```
function dydt = myode(t,y)

% acceleration sample values as a function of time
a = [[0:0.1:3] 3*ones(1,30) zeros(1,30)];

% time sample locations of "a" function
atimes = [0:0.1:9];

% interpolate "a" function to find value at time t between sample
times
at = interp1(atimes,a,t);

% y(1) is displacement, y(2) is velocity
% dydt(1) is velocity (derivative of displacement)
% dydt(2) is acceleration (derivative of velocity)
dydt = [y(2); at];
```

Once the odefile is saved, we can solve the differential equation for velocity and displacement using ode23:

```
» [T,Y] = ode23('myode',[0 9],[0; 0]);
```

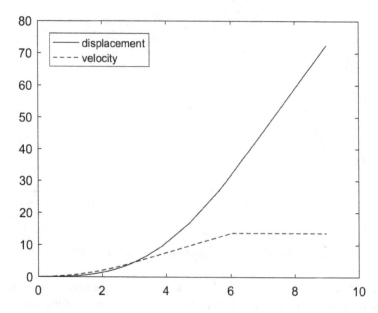

Fig. 8 Displacement and velocity of the car as a function of time.

The first argument is the odefile function name as a string, the second argument is the time range of the requested solution, and the third argument is the initial conditions of the two unknown quantities to be solved. In this case, we assume that the car begins with a displacement of zero and at rest (zero velocity). The numerical solution of the equation is contained in Y. The first column Y(:,1) is the displacement at the times indicated in T, and the second column Y(:,2) is the velocity at the times indicated in T. The displacement (solid line) and velocity (dotted line) are plotted below:

Many other options are available with ode23; quite a few other tools for solving ordinary differential equations are also included in MATLAB. See odeset to review the available tools.

11.6.2 Audio

Many engineering problems require that certain quantities be tracked as a function of time. Examples are chemical concentration levels in a manufacturing process, vibrations of a mechanical structure, and voltage levels in a cell phone receiver. A signal sampled every T seconds can be represented as a sequence of numbers, which can be stored as a vector in MATLAB.

MATLAB provides a variety of tools for manipulating signals. For instance, filters can be defined to smooth out noise (random disturbances) in the data using the command — filter. Approximate derivatives (rates of change as a function of time) can be calculated using diff. The frequency components of a signal can be separated using fft. The details on how to implement and interpret these operations are covered in later courses.

One of the more common signal types of interest both in engineering and in consumer applications is audio. MATLAB is capable of reading, writing, and playing standard audio formats. The most popular of these is the Microsoft WAVE or WAV format (audio files with a .wav extension). To read a standard audio file (such as WAV) into MATLAB, use the following command:

```
» d = audioread('d.wav');
```

A graph of an audio signal of the sound "dee" is shown in Fig. 9. This sound was sampled at 22,050 Hz (hertz, or cycles/second) and is displayed as a function of its index value.

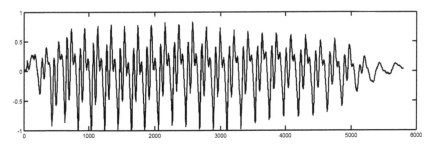

Fig. 9 A graph of an audio signal of the sound "dee."

Signals can also be saved to standard audio file formats. You can create a pure tone of 200 Hz with a duration of 5 seconds and a sampling period of 1/8000 seconds as follows:

```
» f0 = 200;
» nT = [0:1/8000:5];
» x = sin(2*pi*f0*nT);
```

This signal can then be saved to a WAV file:

```
» wavwrite(x,8000,'tone.wav');
```

Note that the signal, the sampling frequency, and the file name must be specified. Other combinations of options are also possible. See the wavwrite help for details.

Signals can also be played as audio from within MATLAB. To play the pure tone generated above:

```
» sound(x,8000);
```

This sends the sound to the sound card on your PC.

The sound function assumes that the vector x to be played has values between −1 and +1. Values outside that range are truncated. The soundsc function has the same functionality as sound, but it first scales the values of x so that they are between −1 and +1 and then plays the result.

Stereo audio (separate left and right signals) can be played on systems that support stereo sound by creating signals having two columns. The left column plays in the left speaker, and the right column in the right.

```
» nT = [0:1/8000:5];
» fl = 200;
» fr = 251;
» x = sin(2*pi*fl*nT);
» y = sin(2*pi*fr*nT);
```

```
» sound([x' y'],8000)
```

11.7 USING MORE OF MATLAB'S CAPABILITIES

MATLAB has a number of important capabilities that have not been covered here. **Simulink** is a companion system to MATLAB that allows the user to model and design dynamic systems in block-diagram form using a graphical interface. These systems can be built and modified interactively and can simulate a wide variety of systems.

The *Symbolic Toolbox* provides the capability of entering mathematical expressions as strings and returning solutions or simplified expressions. The Symbolic Toolbox is actually a front end to Maple®, a symbolic math program that provides analytical solutions to math problems. The Symbolic Toolbox can solve many integrals, derivatives, and differential equations.

Extensive documentation is available, both at the command line and through the MATLAB Help Desk. The Help Desk can be accessed by typing "helpdesk" at the MATLAB prompt. From there you can find documentation on every aspect of MATLAB. A good, detailed introduction to MATLAB is the "Getting Started" link. It expands on many of the concepts found in this chapter.

Other products, such as a number of application-specific toolboxes, a C compiler, C and C++ math libraries are all available from The Math Works. A number of third-party products that operate with MATLAB are also available from other vendors.

EXERCISES

11.1. Create a 9×5 matrix of ones called A. Then create a 3×3 matrix of ones from A using only range notation.

11.2. Create and display a 4×5 matrix A with rand() and then display a 2×3 block consisting of the first two rows and the last three columns.

11.3. Using a single MATLAB expression, calculate the average weight for the heights 155 cm, 179 cm, and 184 cm using the formula weight (kg) = 0.895*[height (cm) – 155] + 48.

11.4. Plot weight as a function of height from the previous exercise and include a legend and x and y labels.

11.5. Write a MATLAB function that returns a square wave (a waveform that switches between +1 and – 1) with a specified period in samples and specified length in samples. (One period is the total number of samples the waveform is at + 1 and then – 1 before starting over.) Assume that there are an equal number of positive and negative samples in each period.

11.6. Solve the following equations in MATLAB: $x + 3y = 2$; $3x + 2y = – 5$.

11.7. A car is traveling at 30 m/s initially. Brakes are applied for 5 seconds beginning at time t=0 to create a deceleration of 2 m/s^2. (Deceleration is negative acceleration.) Plot the velocity for 10 seconds beginning with t = 0.

11.8. Mary Jo contributed $25 to her team's design project. Junior contributed $20, Bubba contributed $35, and Jethro $22. Using a single MATLAB expression, find the percentage of the total contributed by Bubba.

11.9. Create a 5-second sine wave whose frequency is 500 Hz sampled at 44.1 kHz. Create a second waveform that tapers the amplitude down from 1 to 0 in 5 seconds in a straight line.

CHAPTER 12

BASIC ELECTRIC CIRCUITS

12.1 INTRODUCTION

This chapter introduces you to the basic quantities, concepts, and physical devices used in electric circuits. The discussion begins with definitions of charge, current, voltage, and other basic quantities used to describe circuit characteristics. Then, key passive components are introduced, including resistors, capacitors, inductors and switches. Next, the equations needed to solve circuit problems are provided. A basic description of important electronic (semiconductor) devices, such as diodes, transistors, voltage regulators and operational amplifiers, follows. The chapter concludes with a number of circuit examples that have application to the robot challenges.

12.2 BASIC ELECTRICAL QUANTITIES AND CONCEPTS

Atoms have a positively charged nucleus surrounded by orbiting negatively charged electrons – akin to a miniature solar system. When atoms come together to form a solid, some electrons can get detached from their respective nuclei and become free to move through the material. It is the motion of these electrons that forms the basis of electricity. In certain materials, called semiconductors, the motion of the electrons is influenced by the motion of the other electrons, so it appears as if a missing electron is moving in the opposite direction to the rest of the electrons. The moving charge appears to have become positive, and we call this a "hole."

Electricity can be understood most simply by drawing an analogy between charge flowing through a circuit and water flowing through a pipe. The analogy is not perfect, but it provides a great deal of insight into the basic operation of electricity.

12.2.1 Charge

The fundamental unit of charge is the charge on a proton, e, which has a value of 1.6×10^{-19} coulombs (C). An electron has charge $-e$. The total amount of charge in a region is denoted q(t), which is measured in coulombs. It may change with time if charge flows into or out of a region. . We can think of charge as a substance (a large number of "free" electrons) that is contained within an object such as a wire. This is analogous to a hose full of water. When charge moves through a material we call this a current. . If there is an excess of positive or negative charge in a region, these charges can exert forces on other charges even if they do not move, and we call this effect electrostatics.

12.2.2 Current

Current, denoted as I(t), represents charge in motion. The amount of current flow is measured in amperes (A), which is the same as coulombs per second. Just as we speak of the rate at which a volume of water moves through a pipe, we can think of current as a flow rate. The higher the current, the higher the rate at which charge is moving past a point in a conductor. This concept is expressed in equation form as:

$$I(t) = dq/dt \qquad (12.1)$$

In other words, the current in a specific wire is the (instantaneous) rate of movement of charge passing through that wire per unit time.

Regardless of the fact that negatively charged electrons are the carriers of current in wires, it is customary to think of the current in a circuit as the flow of positive charges. All that one needs to remember is that the effect of electrons moving in one direction is equivalent to the flow of positive charges of identical magnitude flowing in the opposite direction.

12.2.3 Voltage

Voltage, denoted V(t), is another fundamental electrical quantity. The unit of voltage is the volt (V). Voltage is always measured between two points and represents how much energy a unit charge will gain or lose as it travels between those two points. As such, it is the electrical equivalent of pressure. Just as a pump creates pressure to force water through a pipe, a voltage source creates pressure to push charge through a circuit.

An ideal voltage source is a circuit element that delivers a specified voltage (electrical pressure) regardless of the current flowing. The most basic *real* voltage source is a battery. A flashlight battery pushes charge through a tungsten wire that comprises the incandescent light bulb filament, creating a kind of friction within the filament, and causing the bulb to heat up and glow. A 6-V battery pushes four times as hard as a 1.5-V battery. The symbol for a constant voltage source (called a **direct current**, or dc, source) such as a battery is shown in Fig. 12.1a. A more general voltage source symbol is shown in Fig. 12.1b. A positive charge that enters the negative terminal of the battery (in Fig. 12.1a) leaves the positive terminal after it gains energy proportional to V_1. A positive charge traveling in the other direction gives up V_1 volts.

Sometimes we refer to voltage as the potential difference between two points. Just as the water pressure is reduced on either side of a kink in the hose that restricts flow, the voltage drops across any part of the circuit that presents a resistance to current. This is a key concept that will be explored more in Section 12.5.

Fig. 12.1 (a) dc voltage source symbol
(b) General voltage source symbol.

12.2.4 Electric Energy and Power

Electric energy, denoted w(t), describes the potential of a circuit or component to do work. The units are the same as for all types of energy: Joules (J). Some components, like batteries, convert other forms of energy to electric energy, thereby "generating" electric energy. Other components, like capacitors and inductors, can take in and store electric energy, to be released at some later time, but they cannot generate energy on their own. Still other components, like resistors, motors, and light emitting diodes (LEDs) can only dissipate electric energy, which is to say that they convert electric energy to other forms of energy, such as heat, kinetic energy, or light. Any component that does not generate electric power is called a "passive" component, in contrast to the sources that are represented in Fig. 12.1.

Electric Power, denoted P(t), is the instantaneous rate of change of energy in a given object:

$$P(t) = dw/dt \qquad (12.2)$$

As such, electric power describes the rate that electric energy flows in a circuit, i.e. the rate that energy enters or leaves circuit components. The units for power are joules/second or watts (W). The instantaneous power in any given component is equal to the current entering that component times the voltage drop that current experiences before it exits the component. This statement may be expressed as an equation:

$$P(t) = [V(t)] [I(t)] \qquad (12.3)$$

Because of this relationship, electrical engineers almost always concentrate on finding voltages and currents in a circuit before they look to evaluate other physical quantities.

12.2.5 Assembling the Basic Circuit Equations

There are two sets of equations that are used to solve for voltages and currents in a given circuit. The first set of equations, discussed in Section 12.5, is related to the way that electrical components are connected to each other in a particular circuit (i.e. the topology of the circuit) – we call these equations Kirchhoff's Laws, after their "discoverer." These equations are independent of which particular component is placed at a particular location in the circuit. The other type of equations is dependent upon the specific components, because those equations describe the intrinsic interrelation between voltage and current in a particular component. These equations are called terminal relationships and will be discussed in Section 12.4, after we describe the materials used to make electric and electronic components.

12.3 MATERIALS FOR ELECTRIC AND ELECTRONIC DEVICES

12.3.1 Conductors

Conductors are materials that have very many "free" electrons and therefore allow a lot of current to flow in them when a voltage difference is applied across the conductor. To quantify what we mean by "a lot", we describe the conductivity of any material by a quantity called its conductivity, measured in siemens (S). We use the Greek letter σ to represent conductivity. Copper is a standard material used to make wire, and the conductivity of copper is 58 **million** siemens—that is what we mean by a lot of current. If we have a wire of cross sectional area A and length l, then if a voltage V is applied across this wire the current that will flow is

$$I = \frac{\sigma A V}{l}, \qquad (12.4)$$

From this equation we can see that the siemen can be written in terms of fundamental units as S=AV⁻¹ m.

Remember that all of this current in conductors is realized by the movement of electrons – negative charges.

Conductors for the most part are metals, and typical conductors used in electric circuits include copper and aluminum, silver and gold, lead and tin, etc. Most wire is made of copper, because it is a very good, relatively cheap conductor. Aluminum is not as good, but it is cheaper and lighter. It is harder to make good permanent connections between aluminum wires, and that is why, when you walk into an electronics store, you will typically find only copper wire. Silver is the best conductor, but more expensive than copper, and also tarnishes in the atmosphere. Gold, though very expensive, is good for making contacts. Tin and lead have low melting points. Mixed together in an alloy they form solder, which is often used to make permanent connections between wires and components.

The key to why conductors work as they do is wrapped up in the meaning of "free" electrons. Conductors, like all materials, are made up of individual atoms that are normally charge-neutral, which means that they have the same number of electrons as protons. In a metal the outer electrons of the atoms become detached from individual atoms and float freely through the surrounding matrix of nuclei and bound electrons.

The metal remains overall charge neutral but the mobile electrons can move and carry large currents. When a battery drives current through a circuit, electrons actually leave the negative terminal of the battery, flow through the circuit, and then return into the positive terminal of the battery

12.3.2 Insulators

Insulators are materials that contain very few free electrons at room temperature. Examples of insulators used in electric circuits are glasses, plastics, ceramics, and rubber. The conductivity of glass, for example, is about twenty orders of magnitude smaller than the conductivity of copper. Insulators are used in electric circuits to protect people and circuit connections by covering wires and other components. They are also used inside some components, particularly in capacitors, as will be discussed in the next section.

12.3.3 Semiconductors

Semiconductors, as the name suggests, have conductivities that fall between those of conductors and insulators. Typical semiconductor materials include the elements silicon and germanium, as well as compounds like gallium-arsenide, cadmium-sulfide, etc. The intrinsic room-temperature conductivity of pure silicon is about 10 millisiemen, or in other words 9-10 orders of magnitude smaller than copper. For some applications pure silicon is adequate, but by far, most electronic components are made from doped semiconductors— semiconductors in which impurities are deliberately introduced to adjust the properties of the material. The impurities generally come in two classes. If the impurity has more electrons in its outer valence band than the semiconductor material (a standard example is phosphorous implanted in silicon), the resultant material has an excess of free electrons and is called an n-type semiconductor. Hence, electrons carry the bulk of the current in n-type materials. On the other hand, if the impurity has fewer electrons in its outer valence band than the semiconductor material (e.g. boron implanted in silicon), the resultant material has shortage of electrons, or more accurately, an excess of free "holes" and is called a p-type semiconductor. So, in contrast, positively-charged "holes" carry the bulk of the current in p-type materials. Whether an n-type or p-type material, the conductivity increases with increasing number of impurities, allowing more and more current to flow for a given applied voltage. The impurity concentrations in a semiconductor device depend on the application and the required properties (i.e. voltage, current, power, etc.) Devices based on n-type, p-type, and combinations of these two materials are described in Section 12.6 below.

12.4 BASIC ELECTRIC COMPONENTS

12.4.1 Terminals and Reference Directions

Every electric component has two or more places where it can be connected to other components in a circuit. These connection points are called "terminals" and are almost always made of metal. The terminals of each component are connected to the terminals of other components to construct the circuit. Permanent connections are often made via solder, which are metals that melt at low temperatures; temporary connections are often made with "wire nuts" or "circuit breadboards." In modern surface-mount printed circuit boards (PCBs) these terminals are small metal pads, but in most other circuits the terminals are metal wires that extend from the ends of the component. For this reason, we draw a generic two-terminal "passive" component as in Fig. 12.2. The terminal wires are represented by the two lines that extend from the rectangular box, which in turn represents the passive component. Note that the terminal connections for voltage sources were represented by the lines (wires) in Fig. 12.1.

Neither current nor voltage is a simple number. Current refers to a direction of charge flow, so in addition to the current value, we need to indicate the direction of flow, as we have done in Fig. 12.2 with the arrow. Voltage represents a potential energy gain between two different points in a circuit, and so we indicate the point of lower energy with a minus sign and the point of higher voltage with a plus sign (as shown in Fig.

12.2. These voltage symbols are the same as those used for voltage sources. Together, the current direction and voltage polarity are known as reference directions. There are two important concepts to remember about reference directions. First, you usually need to assign reference directions to components in a circuit before you actually can solve for voltages and currents. You normally do this by guessing in which direction a current will flow, etc. If you guess wrong, don't worry. All that will happen is that when you solve for that current, for example, is that the value you calculate will be negative. A negative value for a current or voltage is just fine — it simply means that the current really flows in the opposite direction or that the point of lower voltage corresponds to the plus terminal. Second, when you calculate power, a positive power means that energy is flowing into a device **only if** the current and voltage references are coordinated as shown in Fig. 12.2. In other words, the current is flowing into the positive voltage terminal of the device. This coordination is called the passive sign convention, because it is the standard orientation for passive components. In most passive circuit elements, the current flows from regions of high potential to low potential, just as water falls from higher elevations to lower elevations. Whether or not you chose to adopt the passive sign convention is not important. What is important is that you understand the reference direction scheme you choose, so that you understand when energy is entering components and when it's going out of components.

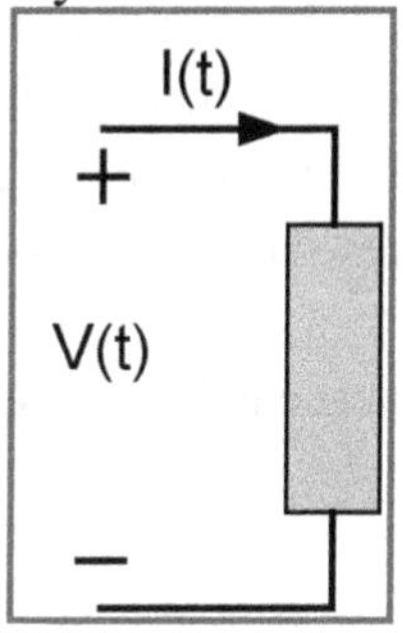

Fig. 12.2 A generic two terminal device showing reference directions.

12.4.2 Resistors

Resistors are passive components that convert electrical power into heat. Resistors are characterized by their resistance, which measures the difficulty of pushing current from one place to another. The unit of resistance is ohms (Ω), which is equal to volts/amps (V/A). The higher the resistance of a device, the more difficult it is to push current through it. Pumping water through a small-diameter pipe requires more pressure to achieve the same flow rate as pumping through a larger-diameter pipe. A high-resistance electrical device is analogous to a small-diameter pipe, resisting more strongly the flow of charge. Likewise, a low-resistance device is analogous to a large-diameter pipe that offers little resistance to flow.

An ideal resistor is a circuit element that has a fixed resistance regardless of the input voltage. The symbol for a resistor is shown in Fig. 12.3, where the resistor is connected to a battery to make a simple circuit. The value of the resistance completes the definition of a resistor and must always be shown on the side of the resistor symbol. The connecting lines between the voltage source terminals and the resistor terminals represent wires, which can be thought of as resistors with nearly zero resistance. In our water analogy, wires are extremely large pipes that offer little resistance to water flow.

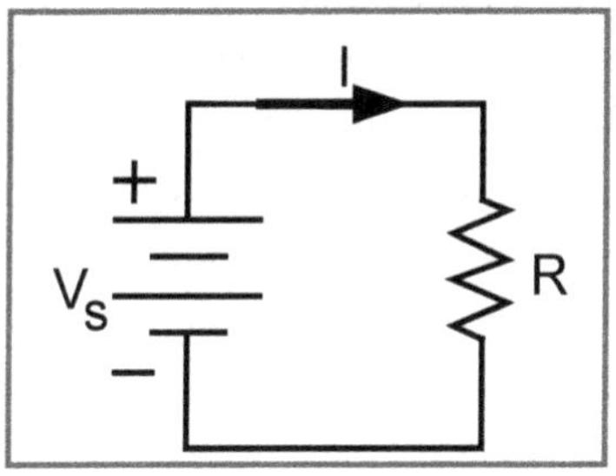

Fig. 12.3 A simple circuit loop containing a dc voltage source and a resistor.

The relation between the applied voltage V_s, the resistance R, and the current flow I is given by:

$$V_s = I\,R \hspace{3cm} (12.5a)$$

$$I = V_s / R \qquad\qquad (12.5b)$$

Equations (12.5a) and (12.5b) are both expressions of Ohm's law. When V_s is expressed in volts and I in amperes, R is given in ohms (Ω).

The resistance R of a conductor is given by:

$$R = \rho L / A \qquad\qquad (12.6)$$

where ρ is the specific resistivity of the metal conductor (Ω-cm); L is the length of the conductor (cm); A is the cross-sectional area of the conductor (cm^2).

Note that the resistivity is the reciprocal of the conductivity, σ, defined in the previous section. We can fabricate resistors by winding very-small-diameter, high-resistance wire on coils, by mixing carbon powder with insulating "filler" material, or by evaporating thin films of metal on ceramic substrates. The relation in Eq. (12.6) allows us to select the metal or carbon percentage, and to size the conductor to provide a specified resistance. Potentiometers are mechanically-tunable variable resistors.

For a resistor connected to a battery (with a constant output voltage) the electric power dissipated as heat can be re-written from Eq. 12.3 as:

$$P = V_d I \qquad\qquad (12.7)$$

where P is the power dissipated and V_d is the voltage drop across the resistor.

Substituting Eqs. (12.5a) and (12.5b) into Eq. (12.7), we obtain useful expressions for power loss in a resistor:

$$P = I^2 R = V_d^2 / R \qquad\qquad (12.8)$$

We have used the subscript "s" with the voltage in Eqs. (12.5a) and (12.5b) and the subscript "d" with the voltage in Eq. (12.7). For the simple circuit shown in Fig 12.3, the supply voltage V_s is equal to the voltage drop across the resistor V_d. In this case, the subscripts can be interchanged. Clearly, it is easy for us to determine the power lost if we know the current flowing through a resistor or the voltage drop across the resistor.

Many applications, perhaps including your robot, use resistors that are cylindrical in shape. You'll note that the size of the resistor has nothing to do with the resistance of the component—the same package size can be used from single digit resistances up to millions of ohms. The package size, instead, is related to the maximum power that the resistor can dissipate without overheating. The smallest packages can only handle ¼ W or less, so it is important to know both the exact resistance and power requirements before purchasing components. One last important item to note is that the resistivity increases (normally) with temperature, so the resistance might change with time in a real circuit, even if the power level is below the maximum allowed. Sometimes resistors that are designed to dissipate a lot of power require physical contact with a heat sink – usually a finned metal block with high heat conductance and heat capacity to keep the temperature to an acceptable level. Sometimes the heat sink needs to be cooled with a fan. (A good example of this is the cooling system for the microprocessor in your desktop computer.)

Note that resistors are not fabricated perfectly, nor are they fabricated with any value that you want. Carbon composition resistors typically have tolerances of ±5% or ±10%. Metal film resistors can have tolerances of ±1% (or less). Resistance values are spaced so that given the tolerances, theoretically all values of resistors could be made. In reality, some resistance values are "less standard" than others and therefore harder to find, but for most applications, obtaining the required resistors is not a difficult problem.

12.4.3 Capacitors

Capacitors are passive components that can store electric energy. They are characterized by their capacitance, which is a bit more difficult to understand than the previous quantities. However, it also has a water-related analogy. It is used to characterize devices that have a degree of "springiness" – devices in which charge can be pumped in but which tend to push charge back when the external pressure is removed. The ease with which charge can be pushed into the device measures its capacity. The unit of capacitance is the farad (F).

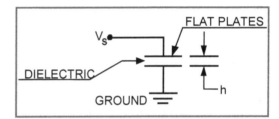

Fig. 12.4 A flat parallel plate capacitor.

An ideal capacitor is a passive electrical component that has a fixed capacitance, regardless of the amount of voltage across its terminals. The circuit symbol for a capacitor is shown in Fig. 12.4. The capacitor ordinarily consists of two flat, parallel-plate electrodes separated by a dielectric, which also serves as an electrical insulator. The approximate capacitance C of a flat-plate capacitor is given in units of picofarad and is determined by:

$$C = \frac{\varepsilon_0 \varepsilon_r A}{d} \qquad (12.9)$$

where ε_r is the relative dielectric constant of the material between the plates, A is the area of the plates, d is the distance between the two plates as shown in Fig. 12.4, and ε_0 is the permittivity of free space (8.854×10^{-12} Fm^{-1}).

When a voltage V is applied across the capacitor, it stores a charge q that is given by:

$$q(t) = C\, V(t) \qquad (12.10)$$

where C is the capacitance in farads and q is the charge on the capacitor given in coulombs.

When a voltage is first applied across the terminals of a capacitor, current flows and it becomes charged. However, when the capacitor is fully charged and maintained at a constant voltage, no current flows through it. The capacitor conducts current only when the voltage across it changes with respect to time. The idea is expressed mathematically as the terminal relationship for a capacitor:

$$I(t) = C\, dV/dt \qquad (12.11)$$

The water analogy for a capacitor is a balloon or membrane stretched across the cross-section of a pipe to block the flow of water. As pressure is applied to attempt to pump water through the pipe, the balloon begins to stretch, which allows a small amount of water to flow. As the balloon stretches, it resists more and more until the backpressure of the balloon resisting the flow equals the forward pressure of the pump. At that point, the water ceases to move.

The electrical energy stored in a capacitor [w(t)] is determined from Eqs. (12.2), (12.3) and (12.11):

$$w(t) = C\,[V(t)]^2\,/2 \qquad\qquad (12.12)$$

where w(t) is the stored energy in joules when C is expressed in Farads.

Similar to resistors, the capacitance is not the only parameter that you need to know before you purchase a capacitor. For capacitors, you also need to know the maximum voltage that can be applied to it.

The shapes, sizes, and construction materials of capacitors vary quite a bit. The size increases both with increasing capacitance and with increasing "working' voltage, i.e. the maximum allowable voltage. Some very high capacitances (tens or even hundreds of microfarads) can be found in small packages. Be careful – these capacitors are probably polarized, which means that one terminal must always be kept at a lower voltage than the other terminal –they **will be destroyed** if even a small voltage is applied with the wrong polarity. The positive terminal is usually marked with a plus sign and sometimes the negative terminal is marked with a minus sign. Capacitor tolerances are generally ±10% or ±20%, but sometimes the tolerances are even wider.

In addition to storing energy, capacitors have another property that gives them an additional important use. Capacitors tend to "absorb" short "spikes" of electrical energy. Without a capacitor, a small spike in energy can cause short-lived, but large, voltage spikes in a circuit that can destroy electronic components. With a capacitor, the energy is "diverted" to the capacitor and the voltage spikes can be greatly reduced (the larger the capacitance, the greater the reduction, of course). For this reason, capacitors are the principle element in the surge protector through which you connect your computer to protect against voltage spikes.

12.4.4 Inductors

Inductors are passive components that can store magnetic energy. They are characterized by their inductance, which relates the flux of magnetic field they generate to the current passing through the inductor. The symbol for the inductor is shown in Fig. 12.5. The inductance of an inductor is hard to calculate in general, but is well-defined when you buy commercial units. One way to produce an inductor is to wind a coil of small-diameter, insulated wire about a magnetic (e.g. iron or ferrite) core. The inductance L is given by:

$$L = \mu N^2 A / l \qquad\qquad (12.13)$$

where   L is the inductance in Henrys (H)
       N is the number of turns of wire in the coil
       μ is the permeability of the magnetic core material
       A is the cross sectional area of the coil
       l is the length of the coil

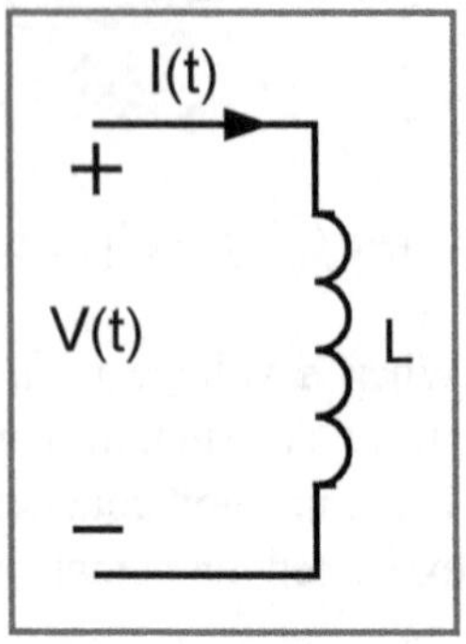

Fig. 12.5 The symbol for an inductor.

When a step voltage is first applied across an inductor, it acts like an open circuit and does not permit current to flow through it. Gradually, current flow through the inductor begins to increase until the inductor offers no resistance at all. In a sense, the inductor has an internal inertia with respect to current. The inertia must be overcome for current to flow. When current begins to flow, the inertia of the inductor tends to keep the current flowing even when the external voltage source is removed. The inductance L is a measure of this electrical inertia. This concept is summarized mathematically as the terminal relationship for an inductor:

$$V(t) = L \, di/dt \qquad (12.14)$$

The water analogy for an inductor is a water wheel in a channel through which water flows. At first, the inertia of the stationary wheel prevents the water from moving through the channel and past the wheel. As the water pushes on the wheel, the wheel begins to turn and speed up, allowing more and more water to pass. Eventually, the wheel is turning at the rate the water would flow without the presence of the wheel. If the water pressure on the upstream of the wheel is turned off, the rotational inertia of the wheel keeps it turning. It pulls water from the upstream side and pushes it downstream. With time, the wheel begins to slow because there is no external pressure, until the wheel and the water flow both cease.

The magnetic energy stored in an inductor (w(t)) is determined from Eqs. (12.2), (12.3) and (12.14):

$$w(t) = L \, [I(t)]^2 /2 \qquad (12.15)$$

where w(t) is the stored energy in joules when L is expressed in henrys.

Similar to resistors, the inductance is not the only parameter that you need to know before you purchase an inductor. For inductors, it is the maximum current that you also need to know. The magnetic core of an inductor is a very nonlinear material, which means that the permeability, μ, is not necessarily a constant, and hence L is not necessarily a constant. The permeability is usually nearly constant for small magnetic fields, and then decreases as the field increases. For an inductor, this means that the inductance is nearly constant for small currents, but then L decreases as the current increases. Companies define differently the maximum allowable current through an inductor, so make sure that you read the product specification carefully. Large inductances usually require lots of wire to make many coil turns, and sometimes, to keep the inductor small, this means that very small wire is used. This wire may have a significant resistance (see the following section), and so this resistance is a third parameter that you need to consider when selecting an inductor. In general, the larger the inductor, the greater the weight, the smaller the resistance and the larger the maximum allowed current. Inductor tolerances generally range from ±10% to ±20%, but higher and lower values exist.

It is unlikely that you will need to deliberately add an inductor to any circuits in your robot, but be aware that motors generally involve coils of wire through which current flows, so they have inductive properties.

12.4.5 Wire

Wire comes in many sizes, shapes and materials. Most wires come surrounded by an insulating sleeve that can withstand the application of a few hundred volts without being destroyed. They are not usually heat resistant and can be accidentally damaged during soldering.

The most common wire is round (circular cross-section) copper wire. This wire comes in many sizes and two configurations: solid and stranded. Stranded wire is more flexible but there are some connections that are easier to make with solid wire. Either style is probably fine for the robot application. The sizes are dictated by the American Wire Gauge number, or AWG # the larger the number, the smaller the diameter of the wire and the larger its resistance per unit length. Each size has a maximum suggested current, above which heating and related problems may occur if the current is applied continuously. Sample wire sizes that may be of use for your applications are given in Table 12.1.

Table 12.1.
A selection of standard wire sizes and properties. 1000 mils = 1 inch

AWG #	Wire diameter (mils)	Weight (oz/foot)	Resistance (mΩ/foot)	Max continuous current (A)
14	65.1	0.20	2.5	15
18	40.3	0.08	6.4	5.8
22	25.4	0.03	16.1	2.3
26	15.9	0.012	40.8	0.92
30	10.0	0.0055	103.2	0.36

12.4.6 Switches

A switch is a device that is used to connect various parts of an assembled circuit together. Simple switches are two terminal passive components which have two states: open and closed. The symbol for an open switch is shown in Fig. 12.6 (a). No current flows between the two terminals. The symbol for a closed switch is shown in Fig. 12.6 (b). Ideally, there is no voltage across the two terminals. There are many different styles of switches. Here we will talk briefly about manual switches, which are useful for connecting batteries to the circuits that they are designed to power. Electronic switches, such as relays and transistors, are discussed in later sections.

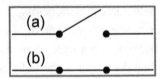

Fig. 12.6 The symbol for a switch: (a) open and (b) closed.

Switches are characterized in terms of how much current they can carry, how many volts they can withstand (when open), how many poles and positions they have, and whether the switch action is permanent or momentary. In a momentary switch, only one of the states is "stable". For example, in a normally-open (NO) switch, the switch is open. When you push the switch button, the switch closes. When you release the button, the switch returns to the open state. There are also normally-closed (NC) switches. Power switches normally are not momentary and the switch action toggles between the two states. Some devices are connected to more than one switch, so that one mechanical action (pushing the button) simultaneously changes the state of several switches. This may be useful, for example, if one needs to connect two different batteries to a circuit at the same time. The number of switches connected in a device is referred to as the number of poles. Sometimes switches have more than two terminals. The most common example of this is the "double throw" switch, which has three terminals. There is a central terminal, and in one state, the central terminal and one other terminal form a closed switch while the central terminal with the third terminal form an open switch. When the switch state is toggled, the roles of the two pairs of switches are reversed. Switches can normally withstand hundreds of volts when in the open state. Maximum currents can be only a few amps or less, so this parameter may be important to consider for the robot application.

12.4.7 Batteries

Batteries are portable devices that convert some form of energy (typically chemical) into electrical energy. Ideal batteries produce a constant potential difference between their two terminals irrespective of the amount of current that passes through the battery and irrespective of the past history and use of the battery. Here, we list and briefly discuss some important electrical properties common to all types of real batteries. **For real batteries:**

(a) **The output voltage decreases with increasing battery current.** A good approximate model for a real battery is shown in Fig. 12. 7. The real battery model has an ideal battery connected to a resistor, called the internal resistance of the battery. This resistor will cause the real voltage to decrease as the current increases. Recall that resistors generate heat – this is why batteries become hot if you try to drain too much current.

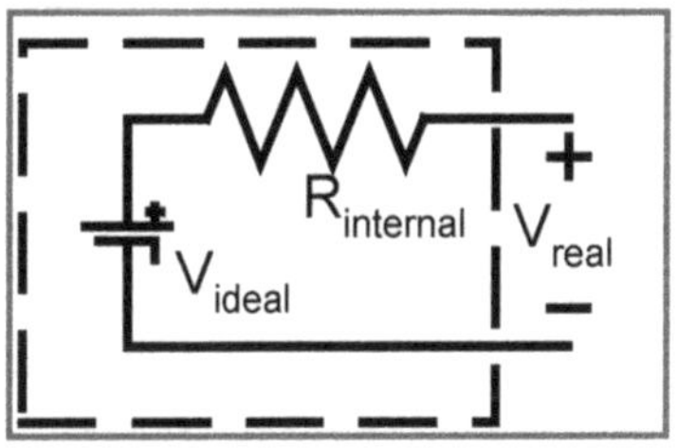

Fig. 12.7 Real battery model

(b) **The output voltage decreases with time during continuous use.** The model for a real battery in Fig. 12.7 is quite simplified. In reality, the ideal (zero current) voltage and the internal resistance of the real battery depend on the current state of the battery. As the battery energy is drained, the output voltage slowly decreases due to the nature of the physical processes occurring in the battery. As the battery becomes nearly completely drained, the output voltage begins to drop off rapidly.

(c) **There is a maximum discharge current that should not be exceeded.** Drawing more current from the battery than it is designed to allow can cause permanent damage related to excess heat. In extreme cases, fire or other potentially dangerous conditions can occur. Some batteries are protected from excessive current, but all batteries should be selected to have performance capabilities that exceed the circuit specifications. The batteries should also always be handled with care, and accidental shorting of the battery terminals should always be avoided.

(d) **The amount of usable stored energy decreases with increased output current.** In reality, the internal losses in a battery are not proportional to the current, so the total amount of energy that can be extracted from a battery decreases with increasing current, so that the battery lifetime decreases more rapidly than corresponding increases in current. This effect is strong in some batteries and weaker in others, so check your particular battery's specifications.

(e) **The amount of usable stored energy decreases with age.** Batteries lose energy just sitting around, without any connections to an electric circuit. Primary batteries, or non-rechargeable batteries, usually have expiration dates labeled on the package as a consequence of this fact. Secondary batteries or rechargeable batteries have a finite lifetime in terms of the number of charging cycles they can handle. The re-charging process does not completely reverse the discharging process, due to damage and other real considerations. This means that advertised energy capacities given for batteries are the maximum that you can expect, and the max stored energy will decrease with time. The best way to minimize this effect is to follow all directions and restrictions with respect to the charge and discharge cycles.

(f) **The recharging cycle depends on the battery chemistry.** Never put a secondary battery in a charger that was not designed to handle it. Never put a primary battery in a charger.

(g) **Problems may occur when the battery is nearly discharged.** Some batteries will experience damage if discharged completely; others work better if completely discharged before a recharge cycle is initiated. If multiple batteries are used in the same circuit, care must be taken that all batteries discharge at about the same rate, or at least that no battery is allowed to remain in the circuit when discharged.

(h) **Electrical combinations of different battery types might lead to problems.** If different types of batteries are used in the same circuit, the danger of having discharged batteries in the circuit increases. If unavoidable, it is important to make sure that none of the specifications are exceeded for any of the batteries and that batteries are replaced or recharged well before they lose all of their stored energy.

12.4.8 Motors

Electric motors are devices that convert electric energy into kinetic energy (and heat, since they are not 100% efficient) via magnetic forces. Normally, the torque required to spin the motor comes from the interaction of a fixed permanent magnet and a rotating electromagnet (i.e. an inductor). The spinning electromagnet has to change the direction of its current periodically so that the torque is always in the same direction. If this change is done mechanically, the motor can be run backwards by reversing the applied voltage; if it is done electronically, the motor might not be reversible.

If it sounds like the electrical properties of a motor are complicated, they are, but there are only a few facts that you need to understand in order to successfully use your motor(s). First, when running at a constant speed, the motors can be modeled as a resistor, with the resistance given by the applied voltage divided by the average current needed. Usually, there is a minimum voltage needed to start a motor and a maximum voltage that will work before it burns out. You can work anywhere between those two limits and adjust the motor's speed by changing the voltage. Also, motors usually require more current to start up than they do to run at their normal operating speed. Finally, the current-switching in motors usually results in electrical "noise", which manifests itself as a high-frequency voltage variation on the motor terminals. These variations can be very large, particularly at moments when power is being disconnected from the motor. Because one main method of controlling motor speed is to periodically "pulse" the voltage applied to the motor, electrical motor noise may be something that you have to reduce.

Figure 12.8 shows oscilloscope (time) traces related to a pulsed-control of the motor speed. The quantitative details of these traces aren't important, and so only a qualitative description is given here. The independent axis represents time. The lower curve shows the voltage applied to the motor. The voltage is normally zero volts, but periodically there are brief 7V pulses. Note that the speed is controlled by changing the width of this pulse — hence the name of this technique: Pulse-width modulation, or PWM.
The upper curve in Fig. 12.8 shows the voltage of a switch that controls the motor. The important thing to note is that when the voltage is removed from the motor, there is always a very large, short voltage spike.

This sharp voltage transient is related to the fact that a typical motor has electrical characteristics that resemble an inductor. This should not be surprising, because the primary electromechanical components of a motor are coils of wire used to produce a magnetic field, similar to what is found in inductors. Recall that in an inductor, the voltage and current are related to one another by the equation: $V = L\, dI/dt$. When the current supplied to an inductor is abruptly stopped (such as when a switch is opened, or a transistor switch is turned off), the current I must immediately become zero. Hence, for this instant in time, we may assume that dI/dt becomes infinite, leading to a very large voltage transient, such as those seen in Fig. 12.8. For this example, the spike is at least seven times larger in voltage than the normal operating voltage, and this spike can damage circuit components if not suppressed.

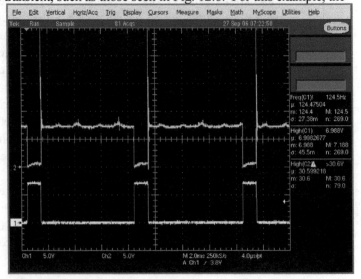

Fig. 12.8. Oscilloscopes traces demonstrating important noise spikes from a PWM-controlled motor.

12.4.9 Relays

Relays are switches that are activated by supplying current to an electromagnet. Relays typically vary from single-pole, single throw (SPST) switches to triple-pole (or even quadruple-pole) double throw (3PDT or 4PDT)) switches. While the electromagnet that controls the switch location(s) is certainly an inductor, the applied voltage is usually a constant (DC – direct current), so it is the resistance of the coil winding that is most important and therefore the quantity normally provided in the specifications. These electromagnets can be energized by a range of voltages; with 5V and 12V being two standard nominal relay voltages. Still, relays have a range of acceptable voltages. There is a minimum voltage, above which the relay is guaranteed to work. There is a maximum voltage above which the relay will overheat and burn out. Usually both values are given in the specifications, so you can judge before buying if the relay will meet your needs. Relays take a finite amount of time, after the electromagnet is energized, before the switch contacts have come to rest in their new locations. This time should also be specified, but normally it is on the order of a few milliseconds or a few tens of milliseconds. While this seems pretty fast to many people, it normally means that relays are good for turning motors on and off, but not very useful for PWM speed control.

12.5 KIRCHHOFF'S LAWS

12.5.1 Topology: Nodes and Loops

The terminal relations given above (Eqs. 12.5, 12.11 and 12.12) provide one-half of the equations needed to solve for all voltages and currents everywhere in a circuit. The other half of the equations comes from Kirchhoff's Laws – rules derived from the connections made between components in a circuit. These rules come from very simple physical principles, yet it is useful to state a few definitions first.

A node is a connection in which two or more components are electrically joined together, either by solder or some non-permanent means. A trivial node has only two components connected. Most people think of nodes as points, but actually nodes in a circuit drawing often look like wires, since they are a means to connect two points together without any voltage drop. Examples of nodes are given in the next section. A loop is any closed path in a circuit that passes through two or more components. If the closed path does not physically enclose any other components (that are not part of the loop), then the loop is also called a mesh. A trivial mesh contains exactly two components. Examples of meshes are given in the next section.

12.5.2 Kirchhoff's Voltage Law

We introduced Ohm's law with a simple circuit loop containing a voltage source and a resistor as shown in Fig. 12.2. This is a good beginning; however, what happens when you begin to insert additional components in the circuit? You will need to develop a few more tools useful in analyzing slightly more complex circuits. Let's begin by adding one additional resistor to the simple circuit loop as shown in Fig. 12.9a. We then determine the relation for the current flowing in this slightly more complex circuit. To approach this problem, we need to use **Kirchhoff's voltage law (KVL),** which is derived from the concept of conservation of energy. This law states:

KVL: The algebraic sum of the voltage drops around any circuit loop is zero.

To understand how to apply this rule, we need to understand the key terms "algebraic" and voltage "drop." Before we can do anything, we must make sure that the reference directions for all components are marked on the loop and that the passive sign convention is used for all passive components. KVL implies that to find the relationship between voltages in a loop, we must "travel" around the loop, summing up the voltages that we encounter along the way. The direction of travel doesn't matter, but to make it simpler to compare answers, let's always agree to travel clockwise around a loop. While traveling around the loop, if we first encounter the positive reference terminal, pass through the component, and exit the negative terminal, we have experienced a voltage drop. If the opposite is true – we enter the negative terminal and leave the positive terminal, we have experienced a voltage rise. The "algebraic sum" requires that a plus sign is placed in front of the voltage drops and a minus sign is placed in front of the voltage rises.

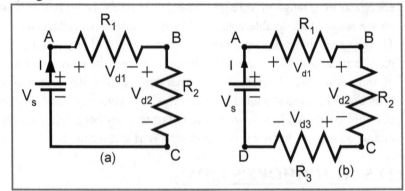

Fig. 12.9 Circuits containing two and three resistors.

Let's use the system above to apply KVL to the mesh in Fig. 12.9a. Let's start at the lower-left corner of the mesh (the only corner not identified by a letter). We go clockwise, so we first pass through the voltage source, entering the negative terminal and leaving the positive one. This means that we have experienced a voltage rise, and we put a minus sign in front of V_s. We next pass through the first resistor, R_1, experiencing a voltage drop along the way. We also see a voltage drop passing through R_2, so both voltages V_{d1} and V_{d2} have plus signs in front. Finally, the wire on the bottom has no voltage drop across it, so we return to the starting point and set the voltages summed up algebraically along the way to zero, to write the following equation:

$$V_s + V_{d1} + V_{d2} = 0 \tag{12.16}$$

Note that this circuit has only one mesh, and so there is only one KVL equation that we can write. It turns out, that no matter how complicated the circuit, you only need to write one KVL equation for every mesh in the circuit… writing a KVL equation for any loop that is not a mesh would not give you any new information.

Note also that this circuit has three nodes – three places where parts are connected together. We have labeled the nodes A, B, and C. Finally, note that node C extends all the way across the bottom of the circuit to connect R_2 to the negative terminal of the voltage source.

As a final note, we have only labeled one current in the circuit, even though there are three components and technically we need three reference currents. Since all of these nodes are trivial, the same current, I, passes through all of the components. Components connected by trivial nodes are said to be in *series*. Using Ohm's Law for the two resistors, we can replace the resistor voltage drops with the current, I, times the resistance, and then solve for the current through the resistors:

$$V_s - I\,R_1 - IR_2 = 0 \tag{12.17}$$

$$I = V_s /(R_1 + R_2) \tag{12.18}$$

If we apply the same analysis techniques to the circuit with the three resistors shown in Fig. 12.9b, it is easy to add another term to the relation shown above and write:

$$V_s - I\,R_1 - IR_2 - IR_3 = 0 \tag{12.19}$$

$$I = V_s / (R_1 + R_2 + R_3) \qquad (12.20)$$

Examine Equations (12.18) and (12.20), and observe that the resistors in a series arrangement in the circuit add together to act like a single resistor with an equivalent resistance R_e given by:

$$R_e = R_1 + R_2 + R_3 = \sum R \qquad (12.21)$$

Equation (12.20) leads to a well-known resistor rule for resistors connected in a series arrangement. The effective resistance of several resistances in series is the sum of the individual resistances. The equivalent resistance is used to simplify the circuit diagram by replacing the three-resistor circuit with a single equivalent resistor as shown in Fig. 12.10.

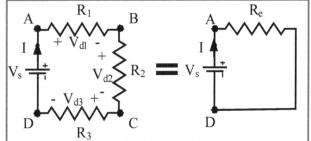

Fig. 12.10 An equivalent circuit with an equivalent resistor replacing a more complex circuit.

12.5.3 Kirchhoff's Current Law

What is the effect if the resistors are not all connected in a series arrangement? Another simple configuration occurs when we have a trivial mesh. The two components in a trivial mesh are said to be in **parallel**. What if we have a parallel arrangement of the resistors such as shown in Figs. 12.11a or 12.11b? To analyze this type of a circuit, we need to introduce **Kirchhoff's current law (KCL)**. This simple law, which comes from the physical concept of charge conservation, states:

KCL: The sum of all currents flowing into a node must be equal to the sum of all currents flowing out of that node.

Let's try KCL out for the circuit in Fig. 12.11a. There are two nodes in the circuit: node A and node B. Node A goes along the entire top of the circuit and the voltage source and both resistors are connected to the node. All three components are also connected to the bottom node. The current through the first resistor, I_1, for example, flows out of node A and into node B. It is easy to see for this example that there will be only one unique KCL equation that we can write for this circuit. Even for more complex circuits, one can show that you need to find one KCL equation less than there are nodes, if you want to find all voltages and currents in the circuit. The current I is flowing into node A, and I_1 and I_2 are flowing out, so an application of KCL for this node yields:

$$I = I_1 + I_2 \qquad (12.22)$$

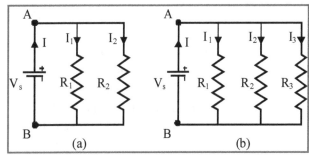

Fig. 12.11 Circuits with parallel arrangements of resistors.

Likewise, KCL for node B in the circuit in Fig. 12.11b yields:

$$I_1 + I_2 + I_3 = I \tag{12.23}$$

KVL, applied to a trivial node, reveals that two components in parallel have the same potential difference across their terminals. Since all resistors in Fig. 12.11 are in parallel with the voltage source, they all have the potential difference of V_s across their terminals. Applying Ohm's Law (Eq. 12.4b) to the resistors in the circuits of Fig. 12.11 yields:

$$I_1 = V_s / R_1, \quad I_2 = V_s / R_2 \quad \text{and} \quad I_3 = V_s / R_3 \tag{12.24}$$

Following the form of Eq. (12.23), we can write:

$$I = V_s / R_e \tag{12.25}$$

where R_e is some equivalent resistor that replaces the two parallel resistors in Fig. 12.11a (or the three resistors in Fig. 12.11b to give the same equivalent circuit as shown in Fig. 12.12.

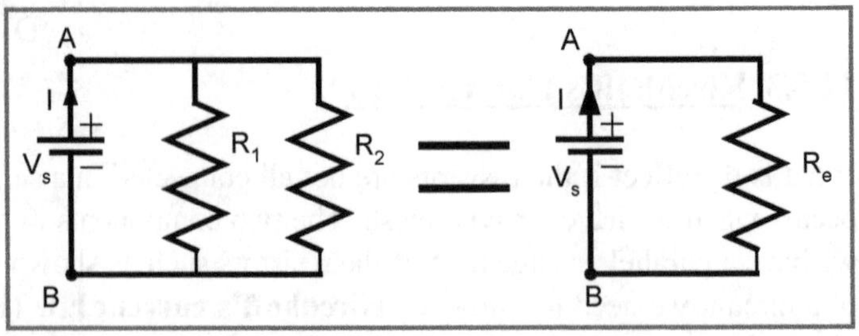

Fig. 12.12 An equivalent circuit contains an equivalent resistor that replaces the two parallel resistors.

Let's substitute Eqs. (12.23) and (12.24) into Eq. (12.21) to obtain:

$$V_s / R_e = V_s [(1/R_1) + (1/R_2)] \tag{12.26}$$

Eliminating V_s from Eq. (12.25) gives:

$$1/ R_e = (1/R_1) + (1/R_2) \tag{12.27a}$$

or

$$R_e = (R_1 R_2) / (R_1 + R_2) \tag{12.27b}$$

If we find three resistors in parallel as in Fig. 12.11b, we follow the same procedure in the analysis of the circuit and show that the equivalent resistance for the three parallel resistors is:

$$1/R_e = (1/R_1) + (1/R_2) + (1/R_3) \tag{12.28}$$

While resistors connected in series were summed to give the equivalent resistance, resistors in parallel follow a different rule. For parallel resistors, the reciprocals are summed and set equal to the reciprocal of the equivalent resistance. **This fact implies that the equivalent resistance for a parallel arrangement of resistors is less than that of the smallest individual resistance.** The conductance, G, is defined to be the reciprocal of the resistance: $G = 1/R$. Given this definition, for parallel resistors, the conductance of the equivalent resistor is equal to the sum of the conductances of the parallel resistors.

12.6 BASIC ELECTRONIC COMPONENTS

12.6.1 Diodes

The diode is another important circuit element. It is constructed by combining a p-type semiconductor material with an n-type semiconductor. While no particular electrical quantity is associated with diodes, they act in some respects like an automatic current switch or a check valve. A diode allows current to flow freely in one direction (ideally, just like a wire) but does not permit current to flow (ideally) in the opposite direction (we call this reverse-biased). Current can flow from the p-region to the n-region but not vice-versa. In a sense, a diode is the electrical equivalent of a check valve in a pipe carrying water. The symbol for a diode is shown in Fig. 12.13. The direction of the arrow indicates the direction in which the diode allows current to flow. We will discuss a common use of diodes later in the chapter.

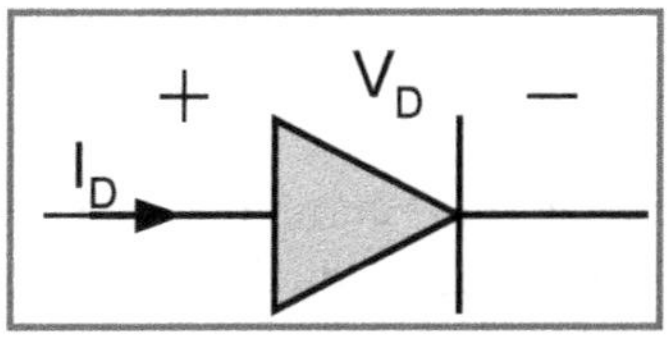

Fig. 12.13 Circuit symbol for a diode.

Real diodes have a voltage drop (V_D) across their terminals when current is passing through the diode. This voltage depends on the particular diode and the exact current (the current increases exponentially with increasing voltage), but it is rarely less than ½ volt and rarely more than 4 V. There is also a limit in terms of how much current can flow through the diode (I_D), as the voltage drop results in the diode dissipating power in the form of heat. For example, if the maximum diode power is 1 W, and the voltage drop is V_D=0.5 V, then the current must be kept below I_D=2 A.

Real diodes also have a maximum negative voltage that can be applied to it. If V_D becomes more negative than this maximum, which is usually tens of volts if not hundreds of volts, current will begin to flow in the opposite direction quickly destroying the diode. This condition is called voltage breakdown. There are some special diodes that are constructed to survive breakdown. The most common diode of this type is the Zener diode, which exploits this property to regulate voltages in a circuit. Real diodes do have a small reverse current when V_D is negative, but below the breakdown threshold. Normally this current is no more than a few micro-amps, and has little impact on most circuits.

Diodes are our first example of a non-linear device. Resistors, capacitors, inductors, and voltage sources are all linear devices. One important consequence of a linear circuit is that if you change the sign of every voltage source (from plus to minus and vice-versa), you can write the solution for all voltages and currents in the "new" circuit just by flipping the signs of those values in the original solution. Another consequence is that if all sources are constant, or vary at the same frequency, **ALL** voltages and currents in the circuit have the same time dependence as the sources. A diode is clearly not linear: you cannot simply change the direction of the current through the diode. As non-linear components, diodes are very useful for converting constant dc circuits to time-varying circuits and vice-versa.

12.6.2 LEDs

LED stands for light-emitting diode, and the name describes well the function of the device. LEDs are found in many commercial devices including laser printers, televisions, and of course, the glowing status lights that are commonplace in consumer electronics. They are becoming more and more common in household lighting applications. Newer high-brightness LEDs are now used in headlights and flashlights. Handheld remote controls use inexpensive infrared LEDs to transmit signals to remote appliances. While infrared LEDs are invisible to the human eye, the built-in camera found on most cellular phones and laptops can easily show infrared signals as a white glow. LEDs are diodes, so current flows only in one direction, but not all of the electric power loss goes into heat – some of it is converted into light. The diodes are usually packaged in transparent or translucent plastic packages to allow the light to illuminate the surrounding area. Their forward diode voltages are typically $V_D > 1.3$ V. The required current is on the order of tens of milliamps. LED performance varies quite a bit, not only with respect to color and brightness, but also with respect to the angular coverage of the light. Viewing angles range from under 10° to over 180°. Thus, some thought should go into exactly what you need the LED to do, before you go searching through the thousands of different models that can be bought online to find the one that's right for your project. Once you have LEDs on hand, they should be tested under realistic operating conditions to make sure that they will perform their intended tasks.

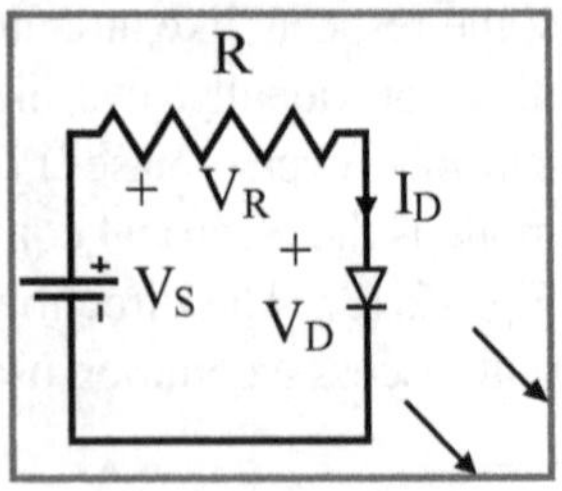

Figure 12.14 A simple LED circuit

Because the current I_D increases exponentially with V_D, some care must be taken when designing the circuit to insure that the diodes don't burn out from too much current. A very simple way to achieve this is shown in Fig. 12.15. The series resistor's value must be selected to obtain the correct LED current, I_D. Using KVL for the only loop in the circuit, we can find an expression for this resistance:

$$R = (V_S - V_D) / I_D \tag{12.29}$$

For a 9.6 V power supply and a diode voltage of $V_D = 2.2V$, for example, if the desired current is 20 mA, the required resistance is:

$$R = (9.6 - 2.2) / 0.02 = 370 \ \Omega \tag{12.30}$$

A 370 Ω resistor is not standard, but either the closest carbon-composition resistance, 360 Ω, or the nearest metal-film resistance, 374 Ω, would be adequate.

12.6.3 Photodiodes and Photoresistors

A photodiode is a semiconductor device that does the opposite of an LED – it converts light into electric energy. A photoresistor is another semiconductor device – often Cadmium-Sulfide or CdS – which acts like a variable resistance resistor. The variation is achieved by changing the amount of light that shines on the device. In Section 12.3, we mentioned that the number of free electrons in a material can depend on the ambient temperature and on the number of impurities. Well, for some materials, like CdS, it also depends on the amount of light, because photons of light can give up their energy to electrons, thereby freeing them from their atomic nuclei. Having more free electrons means more current for the same applied voltage; hence, the resistance of a photoresistor decreases with increasing light. How this property can be exploited in a detector for the robot challenge is described in more detail later in this chapter.

12.6.4 Transistors

Transistors are semiconductor devices that can be used either as amplifiers or as high-speed electronic switches. Unlike the previous two-terminal components that we have discussed up to this point, transistors have three terminals. One of these is used to control current flow between the other two terminals. Bipolar-Junction Transistors (BJTs) use current in the control terminal (called the base) while Field-Effect transistors (FETs) use a potential (voltage) difference between the control terminal (called the gate) and one of the other terminals to control the current flow. Transistors are another example of a non-linear device. An important note for both types of transistors is that current is designed to go in only one direction and voltage is designed to be held off in only one direction. Thus, if you want to use transistors to run a motor backwards, you need more than one transistor to do the job. One possibility is the H-Bridge circuit described later in this chapter.

We recommend the use of field effect transistors in your robotic challenge, because they work as simple voltage controlled switches.

12.6.5 MOSFETS

Although the bipolar junction transistor (BJT) was historically invented sooner , in today's electronic appliances, BJTs have been largely replaced with field-effect transistors, the most common type of which is called a MOSFET. The term MOSFET stands for "metal oxide semiconductor field effect transistor." While the term may seem complicated (and indeed, many use the term MOSFET without ever knowing what it means!) it actually describes concisely both how the device is made and what it does. First: as the name implies, a MOSFET is comprised of a thin oxide region that is sandwiched between a metal (or polysilicon) electrode and the semiconductor underneath. This metal portion is called the gate, because it is used to control the conductivity of the channel underneath, in much the same way that a flood gate regulates the flow of current in a river. When a voltage is applied to the gate, it produces an electric field in the semiconductor region beneath it, similar to the electric field that is produced between the plates of a capacitor. This field causes the electrons (or holes) in the semiconductor to move, creating a conductive path between the two other contacts, called the **source** and the **drain**. Thus, the term "field effect transistor" (FET) thus describes the essential nature of the switching process: an electric field induced by the gate effects the state of conductivity between the remaining two electrodes. In this way, a MOSFET acts as a voltage-controlled switch.

The symbol for an enhancement-mode, n-channel MOSFET is shown in Fig. 12.17, where the MOSFET is inserted into a simple switching circuit. The gate controls the main current flow, which is from the drain to the source. Although it might seem strange that the "source" is on the negative side of the transistor, this is because it is where electrons enter the device and flow towards the "drain." As the gate voltage is increased, the drain current I_D increases and the voltage V_{DS} decreases, Eventually V_{GS} is increased to the point where V_{DS} approaches zero and the MOSFET is saturated. As long as the V_{GS} required for saturation is less than the microcontroller output voltage, one simply needs to hook the gate and source directly to the microcontroller output in order to switch the transistor. **(NOTE: it may be necessary to put a resistor in parallel with the microcontroller output for the switch to work properly, but this requirement is due to the microcontroller properties, not due to MOSFET performance characteristics. It is also advisable to put a resistor, typically around 370 ohm, between the output of an Arduino microcontroller and the gate to protect the Arduino from damage if an accidental short is produced between gate and ground.)**

MOSFET transistors make up the majority of electronic circuits found in computer chips and other integrated circuits. The term "CMOS" stands for complementary metal oxide semiconductor, and describes a technology for integration of n- and p-type MOSFETs on a single semiconductor chip. MOSFETs are also widely used in power electronic circuits, such as power supplies and high-current switches. Power MOSFETs operate according to the same principle as the miniature transistors found in integrated circuits, but the geometry of a power MOSFET is specifically tailored to allow the device to handle much higher currents and voltages required for power applications. As such, power MOSFETs are typically packaged as discrete, three-terminal devices.

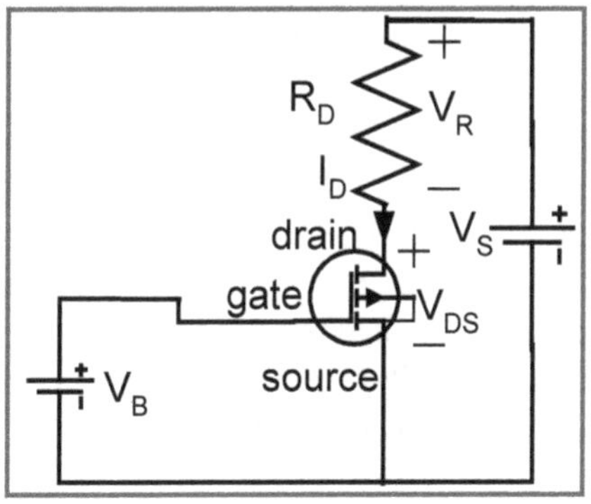

Fig. 12.17 A simple MOSFET circuit

Another possible advantage of the MOSFET is that they usually have smaller voltage drops from the drain to the source, compared to their BJT counterparts, so they act as better switches, with lower power loss and less heat generation. They are more sensitive to damage from electrostatic shocks, but if handled carefully, they are quite robust.

When buying MOSFETs, in addition to making sure it is the right type with adequate gate voltage properties, it must also have sufficient ability to hold off voltage in the off state and sufficient current capability when in the on state. It is important to check that with the available gate voltage (usually 5V from an Arduino) that the MOSFET turns fully on so that the Drain-source voltage drops to a low value.

12.6.6 Transistor Packages and Heat Sinks

An important consideration when buying discrete transistors of either type (BJT or MOSFET) is the package or form-factor of the device. Most transistors are manufactured in a variety of different form-factors, all of which carry the same basic part number. Some form factors are designed for use in surface-mount printed circuit board applications. These devices do not have large, easily accessible leads, but instead have metallic pads or fine-pitched 'gull-wing' leads that are meant to be soldered directly to a corresponding pattern of metallic pads on a printed circuit board. While such devices are economical and ideal for automated manufacturing applications, they are difficult to manually handle, and nearly impossible to incorporate into prototype circuits. Even if you could find a way to make electrical connection to the short, stubby, gull-wing leads of a surface-mount transistor, many rely on intimate contact to the PCB substrate for thermal heat sinking.

Other form-factors are specifically designed for 'through-hole' printed circuit board applications, which indicate that the transistor has protruding leads that can be inserted into a socket, or through holes drilled into a printed circuit board. These form factors are much more suitable for prototyping, because they can be used in solderless breadboards or connected directly with wires. Many of the useful packages with leads have a package name that starts with TO – for Transistor Outline Package. The TO-92 package is very standard for low-power transistors. A better package for high power applications is the TO-220 package. This package comes in styles with two to seven connections (so the three connections needed for a transistor is just fine). They have a metal backing and a hole that can be used to attach the transistor to a heat sink – a metal block (usually with fins) that can be used to draw heat out of the transistor and decrease the operating temperature. Usually a paste-like heat sink compound is applied between the heat sink and the transistor to improve the heat flow. With a heat sink, this package can often dissipate up to 50 W without problem (though often becoming quite hot). There are many other styles of housings as well; the important considerations to reflect on are (1) if it will have the power capability you need and (2) if it will be compatible with your method for assembling your circuit, in terms of its size, connections, and heat sink requirements.

12.6.7 Voltage Regulators

Previously we mentioned that battery voltages are anything but constant. When you buy a 12V battery pack, the voltage could be anywhere from about 11V to 14V and the battery pack would still be considered to be functioning properly. There are many devices, perhaps including your robot, that need to have the battery voltage quite constant in order to function correctly, and the 30% variation typical in batteries can be problematic. The robot's velocity, for example, depends on the motor's speed, which in turn depends on the applied voltage. In previous semesters many hovercrafts have worked well on preliminary trial runs, only to die on their final test run after their batteries have "run down."

Voltage regulators are integrated circuits containing transistors and other electronic components that can maintain the voltage in a circuit extremely constant over the usable life of the battery that powers the circuit. Regulators are generally quite small – many of them come in the TO-220 package so that they can easily mount to heat sinks for temperature reduction.

DC voltage regulators come in two main varieties: linear regulators and switching regulators. Both are fairly cheap. Linear regulators simply "chop off" excess voltage. For example, if your motor needs 12 V at 2 A, but your battery produces 15.4 V, the linear regulator is placed in series with the circuit, and 2.4V is chopped off the battery voltage. This means that the regulator has to dissipate 4.8W all the time. Linear regulators have only three terminals and require at most two small capacitors connected between the terminals to help reduce any voltage "ripple" that occurs with changing circuit conditions (See Fig. 12.18). Most voltage regulators have fixed output voltages, but some have voltages that can be varied via an adjustment to an external resistor.

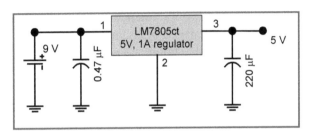

Fig. 12.18 A typical connection for a 5 V output linear regulator.

When selecting a voltage regulator, there are a host of parameters that might be important. After the output voltage, the maximum current that the regulator can handle is critical. The regulator voltage will have a tolerance, which might not be too important, since it is voltage variations that can ruin a good robot performance. The regulator will have limits on the maximum input voltage and the maximum variation in the output voltage with time and/or temperature. The regulator will have fluctuations specified as a function of how much the input voltage changes and how much the output current changes. Usually these variations will be small, but they should always be verified to be adequate. Some regulators will have current protection (so that you cannot draw too much current) and they may have a temperature sensor, which turns the regulator "off" if it overheats. The specifications will have this sort of information, and specifications for almost every electronic part sold can be found on line[1]. Finally, one last parameter that is very important is the dropout voltage. The dropout voltage specifies how much the input voltage has to be above the output voltage to work well. For low dropout voltage regulators, this voltage can be 0.5 – 1 V. For other regulators, this voltage can be as much as two volts or more. Let's continue the example of a 12V motor powered by a battery that outputs 15.4 volts when fully charged and no current drawn. If you have a regulator that has a dropout voltage (like an LM1084) of 1.3 – 1.5V, the output will be regulated at 12V only while the battery voltage is at least 13.3 – 13.5V. When the battery voltage drops below this value, the regulator output will start to drop down and the specifications regarding the voltage regulation no longer apply.

Switching regulators are more complicated than linear regulators. They take in the constant input voltage, then the "chop the voltage" to make a time-varying voltage with a fairly high frequency (above the audible range). They increase or decrease the voltage of the periodic circuit as they convert the signal back to a constant dc voltage. They have several advantages. First, they are more efficient so they usually get less hot. Second, they can be used to lower or raise voltages and may not have a dropout voltage (or at least a smaller dropout voltage). They are somewhat more complicated to connect and usually require a few more extra components, but neither type of regulator is difficult to insert into a circuit and the product specification sheets always contain schematics of the circuits required to use the regulator. A point to note, however, is that regulators dissipate energy. If you want to drive a 9 V motor with a 12 V battery, it might make sense to get a 9V battery instead of using a 12 V battery and the regulator.

---

[1] The authors are impressed with the service received from http://www.digi-key.com

12.6.8 Operational Amplifiers

Operational amplifiers, or **op-amps**, for short, form another class of integrated circuits, each one comprised of dozens of transistors and other passive components. Operational amplifiers are voltage amplifiers that convert supply (dc) power into signal power, which could be dc, sinusoidal, or even non-periodic in nature. They are very useful, very complex devices with a large list of operating characteristics. In this section we will talk only about a small subset of those properties for idealized op-amps.

Real op-amps have at least five connection pins, though some have seven or more. One pin is for the output voltage (V_o), two are for the inputs (V_+ and V_-), and two are for the power supply (or supplies). Op-amps can come in packages with anywhere from 5 to 8 pins. Sometimes op-amps get packaged together – you can fit four op-amps, for example, on a single chip with 14 pins, since all op-amps share the same power source(s). The standard symbol for an op-amp, with power supply connections, is given in Fig. 12.19a. If the op-amp has the rail-to-rail property, the output voltage can range anywhere from the minimum supply voltage to the maximum supply voltage. If an op-amp is NOT rail-to-rail, the minimum output voltage is a volt or so higher than the minimum supply voltage and the maximum output voltage is a volt or so lower than the maximum supply. Single-sided op-amps allow for one power supply with the negative supply terminal simply connected to ground. "Ground" is a node in a circuit (often connected to the negative terminal of a battery) that you agree to call zero volts. If it is not single-sided, the op-amp requires one positive supply and one negative supply.

Fig. 12.19 The symbol for an op amp. All voltages are with respect to ground.

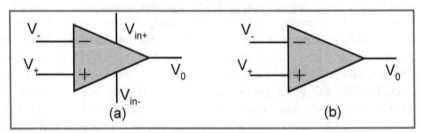

Ideal op-amps have three important properties that make it easy to design and analyze op-amp circuits, and at the same time, make op-amps very useful for many applications. When we consider ideal op-amps, we often ignore the power supply connections, as shown in Fig. 12.19b (don't forget that one must power a real op-amp for the circuit to work). The first ideal assumption is that there is infinite input resistance, which is to say that zero current flows into either input terminal. The second ideal assumption is that the output voltage, which is defined as:

$$V_o = A\,(V_+ - V_-), \tag{12.31}$$

has an open-loop gain, A, which is nearly infinite: $A \rightarrow \infty$. Since the output voltage must be finite, this second assumption means that the two input voltages must be equal:

$$V_+ = V_- \tag{12.32}$$

This assumption is not always true. As long as the output voltage does not try to exceed the power supply limitations, no more than one voltage source is directly connected to the inputs, and there is some feedback (which represents a piece of the output signal fed into the negative input terminal), this assumption is usually not violated. The final assumption is that the output voltage equation (Eq. 12.31) is true irrespective of any component connected to the op-amp output.

12.7 APPLICATION EXAMPLES

12.7.1 Simple Series and Parallel Combinations

Assume there is a requirement for a particular sensor, that you must place a 5.40 kΩ resistor in series with your sensor in order to calibrate it. The sensor will not work correctly if the resistor is less than 5.395 kΩ or greater than 5.405 kΩ. How can this resistance be achieved? Standard resistors available at the local store only come in 5.1 kΩ and 5.6 kΩ packages. The resistors can vary by 5%, so if one has enough 5.6 kΩ resistors, perhaps an adequate resistor could be found. It's far better to build these with series or parallel combinations. With standard resistors, one could use: (a) two 2.7 kΩ resistors in series, (b) a 5.1 kΩ resistor and a 300 Ω resistor in series, (c) a 150 kΩ resistor in parallel with a 5.6 kΩ resistor, (d) a potentiometer, or (e) many, many other possibilities, depending on what component values happen to be available.

12.7.2 Voltage Divider

One consequence of resistors in series is voltage division. Consider the circuit in Fig. 12.20 and determine the voltage across R_2. From Ohm's law (see Eq. 12.5a), we have:

$$I = V_s /(R_1 + R_2) \tag{12.34}$$

$$V_{d2} = IR_2 = V_s R_2/(R_1 + R_2) \tag{12.35}$$

Equation (12.35) shows that the voltage across R_2 is a fraction of V_s. The fraction is $R_2/(R_1 + R_2)$. Another way of expressing this result is:

$$V_{d2}/V_s = R_2/R_{eq} \tag{12.36}$$

where R_{eq} is the equivalent series resistance.

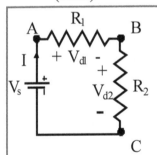

Fig. 12.20 A voltage division circuit

The nice feature about Eq. 12.36 is that it remains correct even if there are more than two resistors in series, as long as it is the voltage across the second resistor that we wish to calculate.

Another consequence of Kirchhoff's laws is that voltage sources in series add just as resistors do as long as the orientation of the source is taken into account. That is why two 1.5 V flashlight batteries are used in an ordinary flashlight; the voltage of the two batteries adds to provide a total voltage source of 3 V. On the other hand, if one of the batteries is inserted upside down, it will act **against** the voltage of the first battery and give a net zero voltage. In that case, the flashlight will not work.

At this point, you should recognize that Ohm's law and Kirchhoff's two laws can be employed to analyze circuits. Also, you should be able to take a complex circuit and use equivalent resistances to reduce the circuit to its most elementary form.

12.7.3 Current Divider

One consequence of resistors in parallel is current division. Consider the circuit in Fig. 12.21 and determine the current through R_2. From Ohm's law [see Eqs. (12.25) and (12.27b)], we have:

$$I = V_s (R_1 + R_2)/ R_1 R_2 \qquad\qquad (12.37)$$

$$I_2 = V_s /R_2 = IR_1/(R_1 + R_2) \qquad\qquad (12.38)$$

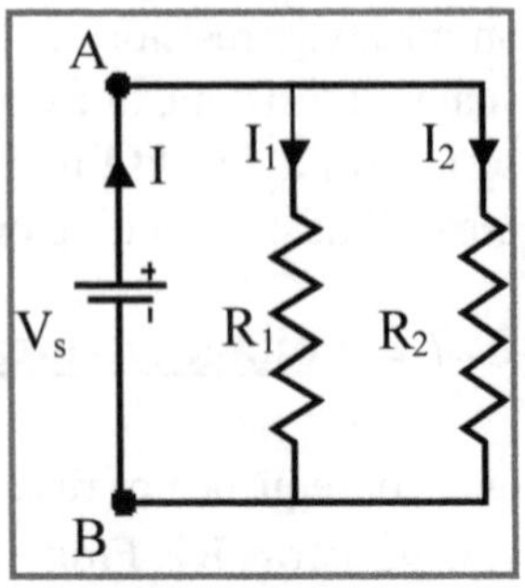

Fig. 12.21 A current divider circuit

Equation (12.38) shows that the current through R_2 is a fraction of I. The fraction is $R_1/(R_1 + R_2)$. If we want a formula that is good for more than two resistors in parallel, we need to write the formulas in terms of the conductances of the components:

$$G_1 = 1/R_1 \qquad G_2 = 1/R_2 \qquad G_{eq} = G_1 + G_2 \qquad\qquad (12.39)$$

For conductances, we get an expression that works for more than two resistors:

$$I_2/I = G_2/G_{eq} \qquad\qquad (12.40)$$

12.7.4 Controlling Motor Speed

A simple motor controller is shown in Fig. 12.24. The microcontroller uses PWM to control motor speed, at a frequency that is too high to switch the power on and off with a typical relay. A MOSFET is shown in the figure since the connection is straightforward. It is advisable to put a resistor R between the output of the microcontroller and the ground (the negative side of the battery in this figure). This is because the MOSFETs have nearly infinite gate resistance and the microcontroller needs a finite resistance to work properly. The exact resistance does not matter: anything from 1 kΩ to 10kΩ should certainly work just fine. A capacitor can be connected across the motor terminals to reduce electrical noise that might affect the operation of the circuit. Typical values are on the order of 0.1-0.3 µF soldered as close to the motor as possible. **DO NOT USE POLARIZED CAPACITORS.** The battery pack which energizes the motor is shown as V_s; if a voltage regulator is used, Fig. 12.18 can be used to synthesize the correct circuit drawing.

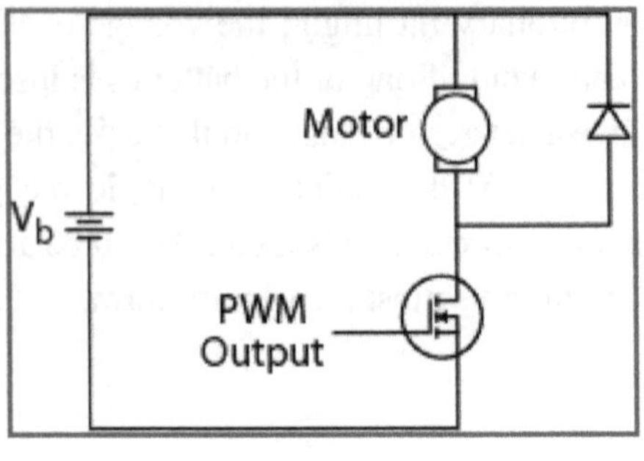

Fig. 12.22 A MOSFET circuit for controlling motor speed via PWM

The diode shown in Fig. 12.22 plays an important and crucial role in helping to mitigate the potentially damaging voltage spikes that would otherwise occur when the MOSFET is switched to the off state. As discussed earlier, a motor is a highly inductive load, for which the current cannot be changed abruptly without producing a large voltage transient. The diode in Fig. 12.22 provides an alternative pathway for current to flow when the MOSFET switch is in the OFF position. When the switch is open, the MOSFET behaves like a conductor that connects the lower terminal of the motor to ground. In this case, the diode is reverse-biased, and therefore draws no current. Essentially, when the switch is ON, the diode does not play a role in the circuit performance. However, when the switch is turned OFF, the negative terminal of the motor gets disconnected. Before the voltage spike can occur, the diode becomes forward biased, allowing the current to circulate back through the motor instead of stopping abruptly. When selecting a diode for this circuit, it is important to ensure that the chosen component can sustain a peak current that is comparable to the maximum current passing through the motor.

12.7.5 <u>Changing the Motor Direction</u>

In addition to adjusting the motor speed, the motor can be in one of three states: forward, off, reversed. If one simply wants to toggle between off and forward (slowly), the circuit shown in Fig. 12.23 can be used with a single SPST relay. When the microcontroller output is zero, the motor is off, the relay inductor is not energized, the switch is open and the motor is not connected. When the control output is on, the switch is closed and the motor runs.

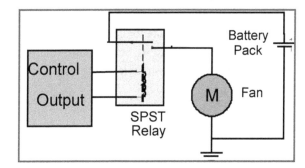

Fig. 12.23 A simple on-off circuit

A double-pole, double throw (DPDT) relay can be used if you want to toggle between forward and reverse (but never be off). This circuit is shown in Fig. 12.24. When the microcontroller output is zero, the center contact of each switch is in contact with its upper contact, causing the top of the motor to be connected to the positive battery terminal and the motor bottom to be connected to the negative battery terminal. When the microcontroller output is high, both switches change positions and the motor connections are reversed. Remember that this will not work if you have a motor that does not allow voltage reversals.

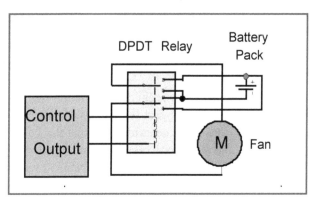

Fig. 12.24 A simple way to reverse motor direction

If you want to be able to access all three states (slowly), two DPST relays (and two diodes) can be used to accomplish this task, since the microcontroller output has three states: off, positive output, and negative output. The required circuit for this is shown in Fig. 12.25. When the microcontroller output is zero, the motor is not connected to anything and does not spin. When the microcontroller output power is on positive, the top microcontroller output (in the figure) is at a higher potential than the lower output. Thus, the upper diode is on and that energizes the upper relay, connecting the motor to the battery. Thus, the positive terminal of the battery is connected to the top of the motor. The lower relay is not energized because the lower diode is reverse-biased. When the microcontroller has the motor direction reversed, the lower microcontroller output is at a higher voltage than the upper output; the upper diode is off and the lower diode is on; the upper relay is off and the lower relay is energized, and the top of the motor is now connected to the negative terminal of the battery reversing the motor direction. The battery would be shorted out if both relays were on for some reason. If identical relays are used, the chance of that would be pretty small, but a good design practice is to always turn the motor off for a short period of time (a small fraction of a second) between motor direction changes.

A good way to both reverse motor direction and control motor speed with a PWM system is to use an H-bridge circuit. An H-bridge is a set of four transistors that are used to reverse the connections to the motor. When all four transistors are off, the motor receives no power. When one pair of transistors is on (and the other pair off), the motor is powered in the forward direction. When the transistor pair roles are reversed, the motor is powered in the reverse direction. The transistor pairs should never be turned on at the same time to prevent shorting the motors. One can certainly build an H-Bridge with individual transistors, but there are off-the-shelf H-Bridges that can handle up to 5A or more. Up to 5A, these chips are fairly cheap and readily available. They have some protections (like avoiding having all transistors on) and they are very compact and reliable. These chips, their wiring diagrams and explanations of circuit operation are all readily available on the web.

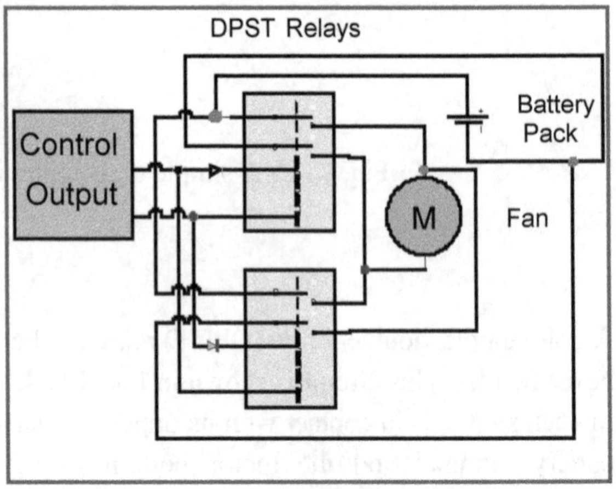

Fig. 12.25 A simple way to drive the motor forward, reverse and stop.

12.7.6 System Requirements Calculation

Assume that you have two 12V, 10 Ω motors that you intend to power with a 15.4 V battery. There are also six sensors that take 100 mA each at 6 V. A 12V voltage regulator is in series between the battery and the motors to insure the correct operating voltage, but a series resistor will be used to drop the 12V regulator output to the 6V needed for the sensors. If you want the batteries to last for 5 trial runs of 15 minutes each, what is the minimum battery capacity (in mAh) that you need? What is the minimum acceptable current specification for the voltage regulator?

The problem may seem daunting, but a single figure will help organize the process (see Fig. 12.26. In Fig. 12.26a, the original layout is shown, with an "M" indicating a motor and an "S" indicating a sensor. The Resistance "R" has to be determined so that the voltage across the sensors is 6V. The capacitors need to be chosen so that the regulator works well, but they don't affect the calculations that we need to do. Likewise, we can replace the battery and regulator combination with an equivalent 12V battery. The current that this battery delivers to the circuit is the current that the voltage regulator must be able to handle. In Fig. 12.26b this simplification has been made, and the motors and sensors have been replaced by their equivalent resistances. The motor resistances were given in the problem definition and the sensor resistances were calculated by Ohm's Law (Eq. 12.5a) given that they use 0.1A when energized to 6V. In Fig. 12.26c the two 10Ω parallel resistors have been replaced by their equivalent resistance and so have the six parallel 60Ω resistors. Using the voltage divider equation (Eq. 12.35), we can find that R = 10Ω. Combining the two 10Ω series resistors, after making one final parallel combination, the simplified circuit in Fig. 12.28d is achieved. Applying Ohm's Law to this circuit, we see that the required regulator current is 3A. Given that the batteries must last 1.25 hours, the battery needs a capacity of 3.75 AH or 3,750 mAh.

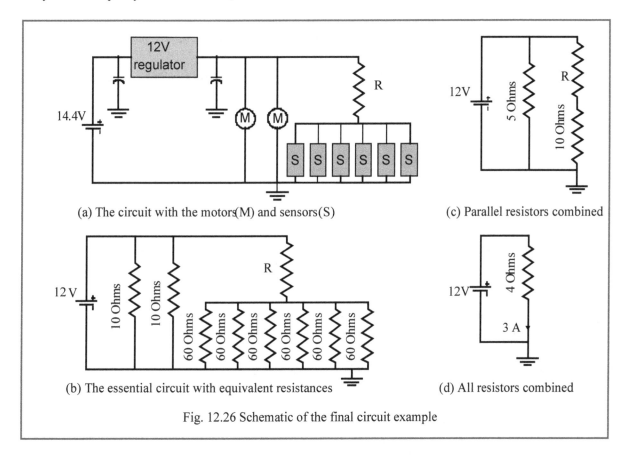

(a) The circuit with the motors(M) and sensors(S)

(c) Parallel resistors combined

(b) The essential circuit with equivalent resistances

(d) All resistors combined

Fig. 12.26 Schematic of the final circuit example

12.8 <u>SUMMARY</u>

In this chapter, we have tried to give a basic, yet thorough, introduction to key electric circuit theory concepts. We started with a discussion of the basic physical quantities and later described the origin of the equations that are needed to solve for all circuit variables. We introduced the material classes that are used to make electronic parts before giving a basic description of a number of different components, such as resistors, transistors, op-amps, regulators, etc. The list of components included the majority of the electronic devices that might prove useful for the design of a small autonomous vehicle. In most instances, we developed the terminal equations necessary to analyze the circuit or gave examples of how the devices can be used.

We also introduced Kirchhoff's Voltage and Current Laws, the two equations which describe how components are connected together to form functioning circuits. We gave a number of examples of circuit analysis, including robot control and detector circuits, as well as classic examples that introduce the useful concepts of equivalent circuits for various resistor combinations, and in particular, series and parallel-connected resistors.

REFERENCES

1. Irwin, J. D., <u>Basic Engineering Circuit Analysis</u>, 8th ed., John Wiley & Sons, New York, 2005.
2. Irwin, J. D. and Kerns, D. V., Jr., <u>Introduction to Electrical Engineering</u>, Prentice-Hall, 1996.
3. Horowitz, P. and Hill, W., <u>The Art of Electronics</u>, 2nd ed., Cambridge University Press, New York, 1989.
4. Dorf, R. C. and Svoboda, J. A., <u>Introduction to Electric Circuits</u>, 6th ed., John Wiley & Sons, New York, 2003.
5. Brindley, K.: <u>Sensors and Transducers</u>, Heinemann, London, 1988.
6. Sedra, A. S. and K. C. Smith: <u>Microelectronic Circuits</u>, 3rd edition, Holt, Rinehart, and Winston, New York, 1991.
7. Shukla, A and Dally, J. W., <u>Instrumentation and Sensors for Engineering Analysis and Process Control</u>, College House Ent., LLC, Knoxville, TN, 2013.

EXERCISES

12.1 Draw the symbols used on drawings to represent a resistor, a capacitor and an inductor. Also, write the relations for the voltage across these components in terms of the current flow.

12.2 If you design a circuit containing an equivalent resistance of 3,600 Ω with a 9-V battery power supply, find the current drained from the battery. If the battery has a capacity of 2.5A-h (ampere-hours), determine the life of the battery. Assume that it cannot be recharged. What is the power dissipated by the resistor?

12.3 A size D, 1.5 V, alkaline battery has a capacity of 1,100 hours when the battery is discharged through a 120 Ω resistor. Determine the current flow through this resistor, and state the capacity in terms of ampere-hours (Ah).

12.4 A 9 V alkaline battery has a capacity of 9 hours when the battery is discharged through a 220 Ω resistor. Determine the current flow through this resistor, and state the capacity in terms of ampere-hours (Ah).

12.5 Write Kirchhoff's two laws. Use circuit diagrams to illustrate these two laws.

12.6 Draw a circuit with four series-connected resistors and give the expression for the equivalent resistance. If these resistors have values of 900, 500, 1,700 and 15,000Ω, find the equivalent resistance. Draw a circuit diagram that contains the equivalent resistor, which is also equivalent to your initial circuit diagram.

12.7 Draw a circuit with four parallel-connected resistors, and give the expression for the equivalent resistance. If these resistors have values of 900, 500, 1,700 and 15,000Ω, find the equivalent resistance. Draw a circuit diagram that contains the equivalent resistor, which is also equivalent to your initial circuit diagram.

12.8 You have a relay that runs off of 6V DC. When energized, the current through the electromagnet wire is 75 mA. Assume that this wire can be modeled by a simple resistor.

   a. What is the resistance of the electromagnet wire?
   b. If the electromagnet wire can dissipate no more than 2 Watts, what is the maximum voltage that you can place across the electromagnet terminals?

12.9 Assume that you have a motor that can be modeled as a 0.38 Ω resistor. You have a FET transistor that can take up to 10A DC. This transistor will be used to energize a motor (in a series configuration).

a. What is the maximum battery voltage that you can use if you want to make sure that you don't exceed the transistor current limit?

b. Let's say that you set the FET gate so that only 6.0 A flows through the motor/transistor combination when the circuit is energized to the maximum battery voltage. How much power (on average) is being dissipated by the transistor?

12.10 Assume that you have three photo-sensors, all of which have resistances of 12.5 kΩ when they are not reflecting light. Assuming that is the case (lights are off):

a. What is the effective (combined) resistance of the three sensors together, if they are connected in series?

b. What is the effective (combined) resistance of the three sensors together, if they are connected in parallel?

12.11 You have a relay that typically operates with an actuation voltage of 12V DC. When energized to this voltage, the current through the electromagnet wire is 32 mA. Assume that this wire can be modeled by a simple resistor.

a. What is the resistance of the electromagnet wire?

b. Let's say that you want to use the output from a port on the Arduino PCB to energize this relay. What would be the power dissipated by the relay coil if the Arduino PCB output were limited to 40 mA?

c. Look up online the characteristics of a real 12V relay. Explain whether or not this relay would not work with the Arduino PCB (in part b) and justify your answer with information from the relay's data sheet! Don't forget to give the part number for the relay!

12.12 You have three motors powering your robot. Two drive motors run at 9.6V, 1A when running continuously and a 12V motor for steering requires 0.45 A. For each of the batteries / battery combinations below, state whether or not they could be used to power your robot for three trial runs on a prescribed course (worst case) and explain your assumptions and reasoning clearly. Support your conclusion with detailed calculations.

a. one 9.6 V, 2,200 mAh Lithium-ion battery

b. one 12 V, 1,500 mAh Ni-Cad battery

c. one 12 V, 700 mAh Ni-Cad battery and one 9.6V, 700 mAh Lithium-ion battery

CHAPTER 13

SENSORS

13.1 INTRODUCTION

Transducers are electromechanical devices that convert a mechanical change such as displacement or force into a change in an electrical signal that can be monitored as a voltage after signal conditioning. A wide variety of transducers are commercially available for use in measuring mechanical quantities. Primarily the sensor that is incorporated into a transducer to produce the electrical output determines the device's characteristics. For example, a set of four strain gages arranged on a tension link provides a transducer that produces a resistance change $\Delta R/R$ in proportion to the load applied along the axis of the link. The strain gages serve as the sensor in this force transducer and play a significant role in establishing the characteristics of the transducer.

Sensors used in transducer design include potentiometers, differential transformers, strain gages, capacitors, piezoelectric and piezoresistive crystals, thermistors, fibers, etc. The important features of these different sensors are described in this chapter.

13.2 POTENTIOMETERS

The simplest type of potentiometer, shown schematically in Fig. 13.1, is the slide-wire resistor. This sensor consists of a length L of resistance wire attached across a voltage source V_i. The relationship between the output voltage V_o and the position x of a wiper, as it moves along the length of the wire, can be expressed as:

$$V_o = (x/L)V_i \qquad x = (V_o/V_i)L \qquad (13.1)$$

Thus, the slide-wire potentiometer can be used to measure a displacement x.

Resistors fabricated from a short length of straight wire are not feasible for most applications, because its resistance is too low. Very low resistance of a sensor imposes excessive power requirements on the voltage source. To alleviate this difficulty, high-resistance, wire-wound potentiometers are obtained by winding the resistance wire around an insulating core, as shown in Fig. 13.2. The potentiometer illustrated in Fig. 13.2a is used for linear displacement measurements. Cylindrically shaped potentiometers, similar to the one illustrated in Fig. 13.2b, are used for angular displacement measurements. The resistance of a wire-wound potentiometer can range between 10 and 10^6 ohms Ω depending upon the alloy and diameter of the wire used and the length of the coil.

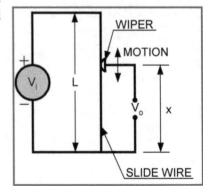

Fig. 13.1 Slide-wire resistance potentiometer.

The resistance of the wire-wound potentiometer increases in a stepwise manner, as the wiper moves from one turn to the adjacent turn. This step change in resistance limits the resolution of the potentiometer to L/n, where n is the number of turns in the length L of the coil. Resolutions ranging from 0.05 to 1% are common, with the lower limit obtained by using many turns of very small diameter wire.

The active length L of the coil controls the range of the potentiometer. Slider potentiometers are available in many lengths up to about 1 m. Producing the coil in the form of a helix can extend the range of the angular-displacement potentiometer. Helical potentiometers are commercially available with as many as 20 turns; therefore, angular displacements as large as 7,200 degrees can be measured quite easily.

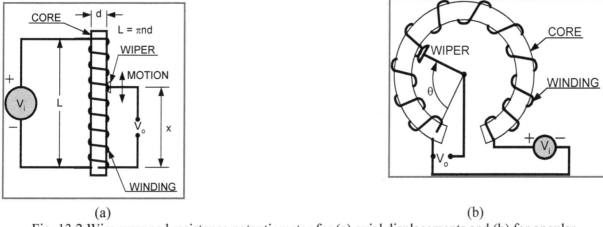

(a)   (b)

Fig. 13.2 Wire-wrapped resistance potentiometer for (a) axial displacements and (b) for angular displacements.

In order to improve resolution, potentiometers have been introduced that utilize thin films of conductive plastic with controlled resistivity instead of wire-wound coils. The film resistance on an insulating substrate exhibits very high resolution together with lower noise and longer life. For example, a resistance of 50 to 100 Ω/mm can be obtained with the conductive plastic films that are used for commercially available potentiometers with a resolution of 0.001 mm or less. The frictional force that must be overcome to move the wiper is depended on the construction details of the particular potentiometer; however, 1 to 2 ounces is common. The life expectancy for potentiometers fabricated with conductive plastics is commonly specified at 20×10^6 strokes.

The dynamic response of both the linear and the angular potentiometer is severely limited by the inertia of the shaft and wiper assembly. Because this inertia is large, the potentiometer is used only for static or quasi-static measurements, where high frequency response is not required.

Potentiometers are used primarily to measure large displacements—10 mm or more for linear motion and 15 degrees or more for angular motion. Potentiometers are relatively inexpensive yet accurate; however, their main advantage is simplicity of operation, because only a voltage source and a digital voltmeter (DVM) to measure voltage comprise the complete instrumentation system. Their primary disadvantages are a limited frequency response that precludes their use for dynamic measurements and relatively high forces required to overcome friction of the wiper.

13.3 DIFFERENTIAL TRANSFORMERS

Differential transformers, based on a variable-inductance principle, are also used to measure displacement. The most popular variable-inductance sensor for linear displacement measurements is the linear variable differential transformer (LVDT). A LVDT, illustrated in Fig 13.3a, consists of three symmetrically spaced coils wound onto an insulated bobbin. A magnetic core, which moves through the bobbin without contact, provides a path for magnetic flux linkage between coils. The position of the magnetic core controls the mutual inductance between the center or primary coil and the two outer or secondary coils.

When an AC voltage is applied to the primary coil, voltages are induced in the two secondary coils. The secondary coils are wired in a series-opposing circuit, as shown in Fig. 13.3b. When the core is centered between the two secondary coils, the voltages induced in the secondary coils are equal but out of phase by 180 degrees. Because the coils are in a series-opposing circuit, the voltages V_1 and V_2 in the two coils cancel and the output voltage is zero. When the core is moved from the center position, an imbalance in mutual inductance between the primary and secondary coils occurs and an output voltage, $V_o = V_2 - V_1$, develops. The output voltage is a linear function of core position, as shown in Fig. 13.4, as long as the motion of the core is within the operating range of the LVDT. The direction of motion can be determined from the phase of the output voltage relative to the input voltage.

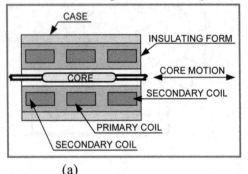
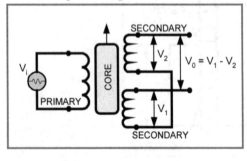

(a)   (b)

Fig. 13.3 (a) Sectional view of a linear variable differential transformer (LVDT).
(b) Schematic circuit diagram for the LVDT coils.

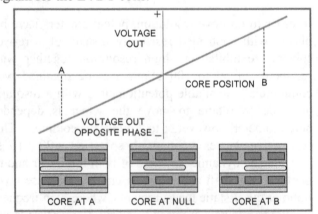

Fig. 13.4 Phase-referenced voltage as a function of LVDT core position.

Because the LVDT, a passive sensor, requires ac excitation at a frequency different from common ac supplies, signal-conditioning circuits are needed for its operation. A typical signal conditioner, shown in Fig. 13.5, provides a power supply, a frequency generator to drive the LVDT, and a demodulator to convert the ac output signal from the LVDT to an analog dc output voltage. Finally, a dc amplifier is incorporated in the signal conditioner to increase the magnitude of the output voltage.

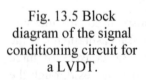

Fig. 13.5 Block diagram of the signal conditioning circuit for a LVDT.

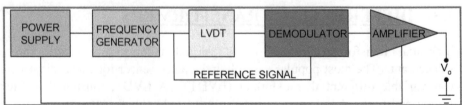

Many different types of LVDTs are commercially available today. They are designed for normal applications, for rugged environments, for high temperature (220 °C) and high-pressure measurements and to monitor motion sensitive mechanisms. Most of the LVDTs are packaged in a stainless steel tube that encases the coils and shields the electronics from noise.

With the development of microelectronics, circuits have been developed that permit miniaturization of the signal conditioner shown in Fig. 13.5. These miniaturized circuits are packaged within the case of an LVDT to produce a small self-contained sensor known as a direct current differential transformer (DCDT). A DCDT operates from a battery or a regulated power supply (± 15VDC nominal) and provides an amplified output signal that can be monitored on either a digital voltmeter (DVM) or an oscilloscope. The output impedance of a DCDT is relatively low (100 Ω).

The LVDT and DCDT have many advantages as sensors for measuring displacement. There is no contact between the core and the coils; therefore, friction and hysteresis are eliminated. Because the output is continuously variable with input, resolution is determined by the characteristics of the voltage recorder. Non-contact also ensures that life will be very long with no significant deterioration of performance over this period. The small core mass and freedom from friction give the sensor a limited capability for dynamic measurements. Finally, the sensors are not damaged by over travel; therefore, they can be employed as feedback transducers in servo-controlled systems, where over travel may occur due to accidental deviations beyond the control band.

In more recent years, the DCDT has been equipped with an analog to digital converter (A/D) to provide a digitized output[1]. Their input voltage may be varied from 8.5 to 30 V, and their output signal has a resolution of at least 15 bits or 1 part in 32,768.

13.4 <u>RESISTANCE STRAIN GAGE</u>S

Electrical resistance strain gages are thin metal-foil grids (see Fig. 13.6) that can be adhesively bonded to the surface of a component or structure. When the component or structure is loaded, strains develop that are transmitted to the foil grid. The resistance of the foil grid changes in proportion to the load-induced strain. The strain sensitivity of metals, first observed in copper and iron by Lord Kelvin in 1856, is explained by the following simple analysis.

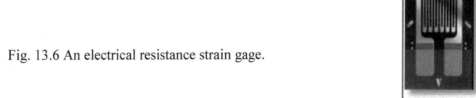

Fig. 13.6 An electrical resistance strain gage.

The alloy sensitivity S_A of the metal or alloy used for the conductor of a strain gage is given by:

$$S_A = \frac{dR/R}{\varepsilon_a} = \frac{d\rho/\rho}{\varepsilon_a} + (1 + 2v) \qquad (13.2)$$

where ρ is the specific resistance of the metal (Ω - cm), v is Poisson's ratio and ε_a is the axial strain.

It is evident from Eq. (13.2) that the strain sensitivity of a metal or alloy is due to the changes in dimensions of the conductor, as expressed by the term $(1 + 2v)$, and the change in specific resistance as represented by the term $(d\rho/\rho)/\varepsilon$. Experimental studies show that the sensitivity S_A ranges between 2 and 4 for most metallic alloys used in strain-gage fabrication. The most commonly used strain gages are fabricated from the copper-nickel alloy known as Advance or Constantan. This alloy is widely used,

---

[1] For example, the Schaevitz Sensor model HC485 displacement sensor has a two wire addressable RS-485 output.

because its response is linear over a wide range of strain (beyond 8%), it has a high specific resistance, and it has excellent electrical stability with changes in temperature.

Most resistance strain gages are of the metal-foil type, where a photoetching process is used to form the grid configuration. Because the process is very versatile, a wide variety of gage sizes and grid shapes can be produced. Standard gage resistances are 120 and 350 Ω; however, special-purpose gages with resistances of 500, 1,000, and 5,000 Ω are also available.

The etched metal-film grids are very fragile and easy to distort, wrinkle, or tear. For this reason, the metal grid is bonded to a thin plastic film that serves as a backing or carrier before photoetching. The carrier film, shown in Fig. 13.6, also provides electrical insulation between the gage and the component after the gage is bonded.

A strain gage exhibits a resistance change $\Delta R/R$ that is related to the strain ε in the direction of the grid by the expression:

$$\Delta R/R = S_g\, \varepsilon \tag{13.3}$$

where S_g is the gage factor or calibration constant for the gage. The gage factor S_g is always less than the sensitivity of the metallic alloy S_A because the grid configuration of the gage with the transverse conductors is less responsive to axial strain than a straight uniform conductor.

The output $\Delta R/R$ of a strain gage is usually converted to a voltage signal with a Wheatstone bridge, as illustrated in Fig. 13.7. If a single gage is used in one arm of the Wheatstone bridge and equal but fixed resistors are used in the other three arms, the output voltage is given by:

$$V_0 = \frac{V_s}{4}\frac{\Delta R_g}{R_g} \tag{13.4}$$

Equation (13.3) into Eq. (13.4) gives:

$$V_0 = \frac{V_s}{4} S_g \varepsilon \tag{13.5}$$

The input voltage is controlled by the gage size (the power it can dissipate) and the initial resistance of the gage. As a result, the sensitivity, $S = V_0/\varepsilon = V_s S_g/4$, usually ranges from 1 to 10 $\mu V/((m/m)$.

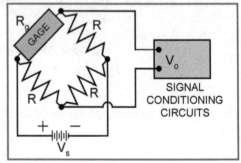

Fig. 13.7 Wheatstone bridge circuit used to convert resistance change $\Delta R/R$ of a strain gage to an output voltage Vo.

13.5 CAPACITANCE SENSORS

The displacement sensor, which is based on changing capacitance and illustrated in Fig. 13.8a, consists of a target plate and a second plate termed as the sensor head. These two plates are separated by an air gap of thickness h and form the two terminals of a capacitor, which exhibits a capacitance C given by:

$$C = \frac{kKA}{h} \tag{13.6}$$

where C is the capacitance in picofarad; A is the area of the sensor head ($\pi D/4$); K is the relative dielectric constant for the medium in the gap (K = 1 for air); k = 0.225 is a proportionality constant for dimensions specified in inches; k = 0.00885 for dimensions given in mm.

If the separation between the head and the target is changed by an amount Δh, then the capacitance C becomes:

$$C + \Delta C = \frac{kKA}{h + \Delta h} \qquad (a)$$

Equation (a) can be written as:

$$\frac{\Delta C}{C} = \frac{\Delta h/h}{1 + (\Delta h/h)} \qquad (13.7)$$

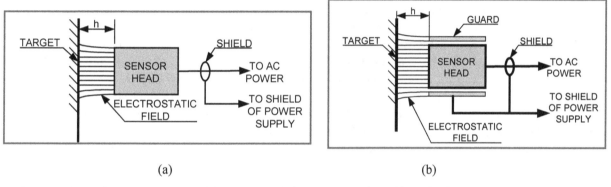

(a)   (b)

Fig. 13.8 Capacitor sensors: (a) without a guard ring, where edge effects in the electrostatic field affect the range of linearity, and (b) with a guard ring to extend the range of linearity.

This result indicates that ($\Delta C/C$) is non-linear, because of the presence of Δ/h in the denominator of Eq. (13.7). To avoid the difficulty of employing a capacitance sensor with a non-linear output, the change in the impedance due to the capacitor is measured. Note that the impedance Z_C of a capacitor is given by:

$$Z_C = -j/(\omega C) \qquad (b)$$

With a capacitance change ΔC it can be shown that:

$$Z_c + \Delta Z_c = -\frac{j}{\omega}\left[\frac{1}{C + \Delta C}\right] \qquad (c)$$

Substituting Eq. (b) into Eq. (c) and solving for $\Delta Z/Z$ gives:

$$\frac{\Delta Z_c}{Z_c} = -\frac{\Delta C/C}{1 + \Delta C/C} \qquad (13.8)$$

Finally, substituting Eq. (13.7) into Eq. (13.8) yields:

$$\frac{\Delta Z_c}{Z_c} = -\frac{\Delta h}{h} \qquad (13.9)$$

From Eq. (13.9) it is clear that the capacitive impedance Z_C is linear in Δh and that methods of measuring ΔZ_C will permit extremely simple plates (the target as ground and the sensor head as the positive terminal) to act as a sensor to measure the displacement Δh. Cylindrical sensor heads are linear and Eq. (13.9) is valid provided $0 < h < D/4$ where D is the diameter of the sensor head.

The sensitivity of the capacitance probe is given by Eqs. (b), (13.6), and (13.9) as:

$$S = \frac{\Delta Z_C}{\Delta h} = \left| \frac{Z_C}{h} \right| = \left| \frac{1}{\omega\, Ch} \right| = \left| \frac{1}{\omega\, KkA} \right| \qquad (13.10)$$

The sensitivity can be improved by reducing the area of the probe; however, as noted previously, the range of the probe is limited by linearity to about D/4. Clearly there is a range-sensitivity trade-off. Of particular importance is the circular frequency ω in Eq. (13.10). Low frequency improves sensitivity, but limits frequency response of the instrument, another trade-off. It is also important to note that the frequency of the ac power supply must remain constant to maintain a stable calibration constant.

The capacitance sensor has several advantages. It is non-contacting and can be used with any target material provided the material exhibits a resistivity less than 100 Ω-cm (any metal). The sensor is extremely rugged and can be subjected to high shock loads (5,000 g's) and intense vibratory environments. Their use as a sensor at high temperature is particularly impressive. They can be constructed to withstand temperatures up to 2,000° F and they exhibit a constant sensitivity S over an extremely wide range of temperature (74 to 1600° F). Examination of the relation for S in Eq. (13.10) shows that the dielectric constant K is the only parameter that can change with temperature. Because K is constant for air over a wide range of temperature, the capacitance sensor has excellent temperature stability.

13.6   EDDY CURRENT SENSORS

A displacement sensor, which uses eddy currents to measure distance between the sensor head and an electrically conducting surface, is illustrated in Fig. 13.9. Sensor operation is based on eddy currents that are induced at the conducting surface, as magnetic flux lines from the sensor intersect with the surface of the conducting material. The magnetic flux lines are generated by the active coil in the sensor, which is driven at a very high frequency (1 MHz). The magnitude of the eddy current produced at the surface of the conducting material is a function of the distance between the active coil and the target's surface. The eddy currents increase as the distance decreases.

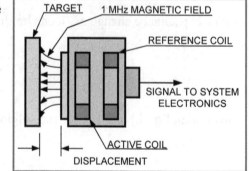

Fig. 13.9 Schematic diagram for an eddy current sensor.

Changes in the eddy currents are sensed with an impedance (inductance) bridge. Two coils in the sensor are used for two arms of the bridge. The other two arms are housed in an associated electronic package illustrated in Fig. 13.10. The first coil in the sensor (active coil), which changes inductance with target movement, is wired into the active arm of the bridge. The second coil is wired into an opposing arm of the same bridge, where it serves as a compensating coil to balance and cancel the effects of temperature change. The output from the impedance bridge is demodulated and becomes the analog signal, which is linearly proportional to distance between the sensor and the target. This signal is then amplified prior to recording on some analog recorder or conversion to a digital signal.

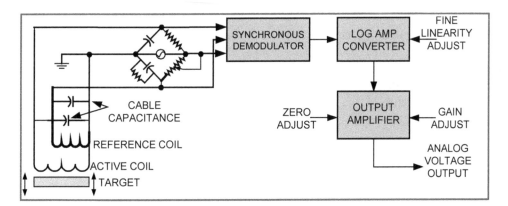

Fig. 13.10 Block diagram of the signal condition circuits used in converting impedance changes due to target displacement to an analog output voltage.

For non-conducting, poorly conducting or magnetic materials, it is possible to bond a thin film of aluminum foil to the surface of the target at the location of the sensor to improve the sensitivity. Because the penetration of the eddy currents into the material is minimal, the thickness of the foil can be as small as 0.2 mm (ordinary kitchen type aluminum foil).

The fact that eddy current sensors do not require contact for measuring displacement is quite important. As a result of this feature, they are often used in transducer systems for automatic control of dimensions in fabrication processes. They are also applied extensively to determine thickness of organic coatings that are nonconducting.

13.7 PIEZOELECTRIC SENSORS

A piezoelectric material, as its name implies, produces an electric charge when it is subjected to a force or pressure. Piezoelectric materials, such as single-crystal quartz, contain molecules with asymmetrical charge distributions. When pressure is applied, the crystal deforms and there is a relative displacement of the positive and negative charges within the crystal. This displacement of internal charges produces external charges of opposite sign on the external surfaces of the crystal. If these surfaces are coated with metallic electrodes, as illustrated in Fig. 13.11, the charge q that develops can be determined from the output voltage V_o because:

$$q = V_o C \hspace{4cm} (13.11)$$

where C is the capacitance of the piezoelectric crystal.

Fig. 13.11 Piezoelectric crystal deforming under the action of an applied pressure.

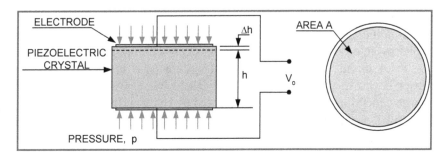

The surface charge q is related to the applied pressure p by:

$$q = S_q A p \qquad (13.12)$$

where S_q is the charge sensitivity of the piezoelectric crystal and A is the area of the electrode.

The charge sensitivity S_q is a function of the orientation of the sensor (usually a cylinder) relative to the axes of the piezoelectric crystal. The output voltage V_o developed by the piezoelectric sensor is obtained by substituting Eqs. (13.6) and (13.11) into Eq. (13.12). Thus:

$$V_0 = \left(\frac{S_q}{kK}\right) h\, p \qquad (a)$$

The voltage sensitivity S_v of the sensor can be expressed as:

$$S_v = \left(\frac{S_q}{kK}\right) \qquad (b)$$

The output voltage v_o of the sensor is then:

$$V_0 = S_v h\, p \qquad (13.13)$$

Again, voltage sensitivity S_v of the sensor is a function of the orientation of the axis of the cylinder relative to the crystallographic axes.

Most piezoelectric transducers are fabricated from single-crystal quartz because it is the most stable of the piezoelectric materials, and is nearly loss free both mechanically and electrically. Its properties are: modulus of elasticity 86 GPa, resistivity 10^{14} Ω-cm, and dielectric constant 40.6 pF/m. It exhibits excellent high-temperature properties and can be operated up to 550 °C.

Most sensors exhibit relatively low output impedance (in the range of 100 to 1,000 Ω). Low input impedance is an advantage, because it is easy to interface the sensor with associated instruments. However, when piezoelectric crystals are used as the sensing elements in transducers, the output impedance is usually extremely high. The output impedance, of a small cylinder of quartz, depends the geometry of the crystal and on the frequency ω associated with the applied pressure. Because the sensor acts like a capacitor, the output impedance is given by:

$$Z_c = \frac{1}{j\omega C} = -\frac{j}{\omega C} \qquad (a)$$

The impedance ranges from infinity for static applications to about 10 kΩ for very high-frequency applications (100 kHz). With this high output impedance, care must be exercised in monitoring the output voltage; otherwise, very serious errors can occur. Recording instruments with an input impedance higher by an order of magnitude than the senor's output impedance are required.

A circuit diagram of a typical system used to measure a voltage produced by a piezoelectric sensor is shown in Fig. 13.12. The piezoelectric sensor acts as a charge generator. In addition to the charge generator, the sensor is represented by two parallel components including a capacitor C_p (about 10 pF) and a leakage resistor R_p (about 10^{14} Ω). The capacitance of the lead wires C_L must also be considered, because even relatively short lead wires have a capacitance larger than the sensor. A charge amplifier with sufficiently high input impedance is often used to isolate the piezoelectric sensor from recording instruments with standard input impedance (one to ten MΩ). If a pressure is applied to the sensor and maintained for a long period of time, the charge q developed by the piezoelectric material leaks, because a small current flows through both R_p and the amplifier resistance R_A.

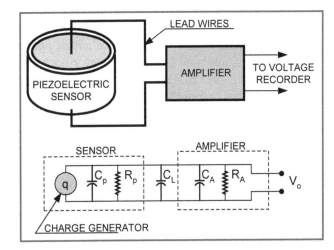

Fig. 13.14 Schematic diagram of a measuring system with a piezoelectric sensor.

The inherent dynamic response of the piezoelectric sensor is very high, because the resonant frequency of the small cylindrical piezoelectric element is so large. The resonant frequency of the transducer depends upon its mechanical design as well as the mass and stiffness of the sensor. However, the most significant advantage of the piezoelectric sensor is its very high-frequency response; hence, they are employed as sensors for many high performance accelerometers.

13.8 <u>PIEZORESISTIVE SENSORS</u>

Piezoresistive sensors, as the name implies, are fabricated from materials that exhibit a change in resistance when subjected to a pressure. The development of piezoresistive materials was an outgrowth of research on semiconductors by the Bell Telephone Laboratories in the early 1950s that eventually led to the transistor. Piezoresistive sensors are fabricated from semi-conductive materials, usually silicon containing trace impurities that are added to ultra-pure single-crystal silicon. The resistivity ρ of a semiconducting material can be expressed as:

$$\rho = 1/(eN\mu) \tag{13.14}$$

where e is the electron charge, which depends on the type of impurity; N is the number of charge carriers, which depends on the concentration of the impurity and μ is the mobility of the charge carriers, which depends upon strain and its direction relative to the crystal axes.

Equation (13.14) shows that the resistivity of the semiconductor can be adjusted to any specified value by controlling the concentration of the trace impurity. The impurity concentrations commonly employed range from 10^{16} to 10^{20} atoms/cm^3, which permits a wide variation in the initial resistivity. For example, the resistivity for P-type silicon with a concentration of 10^{20} atoms/cm^3 is 50,000 $\mu\Omega$-cm, which is about 30,000 times higher than the resistivity of copper. This very high resistivity facilitates the design of miniaturized sensors. Another advantage of using single-crystal semi conducting silicon as a sensor is its high strength and stiffness (modulus). The elastic modulus of silicon is 190 GPA, which is nearly equal to that of steel (207 GPa). The yield strength of silicon—7.0 GPa—is nearly five times as large as 4340 steel with a yield strength of about 1.5 GPa.

The resistivity of a piezoresistive sensor changes when it is subjected to either stress or strain because of the variations in the mobility μ. This change of resistivity is known as the piezoresistive effect and is used to produce piezoresistive strain gages. The sensitivity of a typical piezoresistive stain sensor is high (a gage factor of about 100, while a metal-foil strain gage has a gage factor of approximately 2).

In recent years, considerable progress has been made in applying fabrication techniques used in the microelectronics industry to the development of micro-miniature sensors. These solid-state sensors incorporate silicon as the mechanical element and piezoresistive sensors. In a piezoresistive pressure transducer, a diaphragm is etched from silicon to form the mechanical element of the transducer. Resistances that form the arms of a Wheatstone bridge are diffused or ion implanted directly into the silicon diaphragm to provide the sensing element.

An example of a solid-state sensor that utilizes micromachining single crystal silicon and a number of integrated circuit technologies is shown in Fig. 13.13. This device is a triaxial accelerometer that measures acceleration in the x, y and z directions. It is designed with a square central element (the mass) that is supported by four orthogonally oriented beams. The central mass and the beams are encircled with a square frame. Piezoresistive sensors are implanted in the beams. When subjected to acceleration in any direction, the central mass deflects generating strains in the supporting beams. The piezoresistive sensors respond to these strains and provides a signals that can be interpreted to give the acceleration in the x, y and z directions.

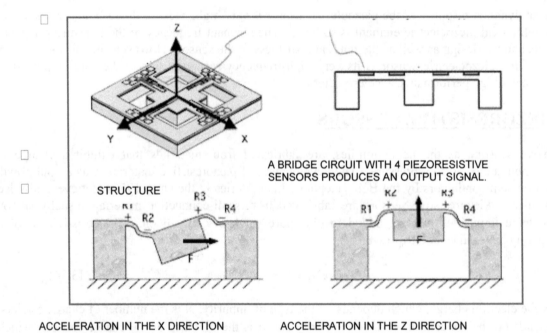

Fig. 13.13 A triaxial accelerometer fabricated from single crystal silicon that utilizes piezoresistive sensors.

13.9 PHOTOELECTRIC SENSORS

In many applications, where contact cannot be made with the object being examined, a photoelectric sensor can often be employed to monitor changes in the intensity of light, which sense the quantity being measured. When light impinges on a photoelectric sensor, it either creates or modulates an electrical signal. Most photoelectric devices employ semiconductor materials and operate by either generating a current or by changing resistivity. These devices are photodetectors that respond quickly to changes in light intensity.

13.9.1 <u>Photoconducting Sensors</u>

Photoconductive cells, illustrated in Fig. 13.14, are fabricated from semiconductor materials, such as cadmium sulfide (CdS) or cadmium selenide (CdSe), which exhibit a strong photoconductive response. When a photon with sufficient energy strikes a molecule of, say CdS, an electron is driven from the valence band to the conduction band and a hole or vacancy remains in the valence band. This hole and the electron both serve as charge carriers, and with continuous exposure to light, the concentration of charge carriers increases and the resistivity decreases. A circuit used to detect the resistance change ΔR of the photoconductor is also shown in Fig. 13.14. The resistance of a typical photoconducting detector changes over about 3 orders of magnitude as the incident radiation varies from dark to very bright.

When a photoconductor is placed in a dark environment, its resistance is high and only a small dark current flows. If the sensor is exposed to light, the resistance decreases significantly (the ratio of maximum to minimum resistance for R_d in Fig. 13.14 ranges from 100 to 10,000 in common commercial sensors); therefore, the output current can be quite large. This change of about two to three orders of magnitude is large, enabling many applications with very simple control circuits.

The photocurrent requires some time to develop after the excitation is applied and some time to decay after the excitation is removed. The rise and fall time for commercially available photoconductors is usually about a second. Because of these delays, the CdS and CdSe photoconductors are not suitable for dynamic measurements. Instead, their simplicity, high sensitivity and low cost lend them to applications involving counting and switching based on a slowly varying light intensity.

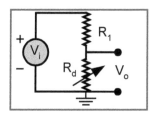
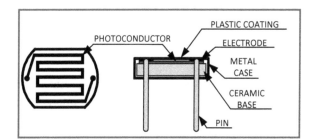

Fig. 13.14 When a photoconductor is used as R_d, the output voltage will vary with the light intensity.

13.9.2 <u>Photodiode Sensors</u>

Photodiodes are semiconductor devices that are responsive to high-energy particles and photons. Photodiodes operate by absorption of photons or charged particles and generate a current that flows in an external circuit. Photodiodes can be used to detect the presence or absence of minute quantities of light and can be calibrated for extremely accurate measurements from intensities below 1 pW/cm^2 to intensities above 100 mW/cm^2. Silicon is the most common semiconductor material used in fabricating planar diffused photodiodes. These photodiodes are employed in such diverse applications as spectroscopy, photography, analytical instrumentation, optical position sensors, beam alignment, surface characterization, laser range finders, optical communications and medical imaging instruments.

Planar diffused silicon photodiodes are P-N junction diodes. A P-N junction can be formed by diffusing either a P-type impurity (anode), such as boron, into a N-type bulk silicon wafer, or a N-type impurity, such as phosphorous, into a P-type bulk silicon wafer. A schematic illustration of a planar diffused photodiode fabricated from N-type silicon is presented in Fig. 13.15.

Semiconductor photodiodes are small, rugged, and inexpensive, and because of these advantages they have replaced the large and expensive vacuum tube detectors in most applications. The photodiodes may be used in either mode (generating either voltage or current) with operational amplifiers to give a responsivity that is exceptionally high.

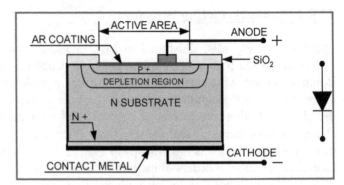

Fig. 13.15 The diffusion areas used to create a
P-N junction in a N-type silicon photodiode.

Circuits showing photodiodes operating in the photoconduction mode and in the photovoltaic mode are
presented in Fig. 13.16.

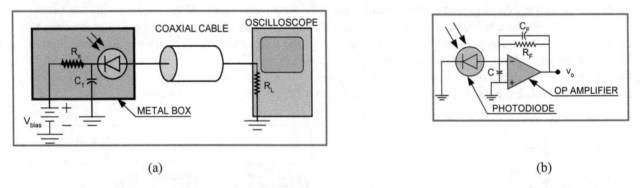

(a)   (b)

Fig. 13.16 Circuits used with photodiodes. (a) High intensity light at high frequency.
(b) Low intensity light with low frequency.

Continued improvements in semiconductor photodiodes expand the sensors that are commercially
available. New products include a line of photodiodes to permit measurement of the variation of light at
discrete points along a line. Also very dense area arrays of photodiodes are used in digital cameras with
high pixel count. These new developments have greatly expanded the number of applications for
photodiodes in the measurement of full field light patterns.

13.10 <u>RESISTANCE TEMPERATURE DETECTORS</u>

The change in resistance of metals with temperature provides the basis for a family of temperature sensors
known as resistance temperature detectors (RTDs). The sensor is simply a conductor fabricated either as
a wire-wound coil encased in a glass or ceramic tube or as a grid formed from a thin film adhered to a
ceramic substrate. The change in resistance of the conductor with temperature is given by the expression:

$$\Delta R/R_o = \gamma_1 (T - T_o) + \gamma_2 (T - T_o)^2 + \ldots\ldots\ldots + \gamma_n (T - T_o)^n \qquad (13.15)$$

where T_o is a reference temperature; R_o is a reference resistance at temperature T_o; $\gamma_1, \gamma_2, \ldots\ldots\gamma_n$ are
temperature coefficients of resistance.

Resistance temperature detectors are often used in ovens and furnaces, where inexpensive but accurate
and stable temperature measurements and controls are required. Platinum is widely used for sensor
fabrication, because it is the most stable of all the metals, is the least sensitive to contamination and is
capable of operating over a very wide range of temperatures (4 °K to 1064 °C).

The performance of platinum RTDs depends strongly on the design and construction of the package containing the sensitive wire coil. The most precise sensors are fabricated with a minimum amount of support; hence, they are fragile and often fail if subjected to rough handling, shock or vibration. Most sensors used in transducers for industrial applications have thin film platinum elements supported on ceramics substrates and encased in glass. These sensors are quite rugged and will withstand shock levels up to 100 g's. An example of a RTD fabricated as described above is presented in Fig. 13.17.

Fig. 13.17 Thin film platinum resistance temperature detectors (Pt-RTD) fabricated with a thin film platinum deposited on a ceramic substrate then encapsulated in glass.
Courtesy of U. S. Sensor Corp.

The dynamic response of an RTD depends almost entirely on construction details. For large coils mounted on heavy ceramic cores and sheathed in stainless steel tubes, the response time may be several seconds or more. For film or foil elements mounted on thin polyimide substrates, the response time can be less than 0.1 s.

13.11 <u>THERMISTORS</u>

A second type of temperature sensor based on resistance change of its sensing element with temperature is known as a thermistor. A thermistor differs from a resistance temperature detector (RTD), because the sensing element is fabricated from a semiconducting material instead of a metal. The semiconducting materials, which include oxides of copper, cobalt, manganese, nickel and titanium, exhibit very large changes in resistance with temperature. As a result, thermistors can be fabricated in the form of extremely small beads as shown in Fig. 13.18.

Resistance change with temperature for a thermistor can be expressed by an equation of the form:

$$\log_e \rho = A_0 + \frac{A_1}{\theta} + \frac{A_2}{\theta^2} + \cdots + \frac{A_n}{\theta^n} \qquad (13.16)$$

where ρ is the specific resistance of the material; A_1, A_2, ... A_n are material constants; θ is the absolute temperature.

The temperature-resistance relationship as expressed by Eq. (13.16) is usually approximated by retaining only the first two terms. The simplified equation is then expressed as:

$$\log_e \rho = A_0 + \frac{\beta}{\theta} \qquad (13.17)$$

Using Eq. (13.17) is convenient and acceptable, when the temperature range is small and the higher-order terms in Eq. (13.16) are negligible.

Thermistors have many advantages over other temperature sensors and are widely used in industry. They are small and, consequently, permit point sensing and rapid response to temperature change. Their high resistance minimizes lead-wire problems and their output is more than 10 times that of a resistance temperature detector (RTD) as shown in Fig. 13.19. Finally, thermistors are very rugged, which permits use in industrial environments, where shock and vibration occur. The disadvantages of thermistors include nonlinear output with temperature, as indicated by Eqs. (13.29) and (13.30) and limited range. Significant advances have been made in thermistor fabrication methods and it is possible to obtain stable, reproducible, interchangeable thermistors that are accurate to 0.5% over a specified temperature range.

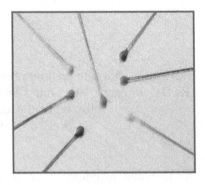

Fig. 13.18 Miniature bead-type thermistors.
Courtesy of U. S. Sensors Corp.

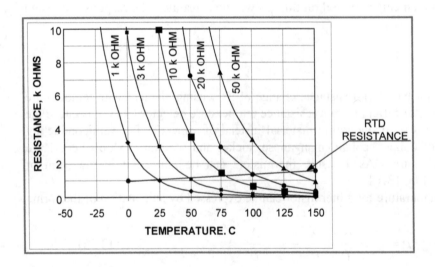

Fig. 13.19 Resistance temperature characteristics of thermistors and a RTD.

13.12 __THERMOCOUPLES__

A thermocouple is a temperature sensor that consists of two dissimilar materials joined together to form both thermal and electrical contact. A potential develops at the interface of the two dissimilar materials as the temperature changes. This thermoelectric phenomenon is known as the **Seebeck effect**. Thermoelectric sensitivities (μV/°C) for a number of different materials in combination with platinum are listed in Table 13.1. Data from Table 13.1 can be used to determine the sensitivity of any thermocouple junction by noting, for example, that:

$$S_{Chromel/Alumel} = S_{Chromel/Platinum} - S_{Alumel/Platinum} = +25.8 - (-13.6) = 39.4 \ \mu V/°C$$

Table 13.1
Thermoelectric sensitivities for different materials in contact with platinum

Material	Sensitivity (μV/°C)	Material	Sensitivity (μV/°C)
Constantan	-35	Copper	$+6.5$
Nickel	-15	Gold	$+6.5$
Alumel	-13.6	Tungsten	$+7.5$
Carbon	$+3$	Iron	$+18.5$
Aluminum	$+3.5$	Chromel	$+25.8$
Silver	$+6.5$	Silicon	$+440$

Common thermocouple material combinations include iron/constantan, chromel/alumel, chromel/constantan, copper/constantan, and platinum/platinum-rhodium.

The output voltage from a thermocouple junction is measured by connecting two identical thermocouples into a circuit as shown in Fig. 13.20. The output voltage V_o from this circuit is related to the temperature at each junction by an expression of the form:

$$V_o = S_{A/B} (T_1 - T_2) \tag{13.18}$$

where $S_{A/B}$ is the sensitivity of material combination A and B; T_1 is the temperature at junction J_1; T_2 is the temperature at junction J_2.

In practice, junction J_2 is a reference junction that is maintained at a carefully controlled reference temperature T_2. Junction J_1 is placed in contact with the body at the point, where temperature is to be measured. When a meter is inserted into the thermocouple circuit, junctions J_3 and J_4 are created at the contact between the wires (material B) and the terminals of the meter. If these terminals are at the same temperature ($T_3 = T_4$), the added junctions (J_3 and J_4) do not affect the output voltage V_o.

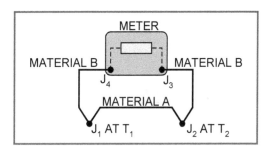

Fig. 13.20 A typical thermocouple circuit.

Designed as a sensor, the thermocouple can be made quite small (0.0005-in. diameter wire is available) to reduce its mass and size. These miniaturized thermocouples have a rapid response time (milliseconds) and provide essentially point measurements of temperature. A thermocouple can cover a wide range of temperatures; however, its output is nonlinear, and linearizing circuits are required to relate the output voltage V_o to the temperature T. In addition to nonlinear output, thermocouple sensors suffer the disadvantage of very low signal output and the need to carefully control the reference temperature at junction J_2.

13.13 COMMERCIAL SENSORS FOR ARDUINO MICROPROCESSOR

Tilt Sensor

A tilt sensor is a device that detects the tilting of an object; however, it does not measure the amount of tilt. This sensor, shown in Fig. 13.21, is an environmental friendly version of a mercury switch. Inside it contains a metallic ball that rolls, when the sensor is tilted, to contact two pins on one side of the device. The ball closes a switch when the sensor tilts by a specific angle. It costs about $8.00.

Fig. 13.21 The tilt sensor is an on/off switch, which is open when level and closed when tilted.

Three Axis Accelerometer

The low cost three axis accelerometer, shown in Fig. 13.22, is produced by Analog Devices. This sensor is a small, thin, low-power and complete 3-axis accelerometer with signal-conditioned voltage outputs. It measures acceleration with a full-scale range of ±3 *g*. It can measure the static acceleration of gravity in tilt sensing applications, as well as dynamic acceleration resulting from motion, shock or vibration.

The bandwidth[2] of the accelerometer is selected to suit the application, with a range of 0.5 Hz to 1,600 Hz for the X and Y axes, and a range of 0.5 Hz to 550 Hz for the Z axis. The accelerometer is supplied with a voltage ranging from 1.8 to 3.6 V and requires 350 µA of current. It is stable with temperature changes and can survive shocks up to 10,000 g. It costs about $25.00 when mounted on a circuit board, as shown in Fig. 13.22.

Fig. 13.22 Three-axis accelerometer capable of measuring ± 3g.

Motion Tracking Sensors

Motion tracking sensors contain several different sensors, which include: a 3-axis gyroscope, a 3-axis accelerometer, a 3-axis magnetometer and three 16-bit analog-to-digital converters (ADCs) for digitizing the gyroscope outputs, three 16-bit ADCs for digitizing the accelerometer outputs and three 13-bit ADCs for digitizing the magnetometer outputs. The sensors are all fabricated using MEMs techniques and are sufficiently miniature to be mounted on a small circuit board, as shown in Fig. 13.23. The sensors are housed in a multi-chip module consisting of two dies integrated into a single package. One die houses the 3-axis gyroscope and the 3-axis accelerometer. The other die houses the 3-axis magnetometer. An on-chip 1024 B buffer helps lower system power consumption by allowing the system processor to read the sensor data in bursts and then enter a low-power mode as the unit collects more data. This unit is priced at $50.

---

[2] Bandwidth defines the range of frequencies that a sensor or recording instrument is capable of sensing or recording without excessive errors.

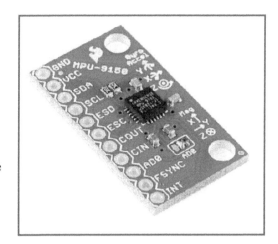

Fig. 13.23 Circuit board supporting the components in the motion tracking device.

For precision tracking of both fast and slow motions, the parts feature a user programmable gyroscope with full-scale range of ±250, ±500, ±1000, and ±2000 °/sec (dps), a user programmable accelerometer with full-scale range of ±2g, ±4g, ±8g, and ±16g, and a magnetometer with full-scale range of ±1200μT. Communication with all registers on the unit is performed using an I2C[3] operating at 400kHz. Additional features include an embedded temperature sensor and an on-chip oscillator with ±1% variation over the operating temperature range. The unit has a shock tolerance of 10,000g, and incorporates programmable low-pass filters for the gyroscopes, accelerometers, magnetometers, and an on-chip temperature sensor.

Force Transducer

The force transducer shown in Fig. 13.24 is similar to that found in a bathroom scale. This load cell is sold by Sparkfun for $10 and is capable of measuring forces up to 500 N. Sparkfun does not provide information regarding the sensors used, but it probably contains strain gages connected to a Wheatstone bridge. The input voltage applied to the bridge may be as high as 10 Vdc with a sensitivity of 1.0 mV per volt of input. The load is applied to the button on the top of the lever arm located above the sensing element. The input resistance is 1,000 Ω and the load cell may be employed at temperatures ranging from 0 to 50 °C. The output is an analog signal and it is necessary to digitize this signal.

Fig. 13.24 Load cell and lead wires.

---

[3] The I2C is a communications interface that allows numerous devices to be connected to the same microcontroller simultaneously.

Infrared Proximity Sensor

Infrared proximity sensor shown in Fig. 13.25 is made by Sharp and is priced at $15. This unit has an analog output that varies from 2.8V at a distance of 150 mm to 0.4V at 1.50 m with a supply voltage between 4.5 and 5.5Vdc. Its power consumption is 33 mA.

Fig. 13.25 An infrared proximity sensor with a LED light source and a IR detector.

This sensor is an integrated combination of a position sensitive detector, an IRED (infrared emitting diode) and signal processing circuit. The different reflectivity of the target, the temperature and the operating time do not markedly influence the distance measurement because of the triangulation method employed enables the voltage output to correspond to the detection distance. The output voltage as a function of distance to the target is presented in Fig. 13.26.

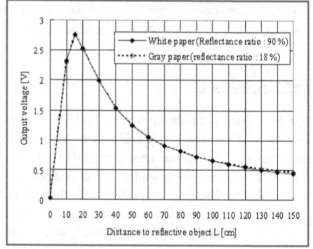

Fig. 13.26 Voltage versus distance for the IR proximity sensor.

CMOS Camera Module

This CMOS camera module, shown in Fig. 13. 72, provides color video images containing 728 by 488 pixels. The unit is equipped with high quality optics, all the on board circuitry to output a RCA signal, and a cable harness. The input voltage ranges from 6 to 20 V and the power consumption at 12 V is 50 mA. The horizontal scanning frequency is 15.7 kHz and the vertical scanning frequency is 60 Hz. The lens is equipped with an electronic shutter and the camera has automatic gain control. The camera can be interfaced with an Arduino microcontroller and operated from its commands. The assembly is priced at $32.

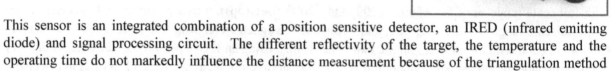

Fig. 13.27 CMOS video camera with lens and associated circuits.

Sound Sensor — MEMs Microphone

This tiny microphone, presented in Fig. 13.28, is a high-quality, high-performance, low-power, analog output, bottom-ported omni-directional MEMS microphone. The module includes a MEMS microphone element, an impedance converter, and an output amplifier. All the components are mounted on a circuit board with the circuitry required for its input and output. The device has a high signal to noise ratio of 62 dBA. Its frequency response is flat (down less than 3Db) from 100 Hz to 15 kHz.

Fig. 13.28 MEMs microphone and amplifier.

The amplifier with a gain of 67 exceeds the bandwidth requirements of the microphone. The amplifier produces a peak-to-peak output of about 200mV when the microphone is held at 0.8 m and exposed to normal conversational speech. The low current consumption of 250 µA enables long battery life for portable applications The module is priced at $10.

Digital Temperature Sensor

This digital temperature sensor, shown in Fig. 13.29, measures temperatures from − 25°C to + 85°C, with a resolution of 0.0625°C, which is achieved with a 12 bit analog to digital converter. Its accuracy is 0.5°C. The sensor requires very low-current 10 µA supplied with voltages that can range from 1.4 to 3.6 V. Communication is achieved through a RS-232 two-wire serial interface. Filtering capacitors and pull-up resistors are included.

Fig. 13.29 Temperature sensor with a 12 bit digital output.

Gas Sensors

There are a series of gas sensors, shown in Fig. 13.30, that provide an analog output that is proportional to gas concentrations ranging from 200 to 10,000 ppm. The sensors detect concentrations of liquefied petroleum gas (LPG), hydrogen H_2, methane CH_4, carbon monoxide CO and alcohol. The sensor and its circuit are simple with a 5 V supply voltage to power a heater coil, a load resistance, and an ADC to convert the analog output to a digital signal.

Fig. 13.30 Carbon monoxide gas sensor.

The CO gas sensor, shown in Fig. 13.30, is fabricated with a stainless steel mesh on its top, which permits the gas to enter the sensor. Inside the container an electrode coated with tin dioxide SnO_2 serves as the sensing element that changes resistance with the concentration of CO in the sensor's chamber. A heating coil maintains the temperature of the chamber. Six pins provide the connections for the signals and the heater supply. A graph showing the resistance change with gas concentration for several different gases is shown in Fig. 13.31.

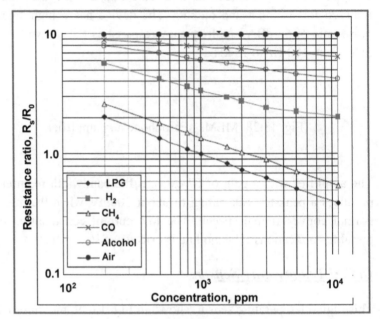

Fig. 13.31 Resistance change with gas concentration for several gas sensors.

13.14 SUMMARY

Several basic sensors have been described in this chapter. These sensors are sometimes used directly to measure an unknown quantity (such as the use of an electrical strain sensor to measure strain); however, in many other instances the sensors are one component of a more involved transducer such as a piezoelectric crystal in an accelerometer or a number of strain gages in a force transducer. Important characteristics of each sensor, which must be considered in the selection process, include:

1. **Size**—with smaller being better, because of enhanced dynamic response and minimum interference with the process or event.
2. **Range**—with extended range being preferred, so as to increase the latitude of operation.
3. **Sensitivity**—with the advantage to higher output signals that require less amplification.
4. **Accuracy**—with the advantage to devices exhibiting errors of 1% or less after considering zero shift, linearity and hysteresis.
5. **Frequency response**—with preference for wide-bandwidth sensors that permit application in both static and dynamic loading situations.
6. **Stability**—with very low drift in output over extended periods of time and with very small changes signal output with variations in temperature and humidity.
7. **Temperature limits**—with the ability to operate from cryogenic to elevated temperatures.
8. **Economy**—with reasonable costs preferred.
9. **Ease of application**—with reliability and simplicity considered significant factors.

Finally, brief descriptions of several low-cost sensors that are compatible with the Arduino microprocessor were provided.

EXERCISES

13.1 Describe the differences between a sensor and a transducer. Give an example of a transducer incorporating a displacement sensor. Give another example of a sensor that is also a transducer.

13.2 A slide-wire potentiometer having a length of 50 mm is fabricated by winding wire with a diameter of 0.10 mm around a cylindrical insulating core. Determine the resolution limit of this potentiometer.

13.3 If the potentiometer of Exercise 13.2 has a resistance of 2,400 Ω and can dissipate 2.5 W of power, determine the voltage required to maximize the sensitivity. What voltage change corresponds to the resolution limit?

13.4 Why are potentiometers limited to static or quasi-static applications?

13.5 List several advantages of the conductive-film type of potentiometer.

13.6 A new elevator must be tested to determine its performance characteristics. Design a displacement transducer that utilizes a 10-turn potentiometer to monitor the position of the elevator over its 120-m range of travel.

13.7 Compare the potentiometer and LVDT as displacement sensors with regard to the following characteristics: range, accuracy, resolution, frequency response, reliability, complexity and cost.

13.8 Prepare a block diagram representing the electronic components in a LVDT. Describe the function of each component.

13.9 Describe the basic differences between an LVDT and a DCDT.

13.10 Design a 50-mm strain extensometer to be used for a simple tension test of mild steel. If the strain extensometer is to be used only in the elastic region and to detect the onset of yielding, specify the maximum range. What is the advantage of limiting the range?

13.11 Compare the cylindrical potentiometer (helipotentiometer) and the RVDT as sensors for measuring angular displacement.

13.12 What two factors are responsible for the resistance change $\Delta R/R$ in an electrical resistance strain gage? Which is the most important for gages fabricated from constantan?

13.13 What function does the thin plastic film serve for an electrical resistance strain gage?

13.14 A strain gage with an initial resistance R_g and a gage factor S_g is subjected to strain ε. Determine ΔR and $\Delta R/R$ for the conditions listed below:

	R_o (Ω)	S_g	ε (μm/m)
(a)	120	2.02	1600
(b)	350	3.47	650
(c)	350	2.07	650
(d)	1000	2.06	200

13.15 For the conditions described in Exercise 13.14, determine the output voltage V_o for an initially balanced bridge if the input voltage V_s is:

(a)	2 V	(c)	7 V
(b)	4 V	(d)	10 V

13.16 From Exercise 13.15, we note that increasing V_i increases V_o. What would happen if we increase V_s to 50 V to improve the output V_o.

13.17 For the gages specified in Exercise 13.14, placed in a single-arm Wheatstone bridge with $V_s = 5$ V, determine the strain ε if the output voltage V_o is:

(a)	1.5 mV	(c)	4.8 mV
(b)	3.3 mV	(d)	5.7 mV

13.18 A short-range displacement transducer utilizes a cantilever beam as the mechanical element and a strain gage as the sensor. Derive an expression for the displacement δ of the end of the beam in terms of the output voltage V_o.

13.19 Prepare a graph of the sensitivity S of a capacitance sensor as a function of frequency ω. Assume the dielectric in the gap is air and consider probe diameters of 1, 2, 5 and 10 mm.

13.20 Write an engineering brief describing the advantages and disadvantages of capacitance sensors.

13.21 Can eddy current sensors be employed with the following target material?

      (a) Magnetic materials
      (b) Polymers
      (c) Non-magnetic metallic foils

    Indicate procedures that permit usage of the sensor in these three cases.

13.22 Compare the characteristics of a piezoresistive sensor with those of a piezoelectric sensor.

13.23 What advantages does the piezoresistive sensor have over the common (metal) electrical resistance strain gage? What are some of the disadvantages?

13.24 Design a circuit to turn on outside lights at your home as it begins to get dark. Use a photoconduction cell in the circuit and provide for an adjustment to control the intensity level for switch activation.

13.25 Compare resistance temperature detectors (RTD) and thermistors as temperature sensors. Give an example that is best suited for each sensor.

13.26 Determine the sensitivity of the following thermocouples.

      (a) chromel-alumel
      (b) copper-constantan
      (c) iron-constantan
      (d) iron-nickel
      (e) gold-silver

13.27 Compute the voltage output (approximate) at the meter in a thermocouple circuit for the five material combinations defined in Exercise 13.26 if:

	T_1 (°C)	T_2 (°C)
(a)	300	0
(b)	200	0
(c)	250	10
(d)	600	100

CHAPTER 14

DYNAMICS AND CONTROL

14.1 INTRODUCTION

After your team has become familiar with your vehicle (robot), its successful operation will depend on how well that the vehicle can be controlled as it negotiates a defined track. The control of the vehicle will depend on the feedback from the sensors and the control algorithm programmed into the Arduino microcontroller. However, before you can prepare a flow diagram for the algorithm, it is essential to understand the underlying mechanics concepts and the basics of control theory. To assist you in this understanding some aspects of kinematics, statics, dynamics, and control theory are introduced in this chapter. In addition, components commonly found on robotic vehicles are described.

14.2 KINEMATICS

Kinematics is the study of the motion of bodies without considering the forces required to produce that motion. For linear motion along a straight line, the equations of kinematics that relate distance s, velocity v, acceleration a and time t are given by:

$$v = \frac{ds}{dt} \qquad (14.1)$$

$$a = \frac{dv}{dt} \qquad (14.2)$$

For angular motion about an axis of rotation, the equations of kinematics take the similar form:

$$\omega = \frac{d\theta}{dt} \qquad (14.3)$$

$$\alpha = \frac{d\omega}{dt} \qquad (14.4)$$

where ω is angular velocity, θ is orientation angle, t is time, and α is angular acceleration.

Integrating Eq. (14.1) yields relations for distance as a function of velocity and time as:

$$s = \int v \, dt . \qquad (14.5)$$

If velocity is constant, Eq. (14.5) becomes

$$s = v \, t + C_1 \qquad (14.6)$$

where C_1 is a constant of integration. If, additionally, s = 0 when t = 0, then:

$$s = v \, t . \qquad\qquad (14.7)$$

Integrating Eq. (14.2) gives velocity as a function of acceleration and time, as:

$$v = \int a \, dt . \qquad\qquad (14.8)$$

If acceleration is constant, Eq. (14.8) becomes:

$$v = a \, t + C_2 \qquad\qquad (14.9)$$

where C_2 is a constant of integration. If, additionally, $v = 0$ when $t = 0$, then:

$$v = a \, t \qquad\qquad (14.10)$$

If the acceleration is constant, the velocity increases linearly with time. We expect a vehicle to reach terminal velocity in a fraction of a second.

EXAMPLE 14.1

Consider a vehicle that is initially at rest. It accelerates for 2.5 s at a constant acceleration of a = 0.8 m/s2. Determine its position and velocity as a function of time during this 1 s interval of time.

Solution:

From Eq. (14.10), write:

$$v = a \, t = (\, 0.8 \text{ m/s2} \,) \, (2.5 \text{ s}) = 2 \text{ m/s} \qquad\qquad \text{(a)}$$

From Eq. (14.7), write

$$s = v \, t = (\, 2 \text{ m/s} \,) \, (2.5 \text{ s}) = 5 \text{ m} \qquad\qquad \text{(b)}$$

At the end of the 2.5 s acceleration period, the vehicle has traveled 5.0 m and achieved a velocity of 2.0 m/s. Is an acceleration of 0.8 m/s2 adequate when you consider the size of the track at about 3 by 3 m or is it excessive? Explain the rational for your answer.

Note that relatively high velocities can be achieved after a short period of time when acceleration is constant.

14.3 STATICS

Statics is the first mechanics course taken by most engineering students. It concerns the forces on stationary bodies or bodies moving with a constant velocity. Equilibrium is the principal concept that enables the student to determine forces acting on a body at rest or one moving at a constant velocity. We will briefly describe the equilibrium of force and moment equations that will be helpful in analyzing the forces acting on your robot.

14.3.1 Equilibrium

Consider a body with no acceleration, i.e., a body that is either stationary or moving at constant velocity. In this case, Newton's second law enables us to write:

$$\Sigma \mathbf{F} = 0, \tag{14.11}$$

where each $\mathbf{F}$ is an external force on the body. These forces are vectors and are denoted here by bold font. The summation includes all the external forces. Another way of representing Eq. (14.11) is:

$$\mathbf{F_1} + \mathbf{F_2} + \ldots\ldots + \mathbf{F_n} = 0, \tag{14.12}$$

where n is the number of external forces.

To illustrate the meaning of the mathematical symbol $\Sigma \mathbf{F}$, consider the arbitrary body in Fig. 14.1. Four coplanar external forces act on this body as shown. When the vector sum of these forces is zero, the body is in equilibrium.

One may recast the vector representation of Eq. (14.11) by writing the equivalent scalar equations as:

$$\Sigma F_x = 0 \qquad \Sigma F_y = 0 \qquad \Sigma F_z = 0 \tag{14.13}$$

$$F_{1x} + F_{2x} + .. + F_{nx} = 0, \quad F_{1y} + F_{2y} + .. + F_{ny} = 0, \quad F_{1z} + F_{2z} + .. + F_{nz} = 0, \tag{14.14}$$

where x, y, and z denote the force components parallel to the x, y, and z axes.

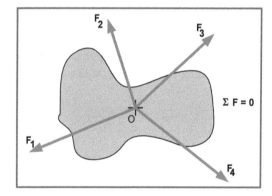

Fig. 14.1 Four coplanar forces acting on a body that is in equilibrium with $\Sigma \mathbf{F} = 0$.

We account for the directions of the forces in Eq. (14.14) by considering only those coplanar forces in either the x, y or z directions. When the forces are constrained to the x, y or z directions, we employ the scalar form of the equilibrium equations. However, when the forces act in directions other than these, the vectors must first be decomposed into their components F_x, F_y and F_z acting in the x, y and z directions, before they can be used in the scalar equilibrium equations.

14.3.2 Moments

The forces shown in Fig. 14.1 are called concurrent because that they all pass through a single point O. Other force systems are less constrained, such as the body in Fig. 14.2. On the vehicle, presented in Fig. 14.3a, we show two propulsion forces, F_p, and two drag forces, F_d. Also shown is a point O that denotes the origin of an x-y coordinate system, which is often chosen to be the center of gravity. These forces are not concurrent because they do not pass through a single point. Consequently, for the vehicle to be in equilibrium an additional equation must be satisfied:

$$\Sigma \mathbf{M} = 0 \qquad\qquad (14.15)$$

where $\Sigma\mathbf{M}$ is the sum of all the moments about any point in the body and $\mathbf{M}$ is, like $\mathbf{F}$, a vector.

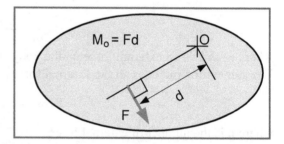

Fig. 14.2 A moment produced by a force depends on the position of the point O.

We have introduced a new term in Eq. (14.15), namely a moment. A moment M_o is produced when a force F, as shown in Fig. 14.2, is applied that causes rotation about some point O. The magnitude of a moment produced by a force is given by:

$$M_o = F\,d \qquad\qquad (14.16)$$

where M_o is a scalar quantity and d is the perpendicular distance from the point O to the line of action of the force F.

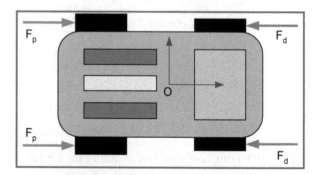

Fig. 14.3a Vehicle with applied propulsion and frictional drag forces (straight line motion $\sum M_O = 0$.

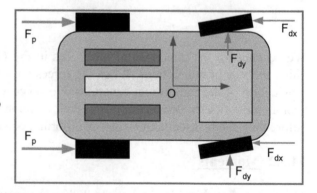

Fig. 14.3b Vehicle propulsion and frictional drag forces. The friction drag force has been resolved into x and y components. These forces produce a moment that will turn the vehicle.

The units of a moment $\mathbf{M_o}$ are typically specified as N-m or ft-lb$_f$. The magnitude M_o is given by Eq. (14.16), and the direction of the vector is perpendicular to the plane in which both F and d lie. As shown in Fig. 14.4, the moment $\mathbf{M_o}$ has direction. In this illustration, the vector $\mathbf{M_o}$ tends to rotate the body in a counterclockwise direction when viewed from above and thus is considered as positive.

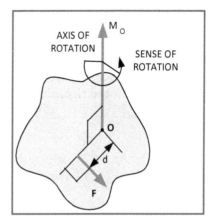

Fig. 14.4 A graphic illustration of the moment $\mathbf{M_o}$ as a vector quantity.

To determine the sign of moment, use the right hand rule. In applying this rule, place the palm of your right hand along the axis of rotation of the moment vector and point your fingers in the direction of the force and rotate your hand. If the direction of the force causes you to rotate counterclockwise, with your thumbs pointing up from the plane in which both F and d lie, then the moment is positive. However, if you must rotate your hand clockwise, with your thumb pointing downward, the moment is negative.

EXAMPLE 14.2

A hexagonal headed bolt is tightened with a wrench as shown in Fig. 14.5. A 160 N force is applied to the handle of the wrench to produce a moment (torque). If the distance from the centerline of the bolt to the point of application of the force is d = 250 mm, find the applied torque.

Fig. 14.5 A force on the handle of a wrench produces a moment (torque) on the bolt head.

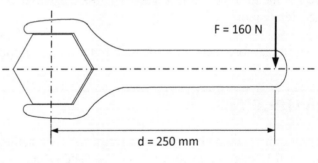

Solution:

From Eq. (14.16), we write:

$$Mo = F\ d = -(160)(0.25) = -40\ \text{N-m.} \qquad (a)$$

This moment is negative since it produces a clockwise rotation about the head of the bolt.

14.3.3 Tipping

The operation of a robotic vehicle is similar to the operation of your automobile. You have a motor or motors to drive the wheels or tracks to propel it in either direction. Steering can be accomplished in a number of different ways depending on the design of the robot. However it is important that the turning radius and velocity are such that the robot does not tip over. Another possible cause of tipping and rolling is the terrain encountered by the robot operating along a defined track. Engaging a slope at the wrong angle of attack can cause the vehicle to tip and then roll. Let's consider both possibilities.

Tipping Due to High-Speed Turning

When a wheeled vehicle turns a centrifugal force F_c develops that tends to tip the vehicle over. The forces acting on a vehicle as it turns at velocity v are shown in Fig. 14.6. The centrifugal force F_C and the weight W of the vehicle, act at its center of gravity. The forces acting on the wheels are depicted as F_L and F_R for the left and right sides respectively.

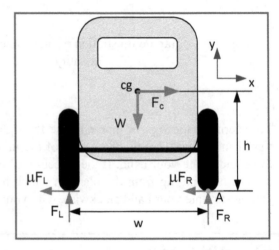

Fig. 14.6 A transverse equilibrium state for a vehicle turning left.

The magnitude of the centrifugal force F_C is given by:

$$F_C = ma = \frac{mv^2}{r} = \frac{Wv^2}{gr} \qquad (14.17)$$

where r is the turn radius, m is mass, v is the tangential velocity, W is the weight of the vehicle and g is the acceleration of gravity.

The centrifugal force acts perpendicular to the vehicle motion and opposite to the turn direction.

EXAMPLE 14.3

Determine the magnitude of the centrifugal force Fc acting on a vehicle traveling around a curve with a radius r = 0.15 m at a velocity v = 0.35 m/s. The vehicle weighs 15 N.

Solution:

$$F_C = \frac{W v^2}{g r} = \frac{15 \text{ N}}{9.807 \text{ m/s}^2} \frac{(0.35 \text{ /s})^2}{0.15 \text{ m}} = 1.249 \text{ N} \qquad (a)$$

In this example the centrifugal force is equal to 8.3% of the vehicle's weight.

Now that we understand centrifugal forces, let's consider the conditions where they become sufficiently large to cause the robot to tip. Examine the illustration in Fig. 14.6 and visualize how the forces acting on the wheels change as the centrifugal force increases. It is evident that with an increase in the centrifugal force the upward force on the right wheel increases while that of the left wheel decreases. The tipping point occurs when the upward force on the left wheel becomes zero. To determine the centrifugal force required to cause $F_L = 0$, let's evaluate the equilibrium equations.

First consider equilibrium of the forces by writing Eq. (14.13) as:

$$\sum F_y = F_L + F_R - W = 0 \qquad \text{(a)}$$

and

$$\sum F_x = F_C - \mu(F_L + F_R) = 0 \qquad \text{(b)}$$

where μ is the coefficient of friction.

Substituting Eq. (a) into Eq. (b) and simplifying yields:

$$F_c = \mu W \qquad \text{(c)}$$

Prior to tipping the centrifugal force is offset by frictional forces.

Next let's consider the tipping condition where F_L and μF_L both vanish, and write the moment equation about the remaining contact point A.

$$\sum M_A = Ww/2 - F_C h = 0 \qquad \text{(d)}$$

or

$$F_C = Ww/(2h), \qquad \text{(14.18)}$$

where w is the distance between the wheels and h is the height of the center of gravity.

Substituting Eq. (14.17) into Eq. (14.18) gives the velocity of the vehicle at the tipping point as:

$$v = \sqrt{\frac{grw}{2h}} \qquad \text{(14.19)}$$

For a curve with a radius r, the velocity can be maximized by decreasing the height of the center of gravity h and increasing the distance between wheels w. This relation implies that you should design your robot to be wide and place your heavy components, especially the batteries, at the base of your chassis.

The center of gravity of a rigid body is the location of a coordinate center O_{xy}, where the two equations shown below are satisfied:

$$\sum M_x = 0 \qquad \sum M_y = 0 \qquad \text{(14.20)}$$

These two relations enable us to write equations for the coordinates of the center of gravity, x_{cg} and y_{cg}, as follows:

$$x_{cg} = \sum_{i=1}^{N} \frac{x_i W_i}{W} \quad \text{and} \quad y_{cg} = \sum_{i=1}^{N} \frac{y_i W_i}{W} \qquad \text{(14.21)}$$

where x_i and y_i are the coordinates of each component, W_i is the weight of each component, W is the total weight of the system and N is the number of components.

While the center of gravity of the robot and all of its components can be determined from Eq. (14.21), the center of gravity can be determined with a simple experiment. Suspend the robot with a piece of string or wire attached somewhere near the center of its front end, as shown in Fig. 14.7. Move the attachment point until the robot is level. This procedure experimentally establishes the distance of the center of gravity (cg) of the vehicle from the contact point of the wheels on the track.

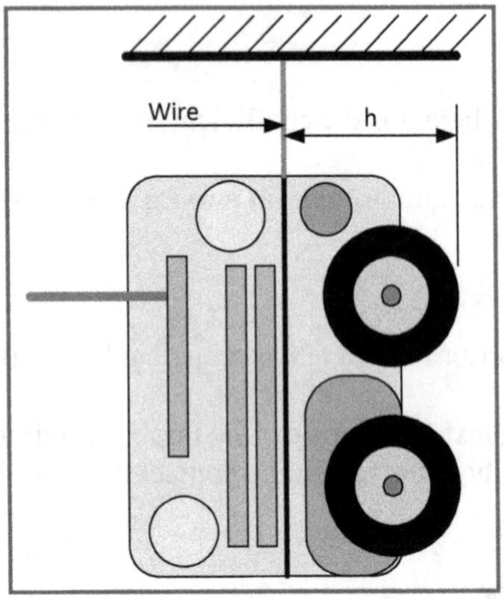

Fig. 14.7 Suspending the robot with a piece wire to experimentally determine the height h of the center of gravity.

Tipping on a slope

Examine the illustration in Fig. 14.8 and visualize how the forces acting on the wheels change as the angle of the slope θ increases. It is evident that as the angle of the slope increases the force on the right wheel normal to the ground increases while that on the left wheel decreases. The tipping point is this force becomes zero for the left wheel.

Next let's consider the tipping condition where F_L and μF_L both vanish, and write the moment equation about the remaining contact point A.

$$\sum M_A = W(w/2) \cos \theta - Wh \sin \theta = 0 \qquad \text{(a)}$$

Solving Eq. (a) for θ yields:

$$\theta = \tan^{-1} w/(2h) \qquad \text{(14.22)}$$

Again it is evident from the results in Eq. (14.22) that the vehicle can negotiate steeper slopes if we maximize w and minimize h.

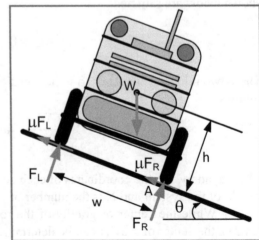

Fig. 14.8 Free body diagram of the vehicle about to tip when navigating a slope.

14.4 DYNAMICS

Dynamics is the study of the effects of external forces on the motion of rigid bodies. For the robot, dynamics is important in assessing your ability to accelerate and steer the vehicle and to overcome friction forces. When the vehicle is in motion, forces in the horizontal direction are required to propel and steer the vehicle. These forces must be controlled to limit acceleration and velocity. Friction forces develop at each wheel (or track) and they act in a direction that opposes motion. The magnitude of the friction forces depend on the geometry of the wheel and the condition of the track over which the vehicle is travelling. With loose dirt, the wheels sink into the surface and rolling friction increases dramatically.

The forces acting on the vehicle for straight line motion are the propulsion forces that act on the driving wheels and the friction forces that act on all of the wheels. If the vehicle chassis contacts the dirt, additional and significant friction forces develop. Wind resistance is negligible for these low velocities.

We introduced friction forces above, but an explanation of friction was deferred because friction did not enter into the analysis of tipping. However, to conduct an analysis of the propulsion force required to achieve a specified acceleration, it is necessary to consider both sliding and rolling friction.

14.4.1 Sliding Friction

Friction occurs when two bodies are in contact and a force is applied to one of the bodies. Sliding frictional forces develop at the contact surfaces that tend to inhibit the movement of one body relative to the other, as illustrated in Fig. 14.9.

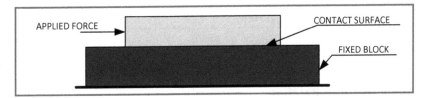

Fig. 14.9 Two bodies in contact with a force applied to the upper block.

If we slowly increase the lateral force applied to the upper block, we observe that it remains stationary. The block remains in equilibrium with a small applied force. If we increase the applied force, the upper block will eventually begin to slide over the surface of the lower block, indicating some upper limit on the frictional forces.

The magnitude of the maximum friction force, F_f, is proportional to the normal force N acting between the bodies as follows:

$$(F_f)_{Max} = \mu N \qquad\qquad (14.23)$$

where μ is the coefficient of friction, a dimensionless property of the contact surface.

The friction force resists motion between the bodies. It is usually independent of the area of the contact surfaces and independent of the velocity except for zero velocity. Representative coefficients of friction are given in Table 14.1 for different material combinations.

The coefficients of friction depend upon whether the body is static or if it is moving. The static coefficients of friction are usually about 20% higher than the dynamic or kinetic coefficients of friction. The static friction force establishes the propulsion force necessary to begin to accelerate the robot. The dynamic or kinetic friction force reduces the trust available to accelerate the robot. Sudden contact of the robot chassis with the surface of the track introduces an undesirable friction force that causes the robot to suddenly decelerate and possibly rotate with subsequent loss of control.

Table 14.1
Friction coefficients for contact between different materials, from www.physlink.com

Materials	Coeff. of Static Friction μ_s	Coeff. of Kinetic Friction μ_k
Steel on Steel	0.74	0.57
Aluminum on Steel	0.61	0.47
Copper on Steel	0.53	0.36
Rubber on Concrete	1.0	0.8
Wood on Wood	0.25-0.5	0.2
Glass on Glass	0.94	0.4
Waxed wood on Wet snow	0.14	0.1
Waxed wood on Dry snow	-	0.04
Metal on Metal (lubricated)	0.15	0.06
Ice on Ice	0.1	0.03
Teflon on Teflon	0.04	0.04
Synovial joints in humans	0.01	0.003

14.4.2 Rolling Friction

Rolling friction occurs when wheels roll. Rolling friction is due to energy losses involved in deformation of the wheel and the surface over which the wheel is rolling. As you drive over a surface your wheels are flattened and the surface is indented over an arc of contact, as shown in Fig. 14.10. For a railroad car with steel wheels rolling on steel rails, the deformation of the wheel and the indentation of the rail are minimized and rolling friction is small. For a robot traveling over soft dirt or sand, the indentation into the track is significant and the rolling friction is a major force to overcome.

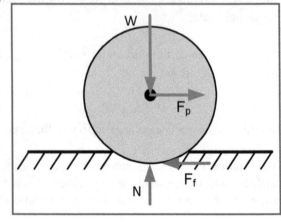

Fig. 14.10 Illustration of a rigid wheel indenting a softer surface and developing a rolling friction force F_f.

Ball and roller bearings are fabricated from smooth hardened-steel and exhibit very low friction coefficients. They minimize the deformation of the balls or rollers as well as the indentation of the contacting surfaces. Another factor contributing to of rolling friction is slipping that may occur between the wheel and the deformed surface, which causes additional energy losses.

Rolling friction is determined in the same manner as sliding friction, as indicated in Eq. (14.24):

$$F_f = \mu_R\, W \tag{14.24}$$

The coefficient of rolling friction μ_R is usually much smaller than the coefficient of sliding friction μ.

14.4.3 Newton's Laws of Motion

Forces acting on a rigid body produce linear and/or angular accelerations. Moments acting on a rigid body produce angular accelerations. Application of the equations from dynamics allows the determination of position, velocity and acceleration (angular or linear) with respect to time. Newton's laws provide the basis for this. Consider Newton's second law, which for linear motion is:

$$\sum \mathbf{F} = \frac{d}{dt}(m\mathbf{v}) \tag{14.25}$$

where $\sum \mathbf{F}$ is the sum of all of the forces acting on the body, m is the mass of the body, and $\mathbf{v}$ is its velocity, and t is time.

The force $\mathbf{F}$ and velocity $\mathbf{v}$ in Eq. (14.25) are vector quantities. The mass m is a scalar quantity:

$$m = \frac{W}{g} \tag{14.26}$$

where W is the weight of the body and g is the acceleration of gravity.

The robot's mass is constant; hence, Eq. (14.25) reduces to:

$$\sum \mathbf{F} = m\frac{d\mathbf{v}}{dt} = m\mathbf{a} \tag{14.27}$$

where $\mathbf{a} = d\mathbf{v}/dt$ is the acceleration vector.

Equation (14.27) is used to define several important fundamental units, as follows:

$$1\ N = 1\ kg\ (\ 1\ m/s^2\ )$$

$$1\ lb_f = 1\ lb_m\ (\ 32.2\ ft/s^2\ ) \tag{14.28}$$

$$1\ lb_f = 1\ slug\ (\ 1\ ft/s^2\ )$$

EXAMPLE 14.4

Determine the acceleration of an robot with a mass of 1.2 kg subjected to a net force of 2 N in the x direction and a force of 3 N in the y direction. The x and y directions form a horizontal plane.

Solution:

To begin, convert Eq. (14.25) from its vector format to its scalar form and write:

$$\sum F_x = m\frac{dv_x}{dt} = ma_x \qquad\qquad \sum F_y = m\frac{dv_y}{dt} = ma_y \qquad\qquad (14.29)$$

Substituting into Eq. (14.29) yields:

$$a_x = F_x/m = 2/1.2 = 1.667 \text{ m/s}^2 \text{ and } a_y = F_y/m = 3/1.2 = 2.50 \text{ m/s}^2 \qquad (a)$$

The magnitude of the total acceleration a is given by the vector sum of the Cartesian values:

$$a = (a_x^2 + a_y^2)0.5 = 3.005 \text{ m/s}^2 \qquad (b)$$

The direction of the acceleration vector is given by:

$$\theta = \tan^{-1}\frac{a_y}{a_x} = \tan^{-1}\frac{2.5}{1.667} = 56.30 \text{ degrees} \qquad (c)$$

14.4.4 Angular Acceleration

In Section 14.4.3, where Newton's second law was introduced, the discussion focused on linear acceleration. However, when moments are applied to a body it rotates. This rotary motion is expressed by a modified version of Newton's second law:

$$\sum M = \frac{d}{dt}(I\omega) \qquad\qquad (14.29)$$

where ΣM is summation of the moments acting on the body and is a vector quantity. Quantity I is the mass moment of inertia and is a scalar quantity, while ω is the angular velocity and is a vector.

For robots the mass moment of inertia I is constant. Thus we can reduce Eq. (14.29) to:

$$\sum M = I\frac{d\omega}{dt} = I\alpha \qquad\qquad (14.30)$$

where α is the angular acceleration of the body.

The mass moment of inertia of a plane body with an area A is determined by:

$$I = \int r^2 \, dm \tag{14.31}$$

Performing this integration on four different elementary geometries gives:

$I = mR^2$ for a ring with a radius R
$I = mR^2 / 2$ for a disk with a radius R
$I = mL^2 / 12$ for a long thin rod of length L rotating about its centerline
$I = mL^2 / 3$ for a long thin rod of length L rotating about its end

---

EXAMPLE 14.5

If the vehicle, shown in Fig. 14.11, was traveling in a straight line at a velocity v = 0.05 m/s when it encounters a loose patch of dirt. At this instant and the friction forces acting on the two front wheels change with Ff1 = 2.4 N and Ff2 = 1.8 N develops. The two propulsion forces do not change and each remain equal to 2.1 N. If the vehicle weighs 15 N and has a mass moment of inertia about O of 0.08544 kg-m2, determine the orientation of the vehicle 1.2 s after it has engaged the area of loose dirt.

Fig. 14.11 Vehicle traveling in a straight line with a velocity v engages a patch of loose dirt and the rolling friction forces differ.

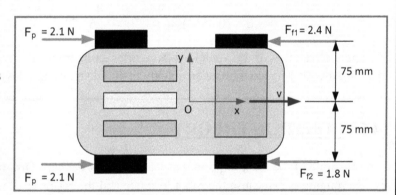

Solution:

The mass of the vehicle is given by:

$$m = \frac{W}{g} = \frac{15.0 \text{ N}}{9.807 \text{ m/s}^2} = 1.530 \text{ kg} \tag{a}$$

Examination of Fig. 14.11 indicates that a moment is produced. The sum of the moments about the center of gravity at point O is written as:

$$\Sigma M_O = F_p (0.075 \text{ m}) - F_p (0.075 \text{ m}) + F_{f1} (0.075 \text{ m}) - F_{f2} (0.075 \text{ m})$$

$$= (0.075 \text{ m})(2.4 \text{ N} - 1.8 \text{ N}) = 0.045 \text{ N-m} = 45 \text{ N-mm} \tag{b}$$

This moment (torque) will impart an angular acceleration of:

$$\alpha = \frac{\Sigma M}{I_C} = \frac{0.045 \text{ N-m}}{0.08544 \text{ N-m-s}^2} = 0.5267 \frac{\text{radian}}{\text{s}^2} \tag{c}$$

The angular velocity ω is obtained by integrating the angular acceleration α with respect to time to obtain at time t = 1.2 s:

$$\omega = \int \alpha \, dt = \alpha \, t = (0.5267)(1.2) = 0.6320 \, \frac{radian}{s} \qquad (d)$$

Next integrate the angular velocity with respect to time to determine the angle of rotation. Assume that the initial angular velocity is $\omega_o = 0$, and the initial orientation angle is $\theta_o = 0$.

$$\theta = \int \omega \, dt = \alpha \, t^2 / 2 \qquad (e)$$

Substituting t = 1.2 s yields:

$$\theta = \frac{(0.6320)(1.2)^2}{2} = 0.4551 \text{ radian} \qquad (f)$$

The angle of rotation is increasing as the square of the time. Converting to degrees yields:

$$\theta = (0.4551)(360°)/(2\pi) = 26.07° \qquad (g)$$

The angle of rotation of the vehicle is large after an interval of only 1.2 seconds. Clearly, corrections by adjusting the steering must be made quickly to avoid significant deviations from the intended path.

14.5 CONTROL THEORY

Control is one of the most important aspects of robot design. Discontinuities in the density of the dirt or surface roughness can cause the robot to rotate and the result is a loss of time as adjustments are made to resume the programmed course. As indicated in Example 14.5, the time available to adjust the direction of the robot is limited. The first question for you to consider is the time required for your navigation system to provide a signal indicating that the robot has deviated from its intended course? Let's assume that you are navigating with a photodiode light sensor that receives signals from navigation beacons provided by your instructor.

The frequency response (bandwidth) of a photodiode in a light sensor is of the order of 10 MHz. The signal from the light sensor is digitized with a 10 bit analog to digital converter operating with a throughput of up to 20 MIP (million instructions per second) requiring less than a microsecond to convert the signal into a form that can be processed in the Arduino. The Amtel® ATmega328 microprocessor in the Arduino UNO is fast and will execute your correction commands in a few microseconds. Thus, it appears that the microcontroller can respond quickly and issue output signals to the steering mechanism to alter course.

It is possible to experimental verify the Arduino PCB response time. You may want to consult with your instructor and obtain the instrumentation necessary to make this measurement.

Clearly, the Arduino PCB and the software program should enable very rapid adjustments of the driving and steering motors to correct for course deviations. This conclusion raises a second question. Why does it take so much time for the motors to respond?

Examine the motors. If you deviate from the intended course, you probably have programmed your Arduino microcontroller to rotate the steering motor to adjust the direction of the vehicle. While the motors are small they have inertia to overcome and they do not achieve their rotational velocity instantaneously. In addition the motors probably drive a set of gears that provide a major reduction in RPM. The gear train adds more inertia and again the system cannot respond instantaneously.

How do you deal with the need to quickly correct the direction of the robot to accommodate deviations due to low density dirt patches, surface irregularities, obstacles, etc. One way is to adapt good driving habits and slow down when you encounter bad road, snow or ice. A lower velocity or even stopping limits the amount of the error in the course and enables the correction to be made more easily. A second approach is to design a control system for steering and propelling that responds rapidly. We will discuss this topic in more detail later when we described design alternatives for propulsion and for steering.

14.5.1 Autonomous Control

A photograph of the Arduino UNO (R3) microcontroller is presented in Fig. 14.12. The Arduino UNO is a microcontroller board based on the ATmega328 microcontroller. UNO means one in Italian and is named to mark the new release of this new Arduino board. The UNO will be the reference version of the USB Arduino boards as revised designs are introduced in the future. Uno is the latest in a series of Arduino microprocessors and the reference model for the Arduino platform.

The Arduino UNO (R3) microcontroller has 14 digital input/output pins (of which six can be used as PWM outputs), six analog inputs, a 16 MHz ceramic resonator, a USB connection, a power jack, an ICSP header and a reset button. The assembly contains everything needed to support the microcontroller. To start the system, connect it to a computer with a USB cable or power it with an AC-to-DC adapter or a battery. The board can operate on an external supply of 6 to 20 volts. However, if it is supplied with less than 7V, the 5V pin may provide less than five volts and the system may become unstable. If you supply more than 12V, the voltage regulator may overheat and damage the board. The recommended range is 7 to 12 volts.

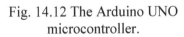
Fig. 14.12 The Arduino UNO
microcontroller.

The Arduino UNO has a number of facilities for communicating with a computer, another Arduino, or other microcontrollers. The ATmega328 provides UART TTL (5V) serial communication, which is available on digital pins 0 (RX) and 1 (TX). An ATmega16U2 on the board channels this serial communication over USB and appears as a virtual com port to software on the computer. The 16U2 firmware uses the standard USB COM drivers, and no external driver is needed. However, for computers using Microsoft Windows a .inf file is required. The Arduino software includes a serial monitor which allows simple textual data to be sent to and from the Arduino board. The RX and TX LEDs on the board will flash when data is being transmitted via the USB-to-serial chip and USB connection to the computer, but not for serial communication on pins 0 and 1.

Up to six sensors may be employed to provide input signals to the Arduino, although efficient control algorithms can be written using the input from fewer sensors. The analog input ports are capable of up to 5 V. These signals are converted from analog to digital with a 10 bit A to D converter. The 14 digital input/output ports provide current up to 40 mA to activate transistors, relays, and other low current devices.

A software serial library allows for serial communication on any of the UNO's digital pins. The ATmega328 also supports I2C (TWI) and SPI communication. For SPI communication, use the SPI library. The Arduino software includes a wire library to simplify use of the I2C bus.

The maximum length and width of the UNO circuit board are 2.7 and 2.1 in. respectively, with the USB connector and power jack extending beyond the length dimension. Four screw holes allow the board to be attached to any flat surface. Note that the distance between digital pins 7 and 8 is 160 mil, not an even multiple of the 100 mil spacing of the other pins.

Programming the Arduino may be accomplished using several different codes. These include Arduino C/C++, avr-gcc, avrdude, AVR Studio, etc. The details of programming in Arduino C/C+ will be described later in Chapter 15.

14.5.2 Open-loop Control

Electronic measurement systems are used in two types of process control: open-loop or operator control and closed-loop or automatic control. Control theory may be applied to processes used to produce chemicals, machine tools used to fabricate parts, vehicles to maintain stability, etc. Control is an important concept that is widely used in several engineering disciplines. Open-loop control, involving a process that is being monitored with several transducers, is illustrated in Fig. 14.13. Data from the transducers are displayed continuously on an instrument panel containing charts, meters and digital displays. An operator observes these quantities and makes the necessary adjustments to the input parameters to maintain control of the process. When you are driving an automobile, you are serving as an open-loop operator, and closing the loop in open-loop control. Open-loop control is effective when decisions regarding adjustments to the control parameters do not need to be made quickly. However, when conditions are changing rapidly, sufficient time is not available for human intervention and closed-loop control becomes mandatory.

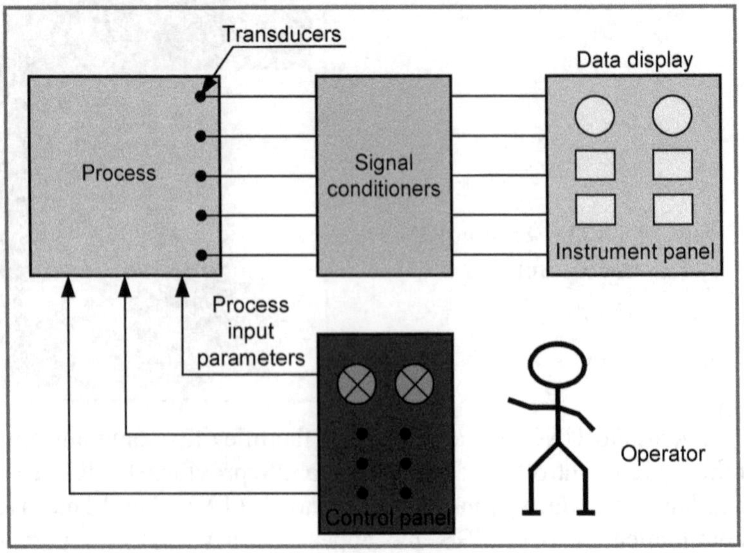

Fig. 14.13 Schematic diagram of open-loop process control that requires an operator to monitor and adjust process parameters.

14.5.3 Closed-loop Control

A second type of process control, known as automatic or closed-loop control, is illustrated in Fig. 14.14. In a closed-loop control system, the operator is not required except for system maintenance. Instead, the signals from an electronic measurement system are compared to command signals that represent voltage-time relationships for the important quantities associated with the process. The first controller measures the difference between the command signal and the transducer (sensor) signal and develops an error or feedback signal. The feedback signal is then transmitted to a second controller where it is amplified and used to drive devices that correct the process. Closed loop control is required for this project.

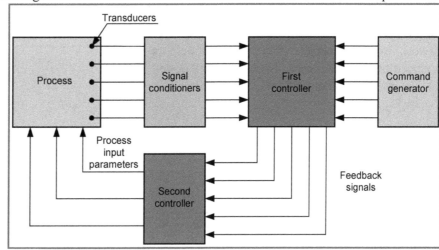

Fig. 14.14 Block diagram of closed-loop process control that does not require operator intervention.

14.5.4 Control Limits

Even with closed-loop control, it is impossible to maintain a process parameter at a precise value. Some deviation from the value specified for the process parameter always occurs. The error bands on the specified value give upper and lower control bands, as illustrated in Fig. 14.15.

Adjustments to the process are made when a control parameter exceeds its upper or lower control limit. In simple control systems, the adjustments are made by turning some quantity on or off (for instance a motor). This on-off adjustment sequence is often called bang-bang control. It produces the response illustrated in Fig. 14.15, where the measurements of the parameter value oscillate between the upper and lower control limits. More efficient control is achieved using proportional control where the adjustments are less drastic. For instance, the speed of the motor is decreased by 10% instead of turning the motor off. With proportional control the measurement of the control parameter will vary from the specified values but the frequency of the variation is smaller and the error encountered is usually smaller.

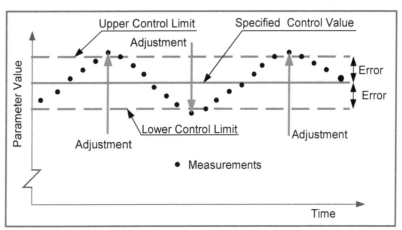

Fig. 14.15 Control limits and adjustments made to maintain the parameter value to within the upper and lower limits.

6.6 MOTORS and GEARS

14.6.1 DC Motors

Most of the motors found in your home are ac devices that operate using power from your local utility at 120 or 240 Vac. Your robot will not have a source for ac power, but will rely on the dc power provided by a battery supply. For this reason we will only describe the operation of two types of dc motors — the classical dc motor and the stepper motor.

The classical dc motor employs axial conductors which move in a magnetic field in a cylindrical gap between two iron cores. The magnetic field is produced by stationary coils that are wound around pole pieces incorporated into the stator as shown in Fig. 14.15.

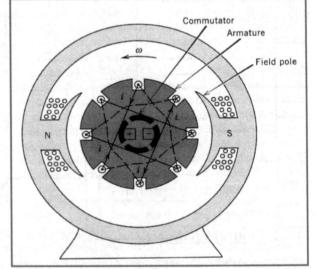

Fig. 14.15 Configuration of armature, field poles and commutator in a dc motor.

In practice, the poles (alternating N and S around the circumference of the stator) are excited by direct current flowing in the field windings. The rotor, or armature, is a cylindrical iron core that carries axial conductors embedded in slots and connected to segments of a commutator. The current to the armature coils is provided through fixed brushes which slide on the rotating commutator. In smaller dc motors the rotor is equipped with permanent magnets eliminating the axial conductors and the commutator. The speed-torque relations for all dc motors are given by:

$$V = K\phi\omega + IR_A \qquad\qquad (14.32)$$

$$T = K\phi I \qquad\qquad (14.33)$$

where V is the terminal voltage; ϕ is the magnetic flux per pole (in Weber); I is the armature current; R_A is the resistance of the armature winding and other resistances; K is a constant for a specific motor and T is the torque produced by the motor.

A small dc motor used in many hobby projects is shown in Fig. 14.16. This motor is operated at voltages than can be varied from 1.5 to 3 Vdc. At no load and maximum voltage input the motor turns at 12,300 RPM. The stall torque is 360 g-mm when it draws a current of 2.1 A. While there are many small dc motors available, they all rotate at high RPM and deliver relatively low torque. The characteristics are not consistent with the requirements for the robot.

Fig. 14.16 A small low cost dc motor for hobbyists.

Typical speed-torque characteristic of these small dc motors is shown in Fig. 14.17.

Fig. 14.17 Speed torque curves for a typical small dc motor.

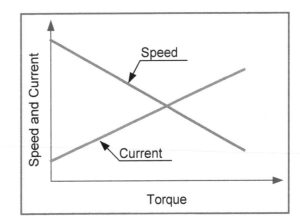

In model vehicles, dc motors can be employed to drive both the steering mechanism and the drive train. The motor speed and output torque can be controlled by adjusting their input voltage. However, the motor RPM is excessive and the torque output is too low. To resolve these two issues, the dc motors are used to drive a gearbox that reduces the RPM while increasing the output torque.

Stepper Motors

Stepper motors differ in several ways from the dc motors described previously. As shown in Fig. 14.18, the stator incorporates a large number of small field poles deployed around its circumference. Moreover, each pole piece is fabricated with several teeth (5 in this example). The armature is constructed with one or two rows of teeth uniformly spaced around its periphery, as shown in Fig. 14.18c. Fabrication of the armature core from permanent magnet materials eliminates the need for the armature coils, the commutator and the brushes.

The stepper motor is driven by a train of pulses delivered to the windings of the field coils. The pulse characteristics (rise time, duration, decay time and amplitude) are matched to the inertia of the motor so that the armature rotates one step for each pulse. Angular position control is accomplished with stepper motors simply by counting the number of input pulses applied to each set of field coils. The speed of a stepper motor is controlled by adjusting the rate of these pulses.

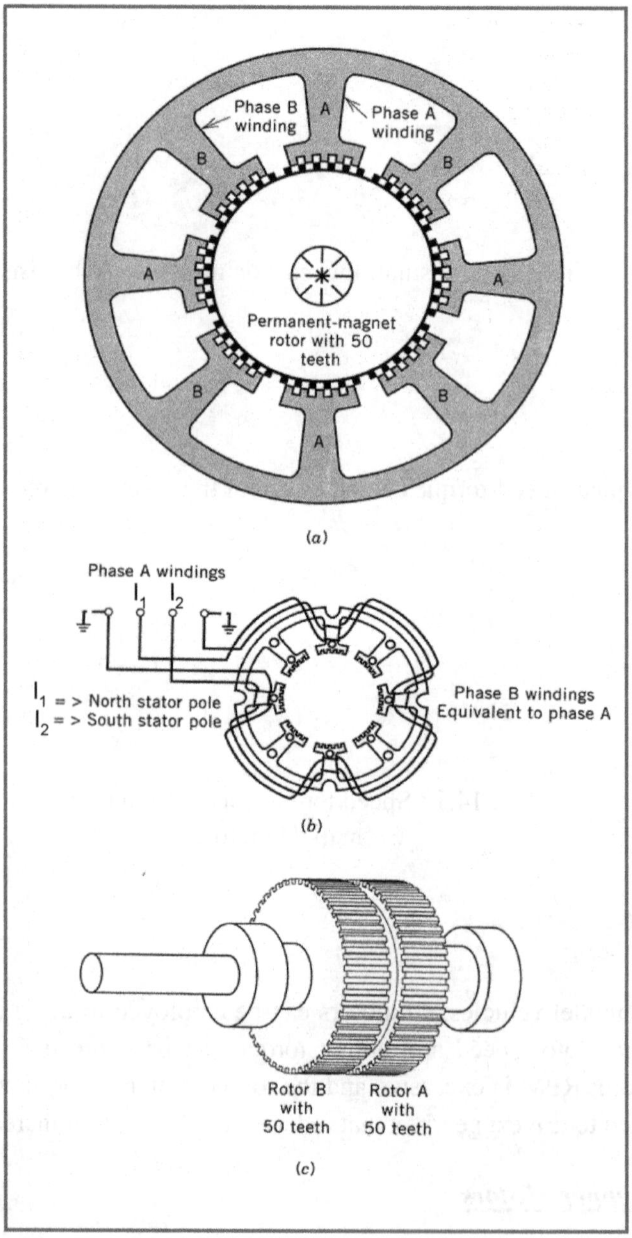

Fig. 14.18 Construction details of a stepper motor.
(a) stator.
(b) field coil windings.
(c) permanent magnet armature.

A typical permanent-magnet stepper motor with a 50 tooth pitch and a four-pole single phase stator produces $50 \times 4 = 200$ steps per revolution. This construction provides for a rotation of 1.8 degrees per step or per pulse. When a second row of teeth is used in constructing the armature, (see Fig. 14.18c) that row is displaced by 1/2 pitch with respect to the first row; hence, the resolution of the stepper motor is improved to 0.9 degrees per step. The precision of a stepper motor is excellent if its torque capacity is not exceeded. Regardless of the number of steps (N), the position of the armature will be 1.8 N degrees with an error of less the 3% of 1 step or 3.24 minutes.

A small uni-polar stepper motor, presented in Fig. 14.19, provides 512 steps per revolution, which is sufficient for steering. The design is not typical as the stepper motor has only 8 steps per revolution, but its output shaft drives a 1/64 gear reducer, producing $8 \times 64 = 512$ steps per revolution. This stepper motor is compatible with the motor shield for Arduino.

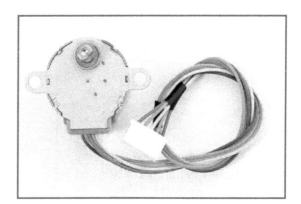

Fig. 14.19 A small stepper motor with a gear reducer
that is used to improve its resolution.

14.6.2 Gearing

Gears are employed to change the speed of rotation and the torque delivered by a motor or engine. The transmission on your automobile is a gear box with several gear sets to progressively increase the auto's speed as the torque required decreases. Because electric motors (excluding stepper motors) turn at high speed, you will be using a gear box to reduce the speed of your robot.

Let's review both gears and belts as methods for changing shaft speed and direction of rotation and output torque.

Simple Spur Gear Reducers

Consider a small spur gear[1] with 8 teeth that is attached to the shaft of a small motor, as shown in Fig. 14.20. The motor is mounted on a frame that also supports the second gear in the drive, which is a large spur gear (40 teeth). The 40-tooth spur gear is attached to an output shaft. The supporting frame for the motor and the output shaft is constructed with a center-to-center distance that enables the pinion and large gear to mesh without interference or an excessive clearance. The output shaft from the large gear drives a pulley or a wheel at a speed that is much lower than the speed of the motor.

     The drive train illustrated in Fig. 14.20 is a simple example of a gear reducer. The shaft from the large gear will rotate at 1/5 the speed of the motor shaft. How do we know the speed reduction? It is as simple as counting teeth. To show this fact, we will introduce a few equations that are useful in designing gear trains.

The relation among power P, torque T and angular velocity ω for any given shaft is given by:

$$P = T \omega \qquad\qquad (14.34)$$

where P the power is measured in watts, W; T the torque is measured in newton-meters, N-m; and ω the angular velocity (speed) is measured in radians.

When using the gearbox, shown in Fig. 14.20, to **reduce** the angular velocity by a factor of five, we **increase** the torque by the same amount[2]. Because gear reduction is often necessary, many suppliers

---

[1] The smallest gear is called the pinion.
[2] The effects of friction are neglected in this conversion. In practice, friction decreases the available torque in the speed-torque tradeoff.

market dc motors equipped with an integral gearbox. These motors rotate at lower speeds, but they deliver much more torque than motors without gear reduction.

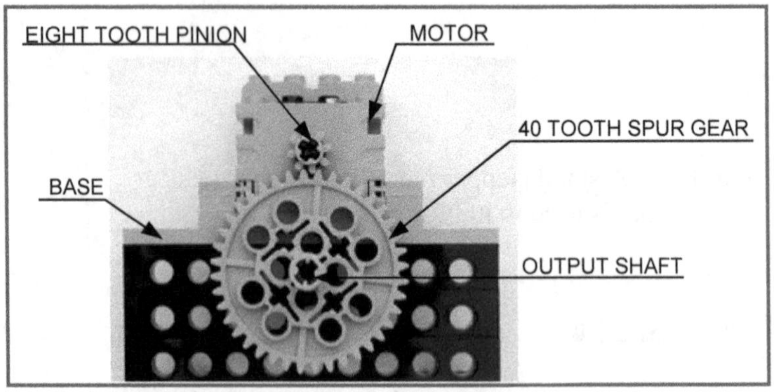

Fig. 14.20 A simple gear reducer consisting of a motor, two shafts and two gears.

Speed reduction is achieved by meshing together two or more gears of different diameters. The pinion, the smaller of the two gears, is attached to the motor shaft and the larger gear is attached to the output shaft. The angular velocity of the output shaft is given by:

$$\omega_O = (N_P / N_G)\omega_I \qquad (14.35)$$

where ω_I and ω_O represent the angular velocities of the input (motor) and output shafts respectively and N_P and N_G are the number of teeth in the pinion and gear respectively.

The torque from the output shaft is given by:

$$T_O = (N_G / N_P)T_I \qquad (14.36)$$

where T_O and T_I are the output and input torque, respectively.

Equations (14.35) and (14.36) indicate the tradeoff between torque and angular velocity that a pair of meshing gears provides. In designing the drive train in Fig. 14.24, an 8-tooth gear was meshed with a 40-tooth gear. The reduction in the angular velocity was by a factor of $40/8 = 5$, and the output torque increased by a factor of five.

The directions of rotation of the input and output shaft of the simple drive train shown in Fig. 14.24 are opposite hand. In other words—if the motor shaft turns counterclockwise, the output shaft from the gearbox turns clockwise. We show this change in rotary direction in Fig. 14.21.

Fig. 14.21 Mating gears rotate in opposite directions.

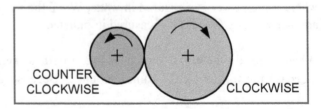

Bevel Gears

Bevel gears are similar to spur gears because they are employed to tradeoff angular velocity with torque. Bevel gears have teeth that are inclined (beveled) relative to the axis of the shaft as shown in Fig. 14.22. The inclination of the teeth enables bevel gears to change the orientation of the axis of rotation by 90°. The shafts of two mating spur gears are parallel; however, the shafts of two mating bevel gears are perpendicular. Equations (14.34), (14.35) and (14.36) also apply to bevel gears.

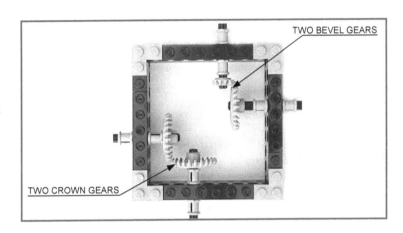

Fig. 14.22 Bevel and crown gears used to change orientation of shafts and to reduce speed and increase torque.

Compound Spur Gear Reducers

In some applications, it is necessary to reduce the shaft speed by amounts larger than feasible with a single pair of spur gears. Very large speed reductions are achieved with compound gear drives that incorporate additional shafts and gears. For example, a compound gear reducer with three shafts and four gears is illustrated in Fig. 14.23. The first shaft, extending from the motor, supports an 8-tooth pinion gear. The second shaft is an intermediate, supports two spur gears—the first is a 40-tooth gear that meshes with the motor pinion and the second is another 8-tooth pinion. The third shaft is the output, which supports a 40-tooth spur gear that meshes with the pinion on the intermediate (second) shaft.

Let's use the previous equations to determine the angular velocity and torque of the output shaft in terms of the speed and torque of the motor. The angular velocity of the intermediate shaft is given by Eq. (14.35) as:

$$\omega_{Int} = (N_{P1} / N_{G2})\omega_I \qquad\qquad (14.37)$$

where ω_I and ω_{Int} are the angular velocities of the input and intermediate shafts, respectively and N_{P1} and N_{G2} are the number of teeth on the pinion and intermediate shaft gear, respectively.

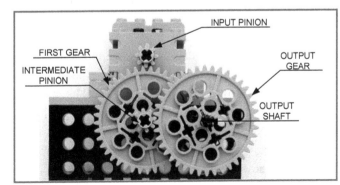

Fig. 14.23 A compound gear reducer with a motor, three shafts and four spur gears.

Next, we apply Eq. (14.35) considering the intermediate shaft to be the input shaft to the third output final shaft.

$$\omega_O = (N_{P2} / N_{G3})\omega_{Int} \qquad (14.38)$$

where ω_{Int} and ω_0 represent the angular velocities of the intermediate and output shafts, respectively and N_{P2} and N_{G3} are the number of teeth on the intermediate shaft pinion and the output shaft gear, respectively.

Substituting Eq. (14.37) into Eq. (14.38) yields:

$$\omega_O = (N_{P2} / N_{G3})(N_{P1} / N_{G2})\omega_I \qquad (14.39)$$

Following a similar process, we can show that torque from the output shaft is given by:

$$T_O = (N_{G2} / N_{P1})(N_{G3} / N_{P2})T_I \qquad (14.40)$$

Rack and Pinion

Gears are also used to convert rotary motion to linear motion. Meshing a rack with a pinion accomplishes this conversion. A typical rack and pinion is illustrated in Fig. 14.24. As the pinion rotates, the rack translates to produce a linear motion. The distance s traveled by the rack is given by:

$$s = \pi d_p n \qquad (14.41)$$

where d_p is the pitch diameter of the pinion; n is the number of rotations of the pinion.

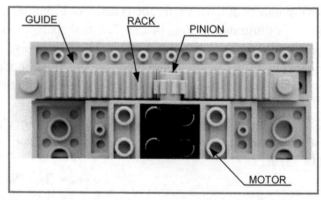

Fig. 14.24 Rack and pinion gears generate linear motion from rotary motion.

One of the difficulties with using the rack and pinion is the high velocity of the rack that is produced if the pinion is mounted directly to the motor as shown in Fig. 14.24. The pinion has 8-teeth and the length of the rack is usually limited. With a motor speed of about 125 RPM, how long do you think it will take the motor to spit out the rack?

To circumvent this difficulty and achieve more significant reductions in motor speed, we employ a worm gear that is attached to the motor's shaft, as shown in Fig 14.25. For every turn of the worm gear, the rack advances by the pitch[3] of the rack. The time required to move the rack a specific distance is increased and it becomes much easier to manage the motion when programming the motor to start and stop within specific time limits.

---

[3] The pitch is the distance between the teeth on the rack.

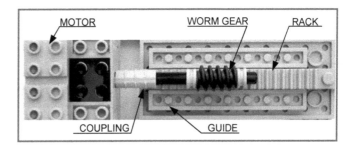

Fig. 14.25 Rack and worm gear to generate
linear motion from rotary motion.

Pulleys and Belts

Pulleys are like gears except they have a groove around their circumference instead of teeth. Two pulleys with a belt can be used to reduce angular velocity and increase torque. A pulley-belt arrangement is shown in Fig. 14.26. The speed change is a function of the diameter of the two pulleys as indicated by:

$$\omega_O = (d_S / d_L)\omega_I \qquad (14.42)$$

where ω_I and ω_O represent the angular velocities of the small pulley (input) and the large pulley (output), respectively and d_S and d_L are the diameters of the small and large pulleys, respectively.

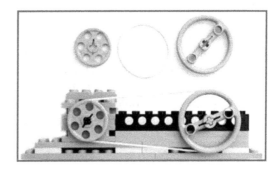

Fig. 14.26 Belt and pulley arrangement with a motor
driving the smaller diameter pulley.

The output torque T_O from the large pulley is related to the input torque T_I on the small pulley by:

$$T_O = (d_L / d_S)T_I \qquad (14.43)$$

The belts are fabricated from reinforced rubber. Belt drives are often preferred over gear drives, because they are much less expensive. Aligning a belt drive is much less critical than aligning meshing gears. However, gear drives are preferred when it is important to maintain accurate rotary position[4]. Belts tend to slip and stretch under load and the angular position of one pulley relative to the other cannot be assured.

Gear Head Motors

There are a large number of motors available, where the gear box and the dc motors have been integrated, such as the one shown in Fig. 14.27. This particular unit is capable of four speeds with gear selection to change the speed. It incorporates a shaft that extends out of both sides of the unit and it can serve as the drive motor for a pair of wheels. With the availability of many different motors and integrated gear box combination, you may be able to procure the unit that provides adequate torque at the speed you seek.

---

[4] An exception is the timing belt on your automobile. In this case, the pulleys have teeth that mesh with the teeth molded on the reinforced rubber belts. While this is a belt drive, it has teeth to insure positive engagement that gives it the characteristics of a gear drive.

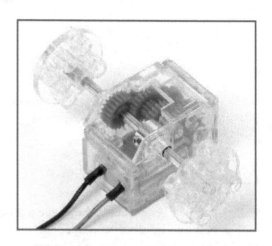

Fig. 14.27 A motor gear box combination that is
capable of four different speeds.

14.6.3 <u>Steering Methods</u>

There are two ways of steering depending on whether you have a tracked or wheeled vehicle. With a tracked vehicle steering is accomplished by stopping the track on one side, while maintaining the rotation of the track on the other side. If you want to turn right you stop or slow the track on the right side and permit the track on the left side to continue at normal speed. After you have executed the change in direction, you resume the speed of the right side track and proceed to travel in a straight line, but in a new direction.

With wheeled vehicles, steering is usually accomplished by rotating a pair of wheels together, as illustrated in Fig. 14.28. The pair of wheels can be on the front or the back of the vehicle. Automobiles are designed with front wheel steering. The rotation of the shaft connecting the two wheels is limited, because a large and abrupt rotation of this shaft could cause excessive stresses imposed on the tire, skidding and loss of vehicle control.

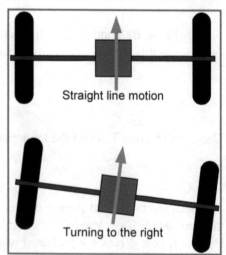

Fig. 14.28 Steering by rotating the two front wheels together.

The turning radius R for a four wheel vehicle can be approximated by:

$$R = \frac{D}{2\sin\theta}\sqrt{\sin^2\theta + 4\cos^2\theta} \qquad (14.44)$$

where D is the distance between the front and rear shafts and θ is the angle of rotation of the shaft.

If the shaft connecting the two front wheels is rotated through an angle $\theta = 30°$, the turning radius is 1.8D. If the distance between the shafts for the front and rear wheels is 250 mm then R = 450 mm. These are typical values for D and θ. Is this turning radius sufficient to enable the maneuvers that you will require of your vehicle?

To avoid sudden rotations of the steering shaft the motor selected to drive the steering mechanism must rotate slowly and be able to reverse slowly. You may want to consider a stepper motor for this application where the rotational speed is controlled by the frequency of the pulses applied, the angle of rotation by the number of pulses and the direction or rotation by the polarity of the pulses.

A more detailed treatment of the dynamics of steering wheeled vehicles is found in reference [4].

14.6.4 Power Supplies

Your robot will be equipped with one or two drive motors, sensors, electronics, etc. All of these devices will require power and the voltage requirements for each device may differ. To meet these requirements, you will need to equip your robot with a suitable battery supply, voltage regulator(s) and motor controls. Let's first consider batteries.

What is the required capacity of this power source? If we select a power source that has excess capacity, we incur a weight penalty and add unnecessary cost to our parts list. If we select one that is too small, we may exhaust the battery before the mission is complete. The current drawn by the drive motors varies with the robot speed, but suppose you have two drive motors each requiring about 2 A maximum at stall, and a stepper motor requiring an average of 0.5 A. Consequently, you will need an auxiliary power supply that can deliver about 4.5 A for sufficient time to complete the mission. This is a crude estimate and the actual current draw will depend on the choice of your components. The voltage requirements will vary with the component, but will probably not exceed about 8 Vdc.

Let's assume that your team has a strategy for completing a challenge in 6 minutes (1/10 h). The capacity C_B of the battery is given as the product of the amperes multiplied by the time that this level of current is delivered.

$$C_B = I\,t \tag{14.45}$$

where C_B is the capacity measured in (Ah) and I is the current draw in A and t is the time in h.

For the 6 minute mission, the capacity of the auxiliary supply is $C_B = 4.5 \times 1/10 = 0.45$ Ah. This value is small compared to the capacity of the battery in your automobile, but you certainly do not want to deal with the weight of an auto battery.

There are four types of batteries that provide this capacity and current drain—Lithium-ion, Nickel Metal Hydride, Nickel Cadmium and Sealed Lead Acid (SLA). All four of these types are rechargeable, which is a requirement to avoid battery replacement after each mission. The sealed lead acid battery is the lowest cost for the capacity, but it is too heavy for the robot. The Lithium-ion battery has the highest energy density and a long life, but is the most expensive. The Lithium-ion battery pack, shown in Fig. 14.29, is made of three balanced 2.2 Ah cells for a total capacity of 6.6 Ah. The cells are connected in parallel and spot-welded to a protection circuit that provides over-voltage, under-voltage and over-current protection. Each cell can provide a current of 4.4 A for a total peak current of 13 Amps. However, to avoid damage by excessive current draw, it is recommended that the current not exceed 6 A in the steady state. This battery pack is priced at about $30.

Fig. 14.29 A lithium-ion battery pack that delivers 6.6 Ah at 3.7 V with a steady state current draw of 6 A.

Nickel-Metal Hydride (NiMH) batteries, shown in Fig. 14.30, are a rechargeable alternative to lithium-ion batteries. NiMH batteries usually have up to three times the capacity of an equivalent sized nickel-cadmium (NiCad) battery. NiMH batteries can be recharged and reused hundreds of times. This type of battery is suited for high drain applications where current draw is significant. It can be recharged at any time as it does not exhibit memory effects. The cost of this battery is about $5.

Fig. 14.30 A Nickel-Metal Hydride battery with a capacity of 0.250 A-h at 8.4 V.

The Nickel-Cadmium battery pack, shown in Fig. 14.31, has a large capacity of 5.0 A-h. This battery pack contains 10 size D NiCd batteries arranged to deliver 12 V. NiCd batteries have a high current capability and this pack can meet a current draw of 0.5 A delivering this current for eight to nine hours. This battery pack is priced at $37.00 and weighs 3 lb. Clearly this is more capacity that you need, but you may want to consider a pack with a smaller number of cells.

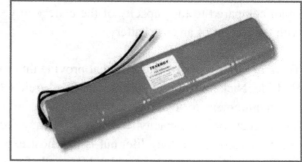

Fig. 14.31 A NiCad battery pack with a capacity of 5.0 A-h at 12 V.

It appears that the Lithium-ion battery will provide sufficient capacity in a small package, but the cost is higher than the Nickel-Metal Hydride battery. The Nickel-Metal Hydride battery is the most cost effective but it will weigh more than the Lithium-ion battery.

What type of battery is in your cell phone? Why? Do you plan on minimizing the weight of your robot or do you want to added weight over the drive wheels to improve traction.

14.6.5 Arduino Shields

Shields are available for use with the Arduino microcontroller that provide the electronics needed for control of the drive motors. The shield, presented in Fig. 14.32, will control two dc motors and conducts up to 2 A per channel. The shield takes its power from the same voltage input line as the Arduino microprocessor. It also includes blue and yellow LEDs to indicate active direction, and all its driver lines are diode protected from back EMF.

Fig. 14.32. A shield for control of two dc motors that is compatible with the Arduino microprocessor.

You may want to explore the website for Sparkfun.com, where several Arduino shields are listed and described.

REFERENCES

1. Shukla, A and J. W. Dally, Instrumentation and Sensors for Engineering Analysis and Process Control, College House Enterprises, Knoxville, 2012.
2. Dally, J. W., and R. J. Bonenberger, Mechanics I: Statics++, College House Enterprises, Knoxville, 2010.
3. Anon, www.robotc.net.
4. Jazar, Reza N., Vehicle Dynamics: Theory and Application, Springer, New York, pp. 379-450, 2008.

EXERCISES

6.1 Measure the torque of your drive motor as a function of the voltage applied. Measure the current draw while you are measuring the torque.

6.2 Measure the thrust developed by your drive motors in the stall condition. Prevent the wheels or tracks from moving when making this measurement.

6.3 Measure the friction forces that are generated when you drive your robot over a hard flat surface. What factors contribute to the friction forces measured.

6.4 Estimate the acceleration of your robot using the measurements made in Exercises 14.1, 14.3 and 14.4.

6.5 You plan to operate your drive motors for four seconds then to shut them down and correct the orientation of the robot in two seconds using the GPS data transmitted to the robot. This cycle will be repeated until your robot completes its mission. Determine the acceleration required to complete a mission involving traveling 20 m in 60 seconds (ten cycles). State the assumptions used in this analysis.

6.6   Your robot is subjected to propulsion forces F_p = 2.4 N and friction forces F_f = 1.3 N, determine the forward velocity v as a function of time from t = 0 to t = 5 s. The mass of your robot is 1.55 kg. Comment on the magnitude of this velocity at t = 5s.

6.7   For the robot in Exercise 14.6 subjected to the forces F_p = 2.4 N and F_f = 1.3 N, determine the distance s traveled as a function of time from t = 0 to t = 5 s. The mass of your robot is 1.55 kg. Comment on the magnitude of this distance at t = 5s.

6.8   Using the results of Exercise 14.7, draw a sketch showing the track of the robot relative to the dirt track marking the position after an elapsed times in one second intervals from one to five seconds.

6.9   Write a one page description of open-loop control.

6.10   Write a one page description of closed-loop control.

6.11   Prepare a strategy for writing a program to control your robot to complete the mission assigned by your instructor. Include in this description your goals for the maximum and average velocities achieved, the type, placement and number of sensors, the technique to handle friction events, the method for navigating to the target, etc.

6.12   Prepare a step by step description of the program you will write to control the robot as it travels over the track to complete its mission.

6.13   Prepare an engineering brief describing the strategy for the design of the steering mechanism. Include a tradeoff discussion on the relative merits of a stepper motor versus a gear head motor for rotating the wheel shaft to change the robot direction.

6.14   Prepare an engineering brief describing the strategy for the design of the robot if your instructor does not place a restriction on its weight. Do you want to develop a heavy robot on one that is light? Does the weight matter?

6.15   Preparing an engineering brief describing the rational for the selection of the chemistry of your battery power supply.

6.16   Prepare an estimate of the capacity required for your power supply and compare this value to the capacity of the power supply you procured. Justify the safety factor involved in your selection.

6.17   If your mission involves finding a candle with a flame that is to be extinguished without tipping over the candle, shown in Fig. 3.10, write an engineering brief describing your strategy for traveling to a position adjacent to the candle.

6.18   If your mission involves finding a candle with a flame that is to be extinguished without disturbing the candle, write an engineering brief describing your strategy for extinguishing the flame.

6.19   If your cargo mission involves finding a cube of wood 50 mm on each side, acquiring it and transporting it back to base, write an engineering brief describing your strategy for executing the three parts of this mission.

6.20   If your construction mission involves transporting rock from Pikes Peak to Death Valley, write an engineering brief describing your strategy for finding Pikes Peak, excavating a quantity of its rock and depositing this rock in Death Valley. See Fig. 3.14 for the location of Pikes Peak and Death Valley.

6.21   If your chemical mission involves locating a toxic chemical source (vinegar), acquiring it and transporting it back to base, write an engineering brief describing your strategy for executing the three parts of this mission. See Fig. 3.11 for the location of the chemical source.

6.22   If your surveying mission involves mapping the contours of Death Valley, write an engineering brief describing your strategy for finding Death Valley and measuring the elevation at different positions. See Fig. 3.13 for the location of Death Valley.

CHAPTER 15

BUILDING AN AUTONOMOUS VEHICLE CONTROL SYSTEM USING THE ARDUINO PLATFORM

15.1 INTRODUCTION: MICROCONTROLLERS

"Look Dave, I can see you're really upset about this. I honestly think you ought to sit down calmly, take a stress pill and think things over..." calmly says the onboard computer HAL9000 to its creator after eliminating the crew of Discovery One in the sci-fi film 2001: A Space Odyssey. In 1968, the year the film was released, a computer with artificial intelligence and with enough sophistication to control an entire spacecraft (including human life support systems) was unbridled fantasy. After all, it was a long way from the Apollo space flight computers that were just slightly more powerful than today's handheld calculators and weighed 32 kilograms. But nowadays the vision of Arthur C. Clarke, the futurist co-author of the screenplay (Stanley Kubrick was the other), does not seem too far-fetched. We have increasingly handed over significant decision-making tasks to computers and made them autonomous. We can only hope that the instructions — programs and codes — we give to our silicon brains do not lead to catastrophic unintended consequences. In this course, we will mimic the experiences of Dave in implementing a computer-based control system that is interfaced with hardware. But we are not going to build a HAL class computer just yet nor will we fly a spaceship. Instead, we will work with small computers – microcontrollers - that are about as powerful as the Apollo flight computers but with ~1/300th the weight, ~1/20th electrical power consumption, and ~1/10^6 in price. And we will navigate around in an engineering building.

Microcontrollers—small and simple computers that collect, process and act on data have become ubiquitous within the last several decades. Everyday appliances, e.g. cell phones, digital clocks, coffee makers and water faucets, are controlled by these tiny brains that execute instructions, faithfully and without complaint. Bigger things such as transportation systems contain a large assortment of these units, which perform a plethora of functions from maintaining the cabin temperature, adjusting the optimal fuel ratio, controlling the roll, pitch and yaw, and controlling the antilock brakes. On an even grander scale, countless numbers of microcontrollers are used to monitor stresses and emergencies in modern buildings and civil structures, automate manufacturing processes, facilitate massive scale sorting and package delivery, gather and process data for health care delivery and security and of course, control robots and bionic limbs. Indeed, all disciplines of engineering have benefited significantly with the advances in microcontrollers.

Microcontrollers are computers' little cousins. They are computers that do not have the usual interface to humans like keyboards and screens, no hard disks to store huge amounts of data indefinitely and incapable of performing multitasking. But what they lack in functionality is more than made up by their low cost and their superb ability to read and write multiple external voltage signals. Despite their sophistication, microcontrollers are surprisingly simple to understand and use. Every microcontroller is composed of input/output (I/O) ports that receive and send signals, a processor that performs logic and arithmetic functions, and memory that stores the instructions and data. In isolation, they cannot do anything. They come to life only with the help of sensors to supply them with actionable information and with actuators to implement the desired actions. The collection of sensors, actuators and the microcontroller and other peripheral components is called hardware, while the computer codes that specify the sequence of commands is called software.

15.2 THE ARDUINO PLATFORM

This project requires you to design the hardware and software that can autonomously control the navigation of a vehicle and other tasks. You will decide on the best approach to monitor the location and trajectory your craft, to steer it in the appropriate direction and in certain cases, design mechanical arms that move when certain conditions are met. You will implement these control systems using a microcontroller platform known as the Arduino, which has been adapted over other systems because of its convenience, cost, and modest learning curve. The Arduino platform was developed by Massimo Banzi, David Cuartielles, Tom Igoe, Gianluca Martino and David Mellis of the Interaction Design Institute Ivrea in Italy. It was inspired, or more precisely, adapted from an earlier platform called Wiring initiated by Hernando Barragan at the Universidad de los Andes, Colombia, (A nice documentary video about its early development is available on line http://vimeo.com/18539129). The motivation was to develop a modern yet inexpensive platform compared with BASIC stamp, the most popular microcontroller at that time. They also made it open source, which allowed many others to contribute to its improvements. We urge you to visit the website www.arduino.cc, visit a few links and return to reading the rest of the chapter. Like any other microcontroller, the Arduino consists of three elements:

- a microcontroller board
- an integrated development environment (IDE) which provides a convenient way to write, compile and run codes,
- programming language, which is similar to C/C++

We will discuss these elements in the following sections.

15.2.1 The Arduino Boards

There are many models of the Arduino board – Uno, Mega, LilyPad, Boardiuno and so on. Being open source, the circuit diagrams themselves are freely available and anyone with electronics skills can make and sell these units, which is why there are so many clones out there. All models are functionally very similar, except for the component layout, memory size and the number of input/output (I/O) ports. They all use ATMEL 8-bit RISC (reduced instruction set computing) microcontroller chips: the Atmega328 for the Uno and clones, and the Atmega 2560 for the Mega and larger boards. Without loss of generality, we will confine our discussion to the UNO, which was introduced in 2010 as a replacement for the highly successful Duemilanove. In geek speak, the UNO operates at 5V, with a clock speed of 16 MHz, and with 32 kB RAM, 2kB of sRAM and 1kB of EEPROM. For the rest of us, suffice it to say that it has the same computing capabilities as the flight guidance computers that went to the moon and back.

The hardware board for the UNO is shown in Fig. 15.1. Initially, it may appear intimidating, but it is much simpler than it looks. For the purposes of clarity, we will discuss the hardware in two passes: the first is a superficial survey of the salient elements of the board and later, more detailed descriptions of its features in conjunction with sample codes.

On the left side of the board, shown in Fig. 15.1, we find two familiar ports: a USB Type B and 2.1 mm power jack. The USB port is used to communicate with the serial port of the programming computer as well as supply electrical power. The 2.1 mm female jack is exclusively used to supply power from an external source with its center pin being positive. There is also a third option labeled "Vin" on the pin on the lower bank. The circuit automatically detects the power source and there is no need to set no electrical connectors or jumpers. One can power the Arduino using an input voltage within the range 9-12 V and capable of delivering at least 300 mA. The Arduino has built-in voltage regulators, which limit the voltage to 5 V to power the chips, but **supply voltages over 20 V, even for short durations**, will destroy the board.

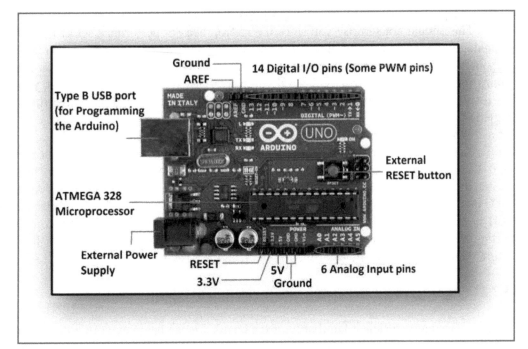

Fig. 15.1 The Arduino Uno Board

Next, we focus on the pin headers – rows of female pin sockets at the top and bottom. The two banks of female ports with 8 pins per bank at the top are labeled from right to left as 0-13, GND (for ground), and AREF. The numbered pins refer to the 14 digital I/O pins. These pins can be designated either as digital input or output ports, meaning that they can be set to read an external voltage (input) or to write (output) a voltage corresponding to a 'HIGH' or 'LOW' state. Generally a HIGH input state is interpreted when the input voltage is from 2.2-5volts and a LOW state is when it is <0.8V. This is consistent with the convention known as the TTL (transistor- transistor logic) levels. A HIGH output is when it is >3.3 V, while a LOW output is <0.2V. These voltages are measured relative to the ground and it is why the ground pin is conveniently placed next to pin 13. Incidentally, because many circuits terminate at ground, there are two other pins (on the lower left bank) that are also designated as GND. The last pin on the top bank is labeled AREF, which is an input port that sets the range of the ranges of the voltages that Arduino is designed to read. Apart from the digital I/O, the Arduino has 6 ports on the lower bank that can read voltage between 0 to 5V. The right half are labeled A0 to A5 and these ports essentially act like voltmeters. In contrast to the digital pins that can only distinguish between 'HIGH' and 'LOW' (or logic 1 or 0),the analog ports under default conditions can distinguish between 0 to 5Volts in 4.8mV increments. These ports report values from 0 to 1023 corresponding to the input voltage range from 0-5V. For instance, a 1.1V input yields the value 225 = 1023 x (1.1V/5V). The resolution of analog input ports can be improved by changing the AREF. For example, if AREF were set to 2.5 volts, then the input voltage can be measured in increments of 2.4mV (4.8mV × (5V/2.5V)), thereby doubling the sensitivity of the voltage reading. A caveat is in order here. Merely placing a voltage on AREF will not suffice to change the analog port accuracy. This will have to be accompanied by a line of code that specifies that the AREF voltage will be used. The last group of pins on the bottom left is the power supply bank. You will find ports labeled 5V and 3V3 which corresponds to 5V and 3.3 V, respectively, along with two extra ground pins and Vin, which was discussed earlier. The RESET pin is on extreme left reboots the processor when momentarily connected to ground in same as pushing the push button on the board.

Now that we are oriented with the pins, we briefly comment on some features of the Arduino UNO. A crystal oscillator operating at 16MHz provides the clock for the processor. Although this corresponds to a short 62.8ns duration per pulse, some instructions take far longer to execute. For sample, the process of reading an analog voltage takes about 100µs (100×10^{-6} s) or nearly 1600 clock cycles! This should not surprise you since the microprocessor performs a set of iterative steps to convert the analog voltage into a 10 bit binary number. Digital data, on the other hand, can be read at higher rates at about 1 ½ the clock period or about 93.7ns. Fortunately, none of the tasks you need to execute will require very high sampling speeds. Even if you were to use the slower analog ports, you can still read data at a respectable rate of 10,000 samples per second. Similarly the rate at which the digital pins can be toggled is about 100 kHz. Using the Arduino library however, you could increase it to as high as 2MHz. But as before, there is probably no reason to require extremely fast cycling of data. More important than read/write speeds, an important parameter is the maximum output current that each port can deliver. For pin 0-12, they can deliver as much as 40mA and pin 13 is somewhat less. Pin 13 is connected to a 1k resistor in parallel to an onboard surface mount LED. 40mA is large enough to blow out most common LEDs that have maximum current limit of 10mA. Therefore it is a good idea to place a resistor (500Ω to 1kΩ) in series with the LED, to limit the current when using pins 0-12. Pin13, owing to the current limiting resistor and built-in LED can be used to drive an LED without a current limiting resistor. This feature comes handy when first starting to play with the Arduino.

15.2.2 The Integrated Development Environment (IDE)

Given that we have sufficient familiarity of the board, we can begin exploring the integrated development environment (IDE) and write a few instructions that the Arduino can execute. The main steps are quite simple:

- Plug in the board using the USB port.
- Write the sketch you want to execute.
- Upload your sketch on the board.

After the last step, you need to wait for a few seconds and your instructions will follow faithfully. If there are errors associated with the syntax or hardware, it will prompt you to correct them before it proceeds. The Arduino's loyalty is wisely conditional. You need to setup the integrated development environment IDE and install the drivers. The procedure is somewhat different for UNIX, Windows and MAC systems, but the Arduino website (**www.arduino.cc/software**) is fairly straightforward to understand. Download the software, plug-in the Arduino and run the software to install the IDE. In most cases, it will recognize your Arduino and install the drivers for it. When the drivers are installed, you can run the Arduino IDE, which will open a window with blue borders and white background as shown in Fig. 15.2.

The format is fairly standard and you might be able to guess the various entries on the menu. "File" and "Edit" perform similar functions as in common windows applications. A "sketch" is synonymous to program codes and the "sketch" menu option provides a dropdown list of actions for sketch compilation, adding libraries and files and opening the sketch folder. "Tools" brings down the options for selecting board type and the serial port selection on the computer connected to the Arduino. The next tier is the toolbar menu. It includes the "compile", "stop" buttons as well as creates a "New" sketch, "Open", "Save" and "Upload" sketches to the board. The last button on the right opens up the new window to display the values written on the serial port. The environment you downloaded contains a collection of Sketches, written by many programmers that illustrate various techniques in using the Arduino. They are found under File>Examples. When you are familiar with the basics of the Arduino, it is instructive to review most of the examples to widen your appreciation of the platform's capabilities.

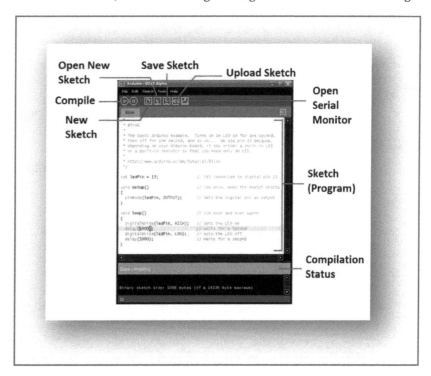

Fig. 15.2. The Arduino
Integrated Development
Environment (IDE)

15.2.3 Programming Your First Sketches: Write, Read and Report

Let's perform our first activity to learn how to:

- turn an LED ON and OFF
- read the state of a pin
- report this state on a monitor screen.

This exercise may look frivolous initially, but imagine the LED as an actuator, say a fan motor: the same sketch can be used to modulate fans use to propel a hovercraft. Shown in Fig. 15.3 is the very simple circuit to connect an LED across pin 13 and ground. Note that diodes have polarities, and we need to make sure that the positive terminal of the LED is connected to pin 13 and the other end to ground (GND). If the LED is new, the longer lead is the positive terminal. If the leads are cut, you can still identify the polarity by examining the base of the diode. The pin closest to the flattened edge of the circular base is the negative terminal.

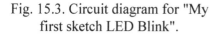

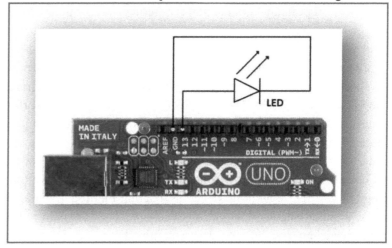

Fig. 15.3. Circuit diagram for "My
first sketch LED Blink".

When the hardware is set, open a blank Arduino file, File>New and write the following program, but ignore line numbers. Alternatively, you could open the File >Examples>Basics>Blink and modify it accordingly.

```
/*My first sketch: LED BLINK*/
1.  int ledpin=3; // connect +LED to pin 13, -LED to GND
2.  void setup()
3.  {
4.  pinMode(ledpin,OUTPUT); //pin 13 is define as output port
5.  }
6.  void loop()
7.  {
8.  digitalwrite(ledpin,HIGH); //turn on LED
9.  delay(1000); //wait a second
10. digitalwrite(ledpin,LOW); //turn LED off
11. delay(1000); //wait a second
12. }
```

Next, compile by pressing the 'Compile' button and then press "Upload" and watch your LED blink once every second. Congratulations, you are now a blinker and the latest esteemed member of the worldwide Arduino users group.

Let's take a closer look at the sketch to understand the general structure of the Arduino code. But first, a few words about the annotations. Any statements inside /*...*/ or after "//" in line will be ignored by the compiler. These are places where texts that explain the code can be written. It is good practice to leave comments not only for others to understand your code, but to help you remember it after some time. It may make perfect sense while you are writing a code, but without any mnemonic, you will have no idea of what it does after week. Also, notice with care that all statements are terminated by a semicolon ";" that is a C language convention.

In all Arduino sketches, two functions have to be executed:

- void setup() {---}, which contains the instructions inside the curly brackets to be run once after RESET. It is similar to the main() function in C/C++ and this is the function where most of the initialization of values take place.
- void loop() {---} which contains the instructions inside the curly brackets to be executed repeatedly until the power is turned OFF or the RESET button is pressed.

Unlike a regular computer, the Arduino cannot end a program or execute multiple programs simultaneously. The setup() program serves to identify constants in the sketch. For example, lines (1) and (4) specify that pin 13 is output pin. Incidentally, you could have omitted line (1), and simply wrote in line (4) as pinMode(13,OUTPUT);. This instruction would work as well. However, it is a good practice to define variables for parameters like ledpin whose value can be assigned in one command line. This is an example of defining a global variable that can be invoked at any part of the program. The declaration "int" before the variable "ledpin" means that we explicitly state that the "ledpin" only take on integer values. Do not be tempted to ignore this seemingly unimportant specification, because without it the program will return an error since the function pinMode(pin,mode) expects an integer for the parameter pin. Computer languages require rigid adherence to syntax in order to reduce ambiguities that cause well-camouflaged bugs in the code. The remaining lines of codes after the setup() routine are obvious:

- digitalWrite(ledpin,HIGH) instructs the ledpin go to HIGH, which sets pin 13 to a value of 5V which turns the LED on.
- delay(1000) is a built-in timer function that keeps the state of the controller unchanged in 1000 milliseconds. When it executes a delay, the Arduino keeps the voltages on all its ports, registers, fixed stacks and pins fixed within the specified time.
- digitalWrite(ledpin,LOW) instructs the ledpin go to LOW, which sets pin 13 to 0 V and turns the LED off.

You can experiment with the delay time to change the tempo, or copy and paste segments of the code to execute other sequences like short-short-short/long-long-long, the universal code for "help". "Compile" and "Upload" after revising the sketch. Save your sketch with a filename "Myfirstsketch" or something similar.

Next, let us make the Arduino 'read' a logic state and report it via the serial monitor. We need to connect a wire to digital pin 2 and for the moment, leave the other end unconnected. As before, start a new sketch, write the following (ignore line numbers), compile and upload.
/*My second sketch: READ & PRINT*/

```
1.   int senspin=2;
2.   boolean pinstate= false;
3.   void setup()
4.   {
5.   pinMode(senspin, INPUT);
6.   Serial.begin(9600);
7.   }
8.   void loop()
9.   {
10.  pinstate = digitalRead(senspin);
11.  delay(300);
12.  if (pinstate==true)
13.  {
14.  Serial.println("HIGH");
15.  }
16.  else
17.  {
18.  Serial.println("LOW");
19.  }
```

When you are done uploading, open the serial monitor window by clicking the last icon on the right on menu bar. Initially you will see the word LOW scroll through the window. In order to get it to say 'HIGH', you need to connect pin 2 to 5V, 3.3 V or Vin. You can toggle between 'HIGH' and 'LOW' by connecting pin 2 to either ground or 5V pin. We introduced 3 new commands in this exercise:

- Serial.begin(9600) in the setup() function instructs the Arduino to open the serial port and communicate at the rate of 9600 baud (bits/sec). This happens to be the common data rate but you can select the baud rate from 1200 to as much as 230400.
- Serial.println(" ") prints the variable inside the quotes. The 'ln' at the end stands for line, which causes printing on a new line. A bare Serial.print() simply continues printing on the same line.
- digitalRead(senspin) reads the voltage level of senspin (pin 2) and returns a value of "1" or "0".

In line (2), we defined "pinstate" as a Boolean variable and set its initial value to false (or 0), and line (5) defined "senspin" (digital pin 2) as an INPUT port. In the loop() line (10), we read the state of pin 2 and assigned it to the variable "pinstate". In lines (12)-(18), we introduced an if-else control structure that prints "HIGH" to the monitor if the pinstate is true, otherwise if prints "LOW". As an exercise, you can combine your earlier sketch with this one, with pin 2 connected to pin 13. (You will have to use your hands or a breadboard to connect the pins). Your goal is print "ON" when the LED is lit, and "OFF", when not.

15.2.4 PWM and Analog Input

Now that we some familiarity with the sketches, it is time to learn to read signals other than binary (1 or 0, TRUE or FALSE) as well as to regulate the amount of power from an output port. In this segment, we will discuss the following concepts:

- Sketch to use Analog input
- Sketch to use PWM to modulate LED intensity
- Illustration of feedback

We have encountered the analog inputs earlier. In the default case, they can read positive voltages in the range 0-5V and assigns them digital code from 0 to 1023. These are useful since in the real world, the signals are not merely binary but can take a range of values. Later on, you will learn that there are transducers that convert physical quantities such as temperature or distance, to electrical quantities. With the Arduino you can simultaneously monitor 6 independent physical quantities.

Close examination of the I/O pins on the board show that pins 3, 5, 6, 9, 10, 11 have special wiggle marking below them. These pins are so called PWM- Pulse Width Modulation pins. Quite simply they can output a square wave signal at approximately 490Hz (2.04 ms period) with variable duty cycle, i.e., the relative durations of the high and low segments of the wave. The command "analogWrite(pin,x)" are specific to those pins, and the range $0 < x < 255$ adjusts the duration of the high and low segments of the wave. For $x = 0$, the wave is low all the time, and the other extreme is $x = 255$ when the wave fixed at 5V. At intermediate values of x, the wave is partly high and partly low. For example, if we set $x = 64$ then value will spend ¼ (64/255) of the time or 0.5 ms at HIGH and ¾ of the time or 1.5 ms at LOW, corresponding to a duty cycle of 25%. This is a very powerful method for controlling the average power delivered to a device (the LED in our case) and in essence effectively provides 256 levels for power control. Additionally, there are also devices, notably the servo motors that operate using PCM- Pulse code modulation. Servo Motors have shafts that rotate to specific angles when supplied with square wave pulses with specific duty cycles.

Let's illustrate these ideas with the following example. The circuit diagram shown in Fig. 15.4 provides the connections to the Arduino. The resistance of the CdS photoresistor changes depending upon the intensity of the light incident upon it. This property exploited by using the voltage divider circuit in which the 5 volts is split between the fixed resistance R and the photoresistor. The voltage across the photoresistor which is read by analog 0 is $= 5[R_{Cds}/(R_{Cds} +R)]$. Thus, the range of voltage that corresponds to dark and bright light levels depends upon the choice of R, which in turn depends upon the characteristics of the photoresistor. A very common range R_{Cds} is 1.5kΩ for bright and 15kΩ for dark, and we choose R = 4kΩ to yield a range between $1.36V = 5[1.5kΩ/(1.5kΩ +4kΩ)]$ for bright to $3.95V= 5[15kΩ/(15kΩ +4kΩ)]$ for dark. Therefore we anticipate the values reported by analogRead() to range from $278 = 1023[1.36V/5V]$ to $808 = 1023[3.95V/5V]$.

/* Using analog write and PWM output*/

/* This is a program to read the light intensity from a photo detector and adjust the intensity level of an LED based on it and the angle of a servomotor */

```
1. int sensorPin = A0;    // input pin for the light sensor
2. int ledPin = 10;      // pin for the LED
3. int servoPin=9;       // pin for the servomotor
4. int sensorValue = 0;  // variable to store the value coming from the sensor
5. int reflightlevel=170+255; //user chosen reference light level, usually close to 170+255
            //light level when exposed to light source
6. int difference=0;  //variable to store difference of reference and current light levels
7. void setup() {
8.    pinMode(ledPin, OUTPUT); // declare the ledPin as an OUTPUT:
9.    Serial.begin(9600); //initialize serial (screen) monitor
10. }
11. void loop() {
12.    sensorValue = analogRead(sensorPin); // read the value from the sensor
13.    int val1=map(sensorValue,0,1023,255,0); // reverse mapping sensor value to variable val1
14.    // Serial.println(val1); //debugging purpose
15.    difference=abs(reflightlevel- val1); //calculate reference-current light levels
16.    analogWrite(ledPin,difference); //write the LED
17.    int val2=map(sensorValue,0,1023,0,255);//mapping light sensor readings to val2 variable
18.    analogWrite(servoPin,val2); //write to servo
19.    delay(20); //delay 20 ms to settle everything
20. }
```

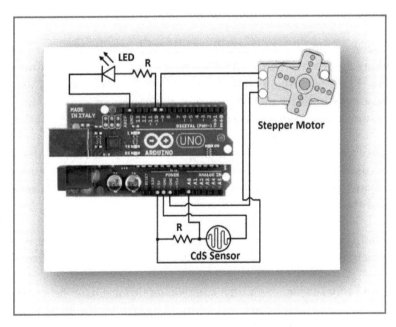

Fig. 15.4. The circuit for analogWrite and PWM output.

This simple program reads the value of the voltage across the photo detector using the analog input port 0 and assigns it to the variable sensorValue in line(12). Since the sensorValue can, in principle, vary from 0-1023, we rescale to fall within the range from 255-0 (using map() function in line 13) and assign it to the variable val1. Note with care that it is reverse mapping, e.g., sensorValue=0 is mapped to val=255, and sensorValue=1023 is mapped to val=0. We use reverse mapping because we want the LED to go bright when the sensor reads low light and dark when the sensor reads bright. In line(15), the difference between a predefined level, reflightlevel and val1 is calculated and in line (16) the resulting number is passed on the analogWrite() function for the LED. A similar procedure is used for val2 which affects the servo. This time, however, we use direct mapping. Incidentally, you will find in line (14), a Serial.println command which can show the value of val1. This instruction, which is later commented out, is used for debugging and for assisting the reflightevel parameter.

15.3 SENSORS AND ACTUATORS

Microprocessors are nothing without the availability of sensors, actuators and other peripheral devices that are interfaced with it. Sensors can be made inexpensively with a little knowledge of electronics or bought from many sources. Store-bought devices come with data sheets or application notes that explain how they are used. It is good practice to read the data sheets and, in particular pay close attention to the sample applications as well as their limitations. In fact, you should read the specifications **before** buying the devices to ensure that they are ideally suited to meet your needs. There is a direct relation between performance and cost. A good engineer understands the requirements for his or her application and will not pay an extra cent for a feature he or she does not need.

Common sensors use the analog input channel, pulse-code modulation or the pulseIn () function that measures a pulse duration. With some exceptions, sensors are typically 3 terminal devices: two connections for Vcc (power) and ground , as well as third for the signal output. Examples of common sensors are:

Light sensor — a sensor returning a value proportional to the amount of light it detects, which can be used to determine the lightness/darkness. (CdS resistor, photodiode and phototransistors).

Reflectance sensor — similar to a light sensor but with a built in light source; it returns a value proportional to the amount of light that is reflected from a surface. (Polulu QTR-1RC/8RC).

Accelerometer — a sensor returning values proportional to the amount of acceleration it detects along 3 axes, which can be used to measure a system's movement in 3 dimensions (acceleration is the 1st derivative of velocity and the 2nd derivative of position). Note that the accelerometer measures both its movements and the effects of gravity, the sensor's constant acceleration in the up/down direction. (Sparkfun ADXL203CE, ADXL320, ADXL3xx)

Gyroscope — a sensor returning values proportional to the angular rotation it detects, which can be used to measure a system's movement in three dimensions, much like an accelerometer. Its advantage is that, unlike the accelerometer, it does not also lump together the effects of movement and the effects of gravity. (Sparkfun ADXRS613, IDG500, MLX90609)

Compass — a sensor returning a heading (which way it is pointing relative to Magnetic North), which can be used to keep a system pointing in a known direction (HMC6352)

Proximity sensor — an ultrasonic or infrared sensor returning a value proportional to the distance to the nearest object along the line of sight, which can be used to sense when approaching obstacles such as people, walls, traffic cones, etc. (Sharp Analog Distance Sensor, Parallax PING)

Temperature sensor — a semiconductor that returns a voltage directly proportional to the ambient temperature. It could be a two-terminal temperature sensitive resistor or a transistor. (thyristor, Analog Devices TMP36)

Pressure sensor — a pressure sensitive sensor that returns a value proportional to the pressure applied to its active area. Could also be a tiny wire that closes a circuit when pressed (poor man's pressure sensor) (Interlink 402 FSR)

Magnetic field sensor — a semiconductor that uses the Hall effect to return a voltage proportional to applied field. There are latching models which keep their voltage once exposed to a high enough field, while non-latching models return to initial state when field is removed. (Latched, Honeywell SS361RT; Unlatched, Allegro Microsystems ATS682LSHTN-T)

Actuators — generally some form of a motor used to induce a mechanical motion. Examples are:

dc motor — an actuator that can be controlled to spin at different speeds (i.e., run at different power levels), which can be used for fans for levitation and propulsion

Servo motor — an actuator that can be controlled to rotate in specified degrees, which can be used to implement wheels for propulsion, wheels for steering, doors that open and close, robotic arms, rudders and vanes, pulley systems, etc. Some models a limited to 180^0 shaft rotation, while others are continuous rotation and can rotate indefinitely. These use pulse code modulation. (Hitec 33065S, Futaba BLS451)

Stepper motor — similar to servo motor but capable of much finer angle positioning accuracy (typically < 1 degree). Uses 4 lines to control. (Parallax 4Phase Unipolar, Panasonic 42SPM24DGYE)

In general, the Arduino signals are not sufficient to power any but the tiniest of these motors, so that intermediate amplifiers are used to bridge between the Arduino and the actuator. In addition, there are also circuits that allow you to circumvent the power limitations of the Arduino. For instance, a dc motor for a levitation fan that requires several amps cannot be run directly by the Arduino output since the maximum current the pin can supply is limited to 40 mA. You could of course use your knowledge of electronics from Chapter 12 to use mechanical relays or transistors to switch high currents. Nonetheless there are two of-the-shelf components are that you may find useful:

H-bridge — an integrated chip that can switch and reverse the direction of current flow using low power signals. (SN754410)

Motor Shield — a board that directly plugs into the Arduino that is capable of controlling 2 servo motors and 4 dc motors or 2 servos and 2 stepper motors (Adafruit Motorshield, Seeestudio Motorshield)

15.4 <u>THE ARDUINO ROVER</u>

To appreciate the interplay between sensors and actuators, let us use the Rover 5.0, an Arduino-Uno compatible platform, which can be procured from Sparkfun Electronics. The Rover 5, shown in Fig. 15.5, uses four independent motors, each with an optical quadrature encoder and gearbox. The entire gearbox assembly can be rotated at 5 degree increments for different clearance configurations. You can replace the treads with traditional wheels. The electric motors rotate at 8,800 RPM prior to speed reduction with the gear heads. The electric motors require 5 to 7.5 VDC with a no load current of 160 mA.

The platform integrates an Arduino Uno controller and a Dagu motor driver board, which is shown in Fig. 15.6. This driver board is designed to accommodate small robotic vehicles with 4-wheel drive. It is equipped with four motor outputs, four encoder inputs and current sensing for each motor.

The motor drivers can be controlled by simply applying a logic 0 or 1 to the direction pin for that motor and a PWM signal to the speed pin. Hence, the speed and direction of four separate motors can be controlled independently using only 8 GPIO pins. The encoder inputs on the driver board mix each pair of encoder inputs using an XOR gate making it possible to read both inputs from a quadrature encoder using only one interrupt pin.

Reading the output from the current sensor pin output is easy, because each pin gives an output of about 1V for each ampere drawn by the associated motor up to 5V. Connecting the current sensor pin to the analog input of the Arduino and you will be able to detect stalls and other motor problems. There are two power connectors on board — one for 5V logic and the other for the motor supply. The board is rated for a maximum motor supply voltage of 12V.

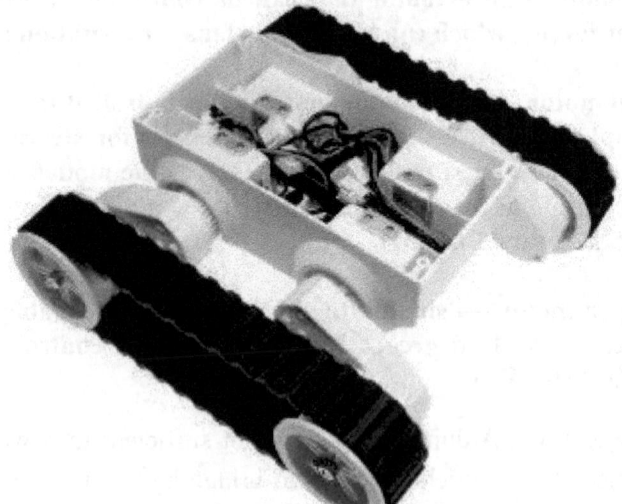

Fig. 15.5 Rover 5 platform designed by Dagu.

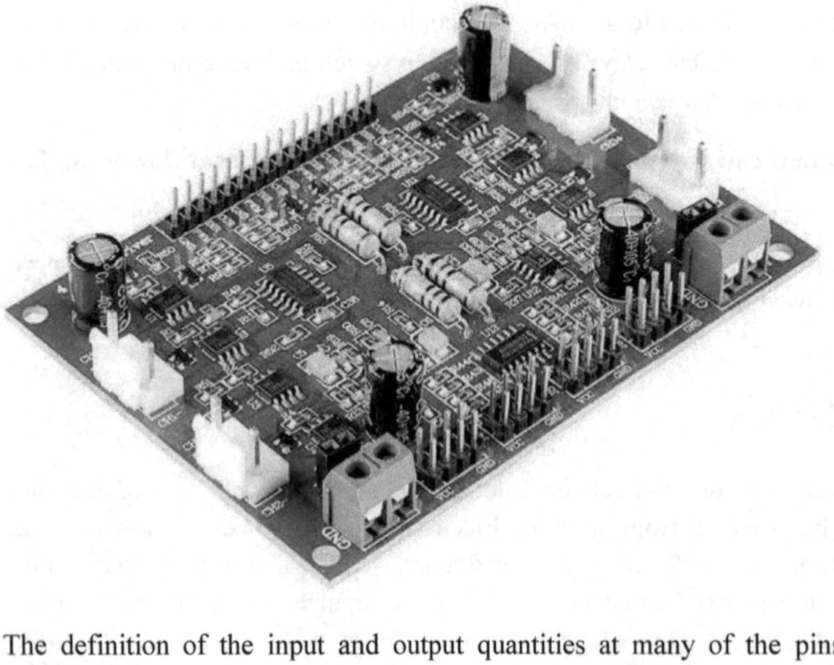

Fig. 15.6 A motor driver board compatible with the Rover 5 platform.

The definition of the input and output quantities at many of the pins on the motor driver board is presented in Fig. 15.7. Wiring the Arduino-Uno with the motor driver board is also shown in Fig. 15.7. Note that only two of the motors are connected to the Arduino in this arrangement. The motor supply voltage, which may vary from 4.5 to 12 VDC is supplied with a 9 V battery. The wiring from the motor driver board to the electric motors is not shown.

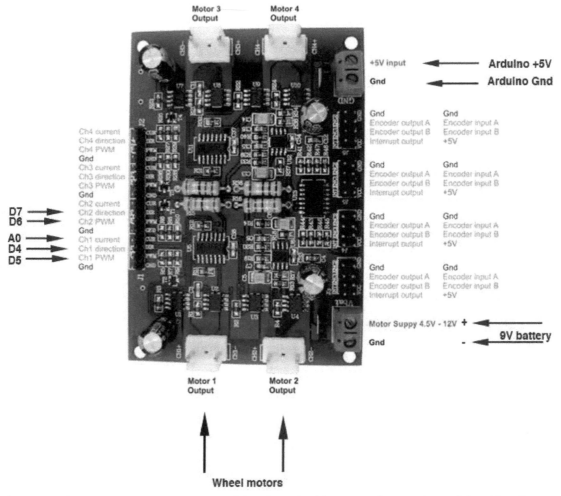

Fig. 15.7 Pin layout on the Dagu motor driver board with connections to the Arduino Uno for two motors.

The code, written by Nick Gammon, found on the Support section of Arduino's website, which is shown below, is designed to drive the robot forwards or backwards for 10 seconds. Then the code attempts to turn the robot by driving the wheels in opposite directions for 1.2 seconds. The current-sense circuit is used to detect if too much current is applied to the motors. If the current is excessive, the code immediately skips to the next phase.

```
// Dagu 5 Chassis example.
// Author: Nick Gammon
// Date: 11th December 2011

volatile int rotaryCount = 0;

#define PINA 8
#define PINB 9
#define INTERRUPT 0 .// that is, pin 2
```

```
#define DIRECTIONA 4
#define MOTORA 5

#define DIRECTIONB 7
#define MOTORB 6

#define TIME_FORWARDS 10000
#define TIME_BACKWARDS 10000
#define TIME_TURN 1200

// Interrupt Service Routine for a change to encoder pin A
void isr ()
{
  boolean up;

  if (digitalRead (PINA))
    up = digitalRead (PINB);
  else
    up = !digitalRead (PINB);

  if (up)
    rotaryCount++;
  else
    rotaryCount--;
} // end of isr


void setup ()
{
  attachInterrupt (INTERRUPT, isr, CHANGE);   // interrupt 0 is pin 2,
interrupt 1 is pin 3
  pinMode (MOTORA, OUTPUT);
  pinMode (DIRECTIONA, OUTPUT);
  pinMode (MOTORB, OUTPUT);
  pinMode (DIRECTIONB, OUTPUT);

}  // end of setup

byte phase;
unsigned long start;
int time_to_go;

void loop ()
{

  analogWrite (MOTORA, 200);
  analogWrite (MOTORB, 200);
  start = millis ();

  // check current drain
  while (millis () - start < time_to_go)
  {
    if (analogRead (0) > 325)  // > 1.46 amps
      break;
    }
```

```
switch (phase++ & 3)
  {
  case 0:
    digitalWrite (DIRECTIONA, 1);
    digitalWrite (DIRECTIONB, 1);
    time_to_go = TIME_FORWARDS;
    break;

  case 1:
    // turn
    digitalWrite (DIRECTIONA, 1);
    digitalWrite (DIRECTIONB, 0);
    time_to_go = TIME_TURN;
    break;

  case 2:
    digitalWrite (DIRECTIONA, 0);
    digitalWrite (DIRECTIONB, 0);
    time_to_go = TIME_BACKWARDS;
    break;

  case 3:
    digitalWrite (DIRECTIONA, 0);
    digitalWrite (DIRECTIONB, 1);
    time_to_go = TIME_TURN;
    break;

  } // end of switch

analogWrite (MOTORA, 0);
analogWrite (MOTORB, 0);
delay (500);

} // end of loop
```

We will show another example of controlling an autonomous vehicle later. But first, it let us develop rudimentary concepts of C programming.

15.5 <u>C PROGRAMMING BASICS</u>

The ARDUINO C/C++ environment implements a subset C programming language that is powerful enough to implement any control system for your craft. The Arduino language supports all standard C constructs and some C++ features. It links against AVR Libc and allows the use of any of its functions. The language reference can be found online at <u>http://arduino.cc/en/Reference</u>, together with links to libraries for interfacing with various hardware and code contributed by others. This section offers a primer on the topic. In general, a complete valid statement is terminated with a semicolon; code snippets without semicolons at the end represent code that would be part of a larger statement. In addition, any characters after "//" in a line, or inside "/*…*/" are treated as comments and ignored by the compiler. These comments are convenient for annotating your code, a practice used by all good programmers and also for "commenting out" a statement without deleting it from the code.

Data Types

The following ARDUINO C/C++ data types are at your disposal, not all of which are classic C types:

Boolean values —Booleans are variables that indicate "true (1)" or "false (0)". For instance, you could use these to indicate whether a reflectance sensor facing the floor is on the black tape or not.

```
boolean sensor_onTape;
if (sensor_onTape) {
        // do something
}
```

Integers —We encountered integers earlier and declared them with "int". They hold 16-bit quantities, or positive or negative integers between –32768 and 32767. Any number outside of this range does not exist, as far as the Arduino is concerned. For instance, if you added say 10 to the value 32767 in a variable, the result will "wrap around" the number scale, and you will be left with a large negative number (-32758). Larger numbers (32-bit), declared as a "long int" can range from -2,147,483,648 to 2,147,483,647.

Floating point numbers —Floating point numbers are used to represent fractions, which are not convenient to write with integers. For instance, if you wrote the following:

```
int a = 9;
int b = 5;
int c = a / b;
Serial.println(c);
```

you would obtain 1, where the decimal part is been truncated because the variable c is incapable of holding a non-integer value. It rounds off to the nearest whole number. The option here is,

```
float a = 9.0;
float b = 5.0;
float c = a / b;
Serial.println(c);
```

This works as expected. Note the float c declaration indicates that the variable is a floating-point number and should be interpreted and printed out as such.

Characters — Characters are 8-bit quantities that range from -128 to 127 which are used to represent ASCII alphanumeric codes. Thus the following declarations are identical

```
char a = "A";
char a= 65;
```

Arrays —An array is a collection of variables that can be accessed using an index. For example, we could declare a set of 6 integers as follows:

```
int a[6] ={1,2,3,4,5};
```

Individual elements are accessed as follows:

```
a[0]     // =1, the first element in the array
a[4]     // =5 the fifth element in the array
a[i]     // the "i_th+1" element (valid for 0 <= i <= 5)
```

Note that the pointer index has to be a positive integer and has to be less than the size of the array to include the index 0. Note too, that although similar, int a[6] and a[6] have completely different meanings.

Strings —A string is a character array is a way of declaring text data and shown as follows:

char Str2[8] = {'a', 'r', 'd', 'u', 'i', 'n', 'o'};
char Str4[ ] = "arduino";
char Str5[8] = "arduino";

Note the double quotes at the beginning and end. Note too that the size of the string array should be at least one more than the data.

Void —This is the declaration of a function that does not return a value, i.e.
```
void loop(){/*do something*/}
```

as opposed to the following, this returns an integer value:

```
int add_two_ints(int a, int b)
{ return a + b;
}
```

Operations on Data

Examples on various ways to operate on data are given below.

Reading/writing data
```
a = 1;          // assigns integer 1 into variable a
a = 1.75;    // assigns the floating-point value 1.75 into the variable a
a = b;          // reads variable b and copies it into variable a
x = list[2]  /*reads the 3rd element in "list" & assigns it to variable x */
list[0] = y; // reads y and writes it to the first item in the array "list"
list[i] = y;   // puts y into the ith slot of the array, e.g.:
                     // if i=1, y is written into the second slot of the array;
                     // if i=32, y is written into the 33rd slot of the array; etc.
a[x] = b[y]; // yes, this is possible, and can be very useful
```

Arithmetic and logic operations work on integers and floating point numbers and produce a result of the same type as the input variables. An integer added to an integer yields an integer; a float

added to a float yields a float. While the Arduino will allow you to operate on different variable types, be careful because its misuse creates unintended bugs. The following are examples:

```
a + b          // addition
a – b          // subtraction
a * b          // multiplication
a / b          // division
a++;           // increments a by 1
a--;           // decrements a by 1
a = b++;       // reads b, puts it into a, and then increments b by 1
a & b          // bit-wise AND function (e.g. 0010 AND 0110 equals 0010)
a | b          // bit-wise OR function (e.g. 0010 OR 0110 equals 0110)
!a             // logical negation (turns a non-zero number into zero, and
               // turns a zero into a 1 – e.g. if a==0110, !a equals 0)
~a             // bit-wise negation (e.g., if a==0110, ~a equals 1001)
a += b;        // short-hand for a = a + b;
a -= b;        // short-hand for a = a – b;
a *= b;        // short-hand for a = a * b;
a /= b;        // short-hand for a = a / b;
a &= b;        // short-hand for a = a & b;
a |= b;        // short-hand for a = a | b;
```

Boolean operations — note that Boolean values are either 0 or 1 (false or true)

```
a == b         // equivalence operation; produces a Boolean result
a != b         // true if a does not equal b
a < b          // true if a is less than b
a <= b         // true if a less than or equal to b
a > b          // greater than
a >= b         // greater than or equal to
a = !b;        // puts the Boolean negation of b into a
a = ~b;        // puts the bit-wise negation of b into a
a && b         // true if a is true (or non-zero) AND b is true (or nonzero)
a || b         // true if a is true (or non-zero) OR b is true (or nonzero)
```

Control Structures

Arduino C/C++ offers the following control structures for your code:

Function — We have seen functions at the very beginning of this chapter. The setup() and loop() are functions predefined by Arduino C/C++. You can also declare your own functions, as exemplified in the code Moving Rover. Parameters can be passed to functions if the function is declared. For example,

```
1. int myAddFunction (int x, int y)
2. {
3. int result;
4. result = x+y;
```

5. return result;
6. }

which can be "called" in the main loop as,

void loop()

7. {
8. int i=2;
9. int j=1;
10. int k;
11. k= myAddFunction (int x, int y); //should contain 3
12. }

When invoked, it is as if you executed the function's code in place.

If/then/else — This control-flow structure allows a programmer to specify one or more code blocks, at most one of which will be executed, based on one or more conditionals. For example,

```
if (SomeSensorValue < 50) {
    // do something
}
```

"does something" if the sensor value satisfies some conditional, so the code body may or may not be executed. The following if-then-else example provides two options, one of which is guaranteed to be executed:

```
if (SensorValue(mySensor)< 50) {
    // turn left
} else {
    // turn right
}
```

Having a simple else statement at the end of an "if" construct is the default case; it specifies code that will be executed if all other conditionals fail. The following is a more complex example:

```
int value = SensorValue(mySensor);
if (value < 25) {
    // turn left
} else if (value < 50) {
    // turn right
} else if (value < 75) {
    // go back
} else {
    // stop
}
```

Note that later conditions in an if-else control block are tested only if all earlier ones fail, and later conditions are not tested if an earlier test succeeds. Hence, only one code block will execute, in particular the first to satisfy its corresponding condition. So in the following example, the third condition would never succeed (turn right).

```
int value = SensorValue(mySensor);
if (value < 25) {
    // turn left
} else if (value < 75) {
    // go back
} else if (value < 50) {
    // turn right
} else {
    // stop
}
```

Note also that the sensor value is read only once and not multiple times, which would have been the case, at least potentially, if we had placed the SensorValue function call in each of the conditionals.

While loop — Formally, it takes as an argument a Boolean statement and conditionally executes whatever code block comes after it. The important thing to remember is this ordering: first the evaluation of the condition statement and then the execution of the following code block. This happens repeatedly until the condition statement evaluates to false, at which point the vehicle moves on to execute the instructions that come after the while loop.

```
while (condition) {
    // code block to be executed (one or more instructions),
    // often called the "loop body"
}
// following instructions
```

This produces a loop format, because the code block will be executed repeatedly until the condition becomes false (the integer value 0). As soon as the condition evaluates to false, the program jumps to the "following instructions" after the loop body.

Note again that the condition test happens before the loop body; this means that it is possible for the loop body to be skipped, if for example on the first test of the condition it evaluates to false. Sometimes, this is not desired, because a programmer might want the loop body to be executed at least once before evaluating the conditional value. This action is what the do-while loop does, as described next.

Do-while loop — As mentioned above, sometimes a programmer needs a loop structure in which the loop body is executed at least once before evaluating the conditional value. For this, the do-while loop is used, as shown below.

```
do {
    // code block to be executed (one or more instructions),
    // often called the "loop body"
} while (condition);
// following instructions
```

This structure is also a loop; the loop body will be executed repeatedly until the microprocessor evaluates the Boolean condition to false. The difference is that the loop body is guaranteed to be executed at least once, even if we know ahead of time that the condition is false, for instance in the following (contrived) case:

```
do {
    // loop body
} while (false);
```

This will execute the loop body once, then test the conditional, which is hard-coded to be false. The condition will fail, and the loop will terminate, passing control to the following instructions. One thing to note: the semicolon at the end of the while(); statement is required for a do-while loop.

For loop — The for loop is generally used for iteration; it takes the following form:

```
for ( /*initialization*/ ; /*condition*/ ; /*update*/ ) {
    // loop body
}
```

The semicolons between the three components (initialization, condition, and update) are required, even if a particular component is left blank. The components function as follows:

- **Initialization** — This happens once, at the start of the loop, before the condition evaluation. It is a good way to set up the variables that will be in use.
- **Condition** — This is identical to the while-loop condition; it is evaluated before the loop body, so if it evaluates to false on the first pass, the loop body is never executed.
- **Update** —This is a statement that is executed after the loop body and before the next evaluation of the condition statement.

Here is a typical use of the for loop:

```
int i, array[1024];
for (i=0; i<1024; i++) {
    array[i] = 0;
}
```

Note that the **initialization** and **update** components can have multiple statements, each separated by commas.

Switch — If you have a control structure similar to the following, where your conditionals are, for the most part, testing for equality as opposed to greater than/less than, you might want to use a switch statement:

```
int value = SensorValue(mySensor);
if (value == 1) {
    // turn left
} else if (value == 2) {
    // go back
} else if (value == 3) {
    // turn right
} else if (value == 4) {
    // spin around
} else if (value == 5) {
    // stop
} else if (value == 6) {
    // spin clockwise
} else if (value == 7) {
    // spin counterclockwise
} else if (value == 0) {
    // shut down
} else {
    // blink lights
}
```

This type of construct can be difficult to read, and because it is difficult to read, it can harbor latent bugs easily. A better construct is the switch statement, which would implement the construct above as follows:

```
int value = SensorValue(mySensor);
switch (value) {
    case 1:
        // turn left
        break;
    case 2:
        // go back
        break;
    case 3:
        // turn right
        break;
    case 4:
        // spin around
        break;
    case 5:
        // stop
        break;
    case 6:
        // spin clockwise
        break;
    case 7:
        // spin counterclockwise
        break;
    case 0:
```

```
                        // shut down
                        break;
            default:
                        // blink lights
  }
```

At the switch (value) statement, the value will be read once, and the microcontroller will immediately jump to which ever code block has the corresponding value expressed with its case statement. The default case is not required but is an extremely good idea to use, even if it is left empty by putting no code after it (perhaps include a comment to the effect that the omission of corresponding code is intentional). The various break; statements are required, except for the very last case in the list. If a particular break; statement is omitted, the program will compile successfully, but at run-time the control flow will "fall through" from one case to the next.

*Continue and break in loops — There are two statements that provide extremely powerful ways to augment the behavior of loops. The continue statement causes a loop to halt mid-loop and go directly to the top of the loop. For example, here is its use in a while loop:

```
while (true) {
        int value = SensorValue(mySensor);
        if (value == 1) {
                // turn left
        } else if (value == 2) {
                // go back
        } else {
                continue;
        }
        // blink lights
        // update state
        // etc.
}
```

In this example, the **continue** statement causes the execution of the code to skip over the lines at the end of the loop (blink lights, update state, etc.) and go straight to the top of the loop, at which point the conditional is re-evaluated, and the sensor value is read.

A **break** statement similarly disrupts control flow, but it instead causes control to exit the loop entirely.

```
while (true) {
    int value = SensorValue(mySensor);
    if (value == 1) {
        // turn left
    } else if (value == 2) {
        // go back
    } else {
        break;
    }
    // blink lights
    // update state
    // etc.
}
```

```
Serial.println("Error situation");
```

This example is just like the previous one, except that the **continue** statement is now a **break** statement. This is one way to exit a loop that would otherwise execute forever. It is usually used in loops to detect anomalous or erroneous situations; normal events should be reflected in the while loop's conditional.

An additional note on loop and if/then conditionals: they are usually Boolean statements, statements that evaluate to either 0 (false) or 1 (true). Because a computer only understands numbers, these are represented with integers, which can have the value 2 or –37, or 1023, and so on. Thus, while it is best practice to use code similar to this:

```
char x, array[1024];
int i;
for (i=1023; i > 0; i--) {
    x = array[i];
}
```

it is just as valid to say the following (the only change to the code is that i > 0 has been replaced by simply i).

```
char x, array[1024];
int i;
for (i=1023; i; i--) {
    x = array[i];
}
```

Why does this code work? Because the Arduino checks to see if the conditional is zero or non-zero; it does not do a strict test of 0 vs. 1. So any number other than 1 is considered equivalent to 1 for the purposes of determining a Boolean result. As soon as the variable *i* reaches the value zero, both of these loops will terminate.

Taking a modular approach tends to facilitate reading, understanding, and thus debugging one's code. This is an extension of our Rover program which illustrates the power of functions; whenever you see the same block of code happening over and over again, with only minor variations, which is a sure bet that you should encapsulate the code in a function call.

15.6 <u>PSEUDOCODE AS A FIRST DRAFT AND AN ACTUAL SKETCH</u>

Thus far, we have studied how to implement control programs using the Arduino. However, one does not start with Arduino programming immediately until some reasonable control flow of one's ideas are conceptualized and written in simple English. This is known as a **pseudocode**, and it is the most commonly used form of sketching out one's programming ideas. Flowchart can also be implemented which is similar to pseudocode and tells you thee approach towards the actual program.

In general, before writing a complex piece of software, one should formulate a picture of what it is the program should do: just as an artist draws sketches before starting to paint, so should an engineer sketch out the overall program before starting to write actual code. Pseudocode is a popular medium because it tends to be self-explanatory, and despite its simplicity, it can still help discover design flaws. For example, here is a rudimentary vehicle-control algorithm for following a dark tape on a light floor, assuming that the vehicle has two downward-pointing light sensors (left_sensor and right_sensor) at the front, with spacing less than the width of the tape. Depending on the conditions on the left and right sensors, the vehicle will move forward, move right, move left or stop.

> do forever:
>   read light sensors (left_sensor, right_sensor)
>   respond to the 4 possible combinations of light sensor values:
>   1. (dark, dark) — both sensors on line
>   move forward
>   2. (dark, light) — vehicle on right of tape
>   turn left
>   3. (light, dark) — vehicle on left tape
>   turn right
>   4. (light, light) — vehicle is completely off the tape
>   move backwards

This is not a particularly robust program and only works when the initial state has at least one sensor on the tape, but it does specify reasonably well what is intended and how it will happen. Because the "code" is mostly human-readable English, it is self-documenting; any engineer reading it should be able to understand what it is intended to do as well as how it expects to go about it.

Note in particular the indentation; here, as in regular programming languages, the nesting of statements indicates their "scope" … e.g., everything below and indented further right than the do forever statement is subservient to that statement (it comprises a loop that will execute until the computer is shut off); the numbered statements that are indented more than the respond to the 4 possible combos statement are the four possible sensor-value combinations, and one would expect that only one could be true, so for any given sensor reading only one will execute. Translating the pseudo code into actual code will invariably look a bit more complicated. The command "move forward" actually means that both wheels of the tank move at the same speed and direction, "move left" means that the left wheel moves in reverse while right wheel moves forward, and so on. Following is an implementation of the pseudocode above for our Rover tank.

/*The Rover Tape Follower*/

```
1.  int E1 = 6; //M1 Speed Control
2.  int E2 = 5; //M2 Speed Control
3.  int M1 = 8; //M1 Direction Control
4.  int M2 = 7; //M2 Direction Control
5.  int right_sensor_pin = A1;// right sensor to Analog input A1
6.  int left_sensor_pin = A0; //left sensor to analog input A0
7.  int val;// initialize variable val as int
8.  int leftspeed = 255; //255 is maximum speed
9.  int rightspeed = 255;

10. void setup()
11. {
12. int i;//intialize variable i as int
13. pinMode(right_sensor_pin,INPUT);// make right_sensor_pin and
14. pinMode(left_sensor_pin,INPUT);//left_sensor_pin as input pins
15. for(i=5;i<=8;i++)//for loop for initializing E1,E2,M1 and M2 as Output pins
16. {
17. pinMode(i, OUTPUT);
18. }
19. Serial.begin(9600);//initiate serial communication at 9600 baud rate
20. }

21. /*Cases: both sensors on black tape, case "craft_on_tape"; right sensor on white and left sensor
    on black, case "craft_on_right"; right sensor on black and left sensor on white, case
    "craft_on_left"; both sensors on white, case "craft_off_line" */
22. void loop()
23. {
24.    char right_sensor = rightsensor();//reading right sensor value
25.    char left_sensor = leftsensor();//reading left sensor value
26.    if (right_sensor == 'D' && left_sensor== 'D') //condition if both sensor values are low
27.      {
28.      forward(leftspeed, rightspeed);//move the craft forward
29.      }
30.    else if (right_sensor =='L' && left_sensor== 'D')/*condition if
31.                         left sensor on the line and right sensor outside the line*/
32.      {
33.      left(leftspeed, rightspeed);// turn the craft left
34.      }
35.    else if (right_sensor == 'D' && left_sensor== 'L')/*condition if
36.                         right sensor on the line and left sensor outside the line */
37.      {
38.      right(leftspeed, rightspeed); //turn the craft right
```

```
39.       }
40.     else if (right_sensor == 'L' && left_sensor== 'L')/* condition if
41.                         both the sensors not on the line */
42.       {
43.         reverse(leftspeed, rightspeed);//if none of the above conditions then stop the craft
44.       }
45. }


46. // Start of function definitions
47. char rightsensor()/* right sensor function defined as char
48.                     as it returns a character value L or D*/
49. {
50.   int sensout;//initialize sensout variable to integer
51.   char sensorstate ='L';//intialize sensorstate to L (Light)
52.   sensout=analogRead(right_sensor_pin);//read right_sensor_pin and store value in sensout
53.   if (sensout>800)//check if sensor value is above 800
54.     {
55.       sensorstate='D'; //if yes change sensorstate to D (Dark)
56.     }
57.   return sensorstate;//return sensorstate value
58. }


59. char leftsensor()/* left sensor function defined as char
60.                     as it returns a character value L or D*/
61. {
62.   int sensout;//initialize sensout variable to integer
63.   char sensorstate ='L';//intialize sensorstate to L (Light)
64.   sensout=analogRead(left_sensor_pin);//read left_sensor_pin and store value in sensout
65.   if (sensout>800)//check if sensor value is above 800
66.     {
67.       sensorstate='D'; //if yes change sensorstate to D (Dark)
68.     }
69.   return sensorstate;//return sensorstate value
70. }


71. void forward(char a, char b)//move craft forward function
72. {
73.   analogWrite(E1,a);//providing speed to motor 1
74.   digitalWrite(M1,LOW);//forward direction to motor 1
75.   analogWrite(E2,b);//providing speed to motor 2
76.   digitalWrite(M2,LOW);//forward direction to motor 2
77. }


78. void reverse (char a, char b)//move craft reverse function
```

```
79.  {
80.    analogWrite (E1,a);//providing speed to motor 1
81.    digitalWrite(M1,HIGH);//reverse direction to motor 2
82.    analogWrite (E2,b);//providing speed to motor 2
83.    digitalWrite(M2,HIGH);//reverse direction to motor 2
84.  }
85.  void left (char a, char b)//move craft left function
86.  {
87.    analogWrite (E1,a);//providing speed to motor 1
88.    digitalWrite(M1,HIGH);//reverse direction to motor 2
89.    analogWrite (E2,b);//providing speed to motor 2
90.    digitalWrite(M2,LOW);//forward direction to motor 2
91.  }

92.  void right (char a, char b)//move craft right function
93.  {
94.    analogWrite (E1,a);//providing speed to motor 1
95.    digitalWrite(M1,LOW);//forward direction to motor 1
96.    analogWrite (E2,b);//providing speed to motor 2
97.    digitalWrite(M2,HIGH);//reverse direction to motor 2
98.  }
```

The first few lines are for initialization: Lines (1) to (4) initializes the motor speed to PWM pins 5 and 6 and motor directions to digital pins 7 and 8, respectively. Lines (5) and (6) initialize the two sensor pins to analog input A0 and analog input A1, line (7) initializes the variable val as integer and (8) and (9) initialize the variables leftspeed and rightspeed, which we set to the maximum motor (PWM) speed of 255. The setup function specifies that the sensor pins are input and that the pins for motor speed and direction are output pins. This is done by using a for loop to reduce the length of the code. Also, Serial communication is initiated at 9,600 baud rate in line (17).

We define the two character variables right_sensor and left_sensor in lines (22) and (23) whose values are returned by the rightsensor() and the leftsensor() functions, respectively. Using if-else if loops, we check for each of four conditions. There can be four conditions possible for (leftsensor, rightsensor), where D=dark and L=light,

      1) Craft on the tape and pointing in the direction of the tape (D,D).
      2) Craft on the tape but pointing to the right of the tape (D,L)
      3) Craft on the tape but pointing to the left of the tape (L, D)
      4) Craft not on the tape (L,L)

and execute the appropriate actions for each case.

Line (45) describes the character function rightsensor. This function is a character function and not a void function because it returns a character value to the line where the function was called in line (22) of the loop function. Here we define two variables, sensout as an integer and sensorstate as the character variable and initialize it to 'L' (Light). We read out the right sensor value on analog input A1 and store the value in the variable sensout in line (49). Further in line (50) we specify a condition if the analog reading is above 800 (this is the threshold determined manually by calibrating the differences in the readings on a white plane and on a black tape) which implies that the sensor is on the black tape and the sensor function should return the character 'D' (Dark). A similar strategy is employed for the left sensor.

The remaining codes define all the functions related to movement, starting with the stop function, which halts the motors and stops the craft. These functions do not return any value and are hence defined as a void functions. To stop, the two pins E1 and E2 that control the motor speed are digitally made LOW by using a digitalWrite() function. The forward, reverse, right and left functions are defined with the appropriate signals for motor directions and speeds.

15.7 SUMMARY

The basic concepts for controlling your craft were described herein by introducing you to the simplicity of an Arduino microcomputer. By performing instructions at high speeds (faster than humans) it can accomplish a wide variety of tasks using two numbers (0 and 1) and very clever circuit and software design. As you design your vehicle control system you will be using your computer and the microcontroller with their memories, sensors and output devices. Using the signals generated by the sensors as feedback, you will design a program in Arduino C/C++ to control the behavior of your vehicle to follow a path and avoid obstacles found along the route.

Snippets of code are introduced early in the chapter to show you how to write sketches for the Arduino board in conjunction with output devices to perform simple tasks. These were short examples that illustrated the use of void setup(), void loop(), curly and semicolon when programming in Arduino C/C++. Then we showed how to convey digital and pulse width modulated data to external devices using output pins, as well as understanding how to read data from external sensors in either binary or analog formats. We also listed some sensors that can be connected to your Arduino board and introduced the Rover, upon which we practiced our simple programming control concepts. Next, we introduced rudimentary yet format concepts in C programming. We focused on data structures and control loops. Finally we gave an example of a pseudocode as a first draft for documenting a program, which was translated into code.

At this point, we have barely scratched the surface. The Arduino community is large and getting larger everyday. There are clever codes and libraries that many users have graciously provided to the community. You should feel free to scour the web for important and current information, and to find solutions to fix bugs and unusual behavior. You may return the favor by writing your own insightful comments.

EXERCISES

8.1   Why would autonomous control of your vehicle be interesting and challenging?

8.2   Describe some behaviors you anticipate in attempting to control your vehicle.

8.3   What sensors are you planning to use in designing your control system?

8.4   What actuators do you plan to use in designing your control system?

8.5   Would it be easier to have a wheeled robot run the course instead of a tracked vehicle? If so why?

8.6   Describe the difference between a microcomputer and a computer.

8.7   Describe the input and output pins of the Arduino.

8.8   Why does the Arduino contain a analog to digital converter (ADC)? How many counts are possible with this 10 bit converter?

8.9   What is the difference between digital pins 3,5,6,9,10,11 and the rest of the Arduino digital pins?

8.10   Why are resistors in series with LED recommended when using all pins except pin digital pin 13?

8.11 Write a code that will write "Good Morning, Dave" on your serial monitor.

8.12 What is the difference between analogWrite and analogRead, are they complements of each other?

8.13 Write a single line of code that would introduce a delay of 4.5 seconds in a program.

8.14 What does delay(), millis() and pulseIn() functions do?

8.15 Write a snippet of code that would record the time of operation of actuators if the output of a light sensor exceeds 55%.

8.16 What is the difference between servo motors and stepper motors?

8.17 Why do you need an amplifier between the Arduino and a large motor?

8.18 What is an Arduino shield? Give two examples.

8.19 Is punctuation important in writing code in Arduino C/C++? Give explanations on when to use { } ( ) ; [ ].

8.20 Why is important to pay attention to numbers designated as integers or floating point?

8.21 What are functions in general? What are void, char and int functions?

8.22 Describe the role of indentation when writing code in Arduino C/C++.

8.23 What are the Boolean values? Why are they important in designing you vehicle control system.

8.24 Prepare a listing of ASCII alphanumeric codes.

8.25 Define the data types used in Arduino C/C++.

8.26 Explain the following read and writing commands:

```
a = 3;
a = 3.5;
a = b;
x = list[2]
list[0] = y;
list[i] = y;
```

8.27 Explain the following Boolean operations:

```
a == b
a != b
a < b
a <= b
a > b
a >= b
a = !b;
a = ~b;
a && b
a || b
```

8.28 How many tasks can you program in Arduino C/C++?

8.29 Write a snippet of code containing a `while` loop.

8.30 Write a snippet of code containing a `do-while` loop.

8.31 Write a snippet of code containing a `for` loop.

8.32 Write a snippet of code containing a `if/then/else` control structure.

8.33 What happens when a `break` statement is inserted in a loop?

8.34 What happens when a `continue` statement is inserted in a loop?

CPSIA information can be obtained
at www.ICGtesting.com
Printed in the USA
BVHW061348110920
588463BV00006B/167